PYRIDINE AND ITS DERIVATIVES

SUPPLEMENT IN FOUR PARTS
PART FOUR

This is the fourteenth volume in the series
THE CHEMISTRY OF HETEROCYCLIC COMPOUNDS

THE CHEMISTRY OF HETEROCYCLIC COMPOUNDS

A SERIES OF MONOGRAPHS

ARNOLD WEISSBERGER and EDWARD C. TAYLOR

Editors

PYRIDINE AND ITS DERIVATIVES

SUPPLEMENT
PART FOUR

Edited by

R. A. Abramovitch
University of Alabama

AN INTERSCIENCE® PUBLICATION

JOHN WILEY & SONS
NEW YORK • LONDON • SYDNEY • TORONTO

An Interscience ® Publication

Copyright © 1975, by John Wiley & Sons, Inc.

Library of Congress Cataloging in Publication Data:

Abramovitch, R. A.   1930–
   Pyridine supplement.

   (The Chemistry of heterocyclic compounds, v. 14)
   "An Interscience publication."
   Supplement to E. Klingsberg's Pyridine and its derivatives.
   Includes bibliographical references.
   1. Pyridine. I. Klingsberg, Erwin, ed.   Pyridine and its derivatives. II. Title.

QD401.A22   547'.593   73–9800

ISBN 0–471–37916–6

Printed in the United States of America

10 9 8 7 6 5 4 3 2 1

Contributors

ELLIS V. BROWN, *University of Kentucky, Lexington, Kentucky*

A. F. CASY, *Faculty of Pharmacy and Pharmaceutical Sciences, University of Alberta, Edmonton, Canada*

R. T. COUTTS, *Faculty of Pharmacy and Pharmaceutical Sciences, University of Alberta, Edmonton, Canada*

RENAT H. MIZZONI, *Ciba Pharmaceutical Company, Division of Ciba-Geigy Corporation, Summit, New Jersey*

HARRY L. YALE, *The Squibb Institute for Medical Research, New Brunswick, New Jersey*

TO THE MEMORY OF

Michael

The Chemistry of Heterocyclic Compounds

The chemistry of heterocyclic compounds is one of the most complex branches of organic chemistry. It is equally interesting for its theoretical implications, for the diversity of its synthetic procedures, and for the physiological and industrial significance of heterocyclic compounds.

A field of such importance and intrinsic difficulty should be made as readily accessible as possible, and the lack of a modern detailed and comprehensive presentation of heterocyclic chemistry is therefore keenly felt. It is the intention of the present series to fill this gap by expert presentations of the various branches of heterocyclic chemistry. The subdivisions have been designed to cover the field in its entirety by monographs which reflect the importance and the interrelations of the various compounds, and accommodate the specific interests of the authors.

In order to continue to make heterocyclic chemistry as readily accessible as possible new editions are planned for those areas where the respective volumes in the first edition have become obsolete by overwhelming progress. If, however, the changes are not too great so that the first editions can be brought up-to-date by supplementary volumes, supplements to the respective volumes will be published in the first edition.

Arnold Weissberger

Research Laboratories
Eastman Kodak Company
Rochester, New York

Edward C. Taylor

Princeton University
Princeton, New Jersey

Preface

Four volumes covering the pyridines were originally published under the editorship of Dr. Erwin Klingsberg over a period of four years, Part I appearing in 1960 and Part IV in 1964. The large growth of research in this specialty is attested to by the fact that a supplement is needed so soon and that the four supplementary volumes are larger than the original ones. Pyridine chemistry is coming of age. The tremendous variations from the properties of benzene achieved by the replacement of an annular carbon atom by a nitrogen atom are being appreciated, understood, and utilized.

Progress has been made in all aspects of the field. New instrumental methods have been applied to the pyridine system at an accelerating pace, and the mechanisms of many of the substitution reactions of pyridine and its derivatives have been studied extensively. This has led to many new reactions being developed and, in particular, to an emphasis on the direct substitution of hydrogen in the parent ring system. Moreover, many new and important pharmaceutical and agricultural chemicals are pyridine derivatives (these are usually ecologically acceptable, whereas benzene derivatives usually are not). The modifications of the properties of heteroaromatic systems by *N*-oxide formation are being exploited extensively.

For the convenience of practitioners in this area of chemistry and of the users of these volumes, essentially the same format and the same order of the supplementary chapters are maintained as in the original. Only a few changes have been made. Chapter I is now divided into two parts, Part A on pyridine derivatives and Part B on reduced pyridine derivatives. A new chapter has been added on pharmacologically active pyridine derivatives. It had been hoped to have a chapter on complexes of pyridine and its derivatives. This chapter was never received and it was felt that Volume IV could not be held back any longer.

The decision to publish these chapters in the original order has required sacrifices on the part of the authors, for while some submitted their chapters on time, others were less prompt. I thank the authors who finished their chapters early for their forebearance and understanding. Coverage of the literature starts as of 1959, though in many cases earlier references are also given to present sufficient background and make the articles more readable. The literature is covered until 1970 and in many cases includes material up to 1972.

I express my gratitude to my co-workers for their patience during the course of this undertaking, and to my family, who saw and talked to me even less than usual during this time. In particular, I acknowledge the inspiration given me by the strength and smiling courage of my son, Michael, who will never know how much the time spent away from him cost me. I hope he understood.

R. A. ABRAMOVITCH

University, Alabama

Contents

Part Four

Part Three

CHAPTER XIII

Pyridine Alcohols

ELLIS V. BROWN

University of Kentucky, Lexington, Kentucky

1

I. Preparation

1. From Nonpyridine Starting Materials

Although pyrones have been heated with ammonia to give pyridines, 3-(1-alkoxyalkyl)pyridines recently have been prepared by heating 2,6-dialkoxy-3-(1-alkoxyalkyl)tetrahydropyrans with ammonia over aluminum oxide impregnated with platinum or palladium (1) **(XIII-1)**. 4-Hydroxy-6-hydroxymethyl-3-methoxy-2-pyridone was synthesized from 2-bromo-6-hydroxymethyl-3-

methoxy-4-oxo-4H-pyran and aqueous ammonia (2) **(XIII-2)**. 1,7-Dioxaindans **(XIII-3)** are converted to 2-(3-pyridyl)ethanols by oxygen and cupric ion in ammoniacal solution (3).

2. Oxidation of Side Chains

The picolines, 2-ethylpyridine and 4-isopropylpyridine, have been oxidized to the corresponding alcohols with air using various catalysts. The reaction is believed to proceed by a radical mechanism (4). The oxidation of alkylpyridines to alcohols by manganese dioxide-containing ores was deemed unprofitable because of the quantity of oxidizing agent needed (5).

3. Hydrolysis of Side-Chain Halides

2,6-Bis(chloromethyl)pyridine on acid hydrolysis produced the bis-alcohol, an intermediate in the preparation of polyurethanes (6).

4. Reduction of Aldehydes and Ketones

Pyridine aldehydes and ketones have been reduced to carbinols but recent papers deal with ketone reduction only. An interesting reducing agent used on a series of 4-pyridylmethyl ketones is tetramethylammonium borohydride (7). 3-Pyridyl 2-thiazolyl ketone was reduced by Raney nickel in acetic acid to the carbinol (8, 9). Lithium aluminum hydride was used to reduce 2-(α-bromophenacyl)pyridine to *threo*-1-phenyl-2-bromo-2-(2-pyridyl)ethanol (10), and the same procedure was used to prepare the corresponding chlorohydrin. 4-(Bromoacetyl)pyridine was also reduced to the halohydrin by sodium borohydride (11). Lithium aluminum hydride was also used to reduce 3-benzoyl-4-phenylpyridine to the carbinol (12). Sodium metal in liquid ammonia followed by the appropriate alkyl halide gave a series of phenylpyridylalkylaminocarbinols (13, 14) (see Table XIII-17). Hydrogenation of 4-acetylpyridine to the carbinol was effected by a palladium catalyst and hydrogen (15). Nitrosobenzyl pyridyl ketones were reduced by Raney nickel and hydrogen to the amino alcohols or by sodium borohydride to the nitroso alcohols (16) **(XIII-4)**.

In vivo reduction of 3-acetylpyridine to 3-pyridinemethylcarbinol was effected by the rat (17). The adrenal glands of several species reduced 2-methyl-1,2-bis

(3-pyridyl)propanone to the alcohol, and the same change was brought about by sodium borohydride (18). Sodium borohydride was also used to prepare phenyl-2-pyridylcarbinol from an ozonolysis product (19).

5. Reduction of Acids and Esters

Pyridine esters were reduced with lithium aluminum hydride to the carbinols. Ethyl nicotinate, ethyl 4-methylnicotinate, ethyl 2,3-di-(2-pyridyl)propane-2-carboxylate, the corresponding butane, and many others were reduced in this way (20–23). Certain esters, such as diethyl pyridine-2,6-dicarboxylate, can be reduced with sodium borohydride (24–26) although there is the danger of nuclear reduction with this agent (26, 27) **(XIII-5)**.

XIII-5

6. Aldol Condensation of Alkylpyridines with Aldehydes

Methyl groups at the 2-, 4-, and 6-positions in pyridine undergo reaction with formaldehyde to give the carbinols and, in some cases, dihydric alcohols. For example, 5-ethyl-2-methylpyridine gives 5-ethyl-2-(2-hydroxyethyl)pyridine (28). 2,4-Lutidine gives a mixture of 2-(2-hydroxyethyl)-4-methylpyridine and 4-(2-hydroxyethyl)-2-methylpyridine, while collidine gives a mixture containing 2,4-dimethyl-6-(2-hydroxyethyl)pyridine (29). 2,6-Lutidine can react either at one or at both methyls (30). A number of industrial preparations of vinylpyridines start with various methylpyridines to give the carbinols (31–35). Arrhenius parameters and rate constants were determined for the formation and dehydration of 2-(2-hydroxyethyl)pyridine (36). Isopropenylpyridines were prepared from 2- and 4-ethylpyridine and formaldehyde through the 2-(2- or 4-pyridyl)propanols (37). A somewhat more novel reaction was the condensation of (2,3-pyrido)cycloparaffins with formaldehyde (38) **(XIII-6)**.

XIII-6

7. From Organometallic Compounds and Pyridine Aldehydes, Ketones, and Esters

2-Pyridine aldehyde has been treated with a number of substituted phenylmagnesium halides to give the corresponding carbinols (39, 40). Numerous 3-pyridyl ketones have been converted to carbinols by Grignard reagents (8, 12, 41-46) as have 4-pyridyl ketones. Several amino alcohols were prepared by treating 2- and 4-pyridyl aminoalkyl ketones with Grignard reagents (47, 48). From 2- and 4-benzoylpyridines with organomagnesium halides various tertiary alcohols were prepared and these were resolved (49). 2-(2-Carbethoxycyclopropyl)pyridine, 5-(2-carbethoxycyclopropyl)pyridine, and 4-(2-carbethoxycyclopropyl)pyridine were converted to carbinols by various Grignard reagents (50-52). Organosodium compounds gave carbinols with pyridine ketones (53). Pyrazinyllithium compounds have been used to prepare pyridylpyrazinylcarbinols (54).

8. From Metallopyridine Compounds

Numerous alcohols have been prepared by the reaction of α-picolyllithium with various ketones (48, 53, 55-61). 2-Ethylpyridine (62), 2,4-lutidine, and 2,4,6-collidine (63) underwent lithiation at the 2-position, and this was followed by carbinol formation (29). 2-Picolyllithium reacted with acrylaldehyde (acrolein) to give the unsaturated carbinol (64). 2-Picolyllithium (65) and various alkylpyridines (66) with epoxides have given carbinols (XIII-7), but with chloromethyl methyl ether ethers were formed (67, 68) (XIII-8).

2-Benzylpyridine was converted to the organolithium compound and then treated with various ketones to form phenyl-2-pyridylcarbinols (69-71).

3-Picolylsodium reacted with formaldehyde to give the alcohol (70). The sodium and potassium derivatives of 3-benzylpyridine have been converted to carbinols by reaction with ketones (71, 72).

Organometallic derivatives of 4-picoline have been reacted with benzoyl chloride (73-75), with cyclohexanoyl chloride (76, 77), various esters (75, 78, 79) and ketones (53, 80) to give the 4-pyridylcarbinols. 4-Benzylpyridine lithium derivative has been reacted with ketones to give 4-pyridylcarbinols (70,

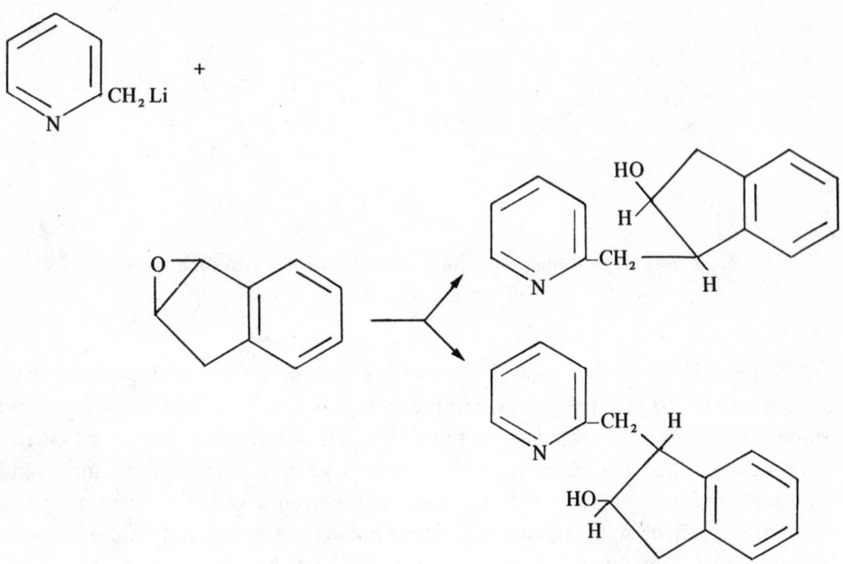

XIII-7

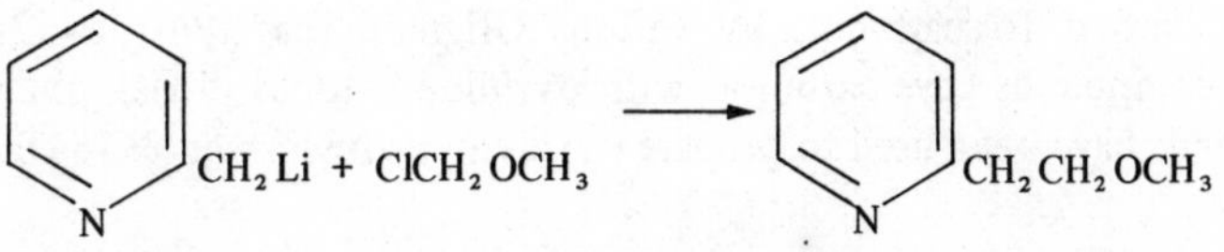

XIII-8

71). 2-Pyridyllithium and 3,4,5-trimethoxybenzophenone afforded the carbinol (81) and the 6-lithium derivative of 4-chloro-3-methylpyridine-1-oxide reacted with cyclohexanone to give the tertiary alcohol (**XIII-9**) (82) (see also Chapter

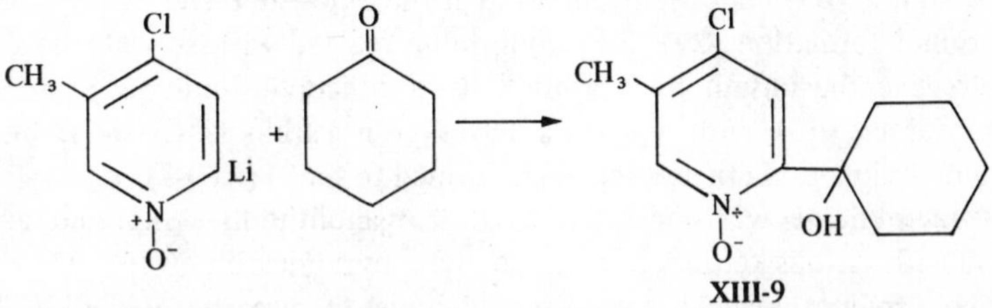

XIII-9

IV). Many ring-lithiated pyridines have given carbinols when treated with ketones (83–86). 5-Ethynyl-2-methylpyridine, when converted to a metallo-acetylene derivative, reacted with ketones to give the ethynylcarbinols (87–89).

2-Pyridyl phenyl ketone reacted with sodium ethylate followed by cyclopentadiene to give α-phenyl-α-[6-phenyl-6-(2-pyridyl)-2-fulvenyl]-2-pyridinemethanol (**XIII-10**), as did the corresponding 3- and 4-pyridyl phenyl ketones (90–93).

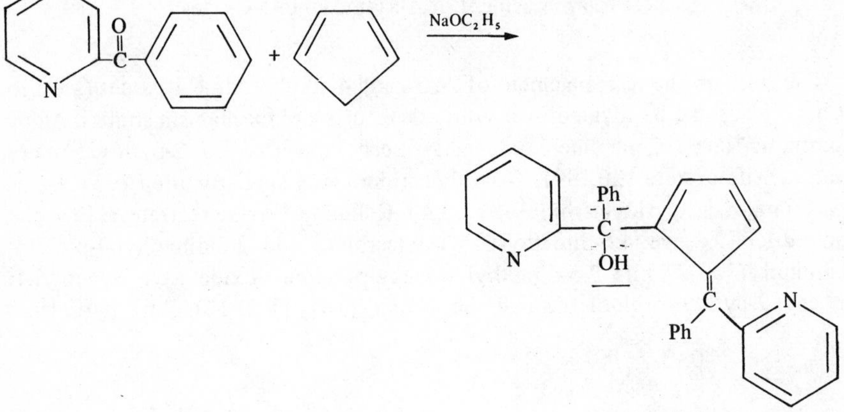

XIII-10

9. Aldol Condensation of Pyridine Aldehydes

Nitroethane reacts with all three pyridine aldehydes in the presence of secondary amines to give the corresponding nitrocarbinols (94). Pyridine aldehydes were condensed with the α-methylene group of several ketones using Amberlite IRS-400 in place of the conventional base (95).

10. Hammick Reaction

In an extension of the Hammick reaction sodium 2-pyridylacetate was decarboxylated in cyclohexanone, acetophenone, and benzaldehyde to give the respective alcohols (96). A biography and bibliography of Dalzill Lewellyn Hammick who died in 1966 has appeared (97). A reaction certainly related to the Hammick reaction is the decarboxylation of homarine chloride in aromatic aldehydes, such as benzaldehyde, to give 2-(α-hydroxybenzyl)-1-methylpyridinium chloride (**XIII-11**) (98).

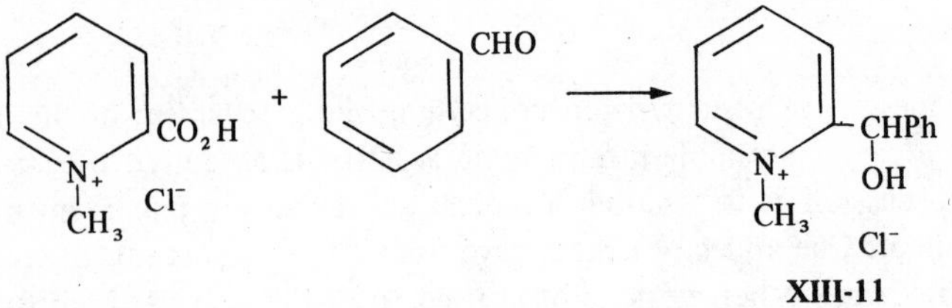

XIII-11

11. Rearrangement of Alkylpyridine-1-Oxides

The work on the rearrangement of 2-, 3-, and 4-alkylpyridine-1-oxides (see also Chapter IV) can be divided into synthetic efforts and mechanism studies. Along synthetic lines, 2-picoline-1-oxide has been converted to 2-pyridylcarbinol through its acetate (20, 99); 4-pyridylcarbinol was similarly prepared (20), as was 6-methyl-2-pyridylcarbinol (97). 2,4,6-Collidine-1-oxide rearranged in acetic anhydride to give 4,6-dimethyl-2-pyridylcarbinol and 2,6-dimethyl-4-pyridyl-carbinol (100), while 2,5-dimethyl-4-phenylpyridine-1-oxide gave 5-methyl-4-phenyl-2-pyridylcarbinol through the acetate (101) **(XIII-12)**. Both pyrindane-

1-oxide (102) and some other (2,3-pyrido)cycloparaffin-1-oxides (38) have been rearranged to the carbinols. 4-Methoxy- and 4-ethoxy-2-picoline-1-oxide were converted to the 2-carbinols (103). 2-[2-(5-Nitro-2-furyl)vinyl]-6-methylpyri-dine-1-oxide was rearranged to the 6-carbinol acetate **(XIII-13)** (104).

XIII-13

Many papers have been written concerning the mechanism of this rearrange-ment and a very reasonable intermediate is **XIII-14**, an anhydro-base structure. There is discussion as to whether a radical pair or an ion pair is involved in the rearrangement. Oae and co-workers have used ^{18}O in the study of this reaction (105, 106). Many other workers have tried to distinguish between the ion pair

XIII-14

and radical pair mechanism (102–112). Rate constants and ^{18}O incorporation have been reported recently (113, 114). The problem has been reviewed in two articles and a book (115–117). Acetyl and benzoyl chlorides have been shown to convert 2-picoline-1-oxide to the corresponding carbinol esters (118) (for a more detailed treatment see Chapter IV).

12. Emmert-Asendorf Reaction

This reaction involves the condensation of a pyridine with a ketone in the presence of magnesium or magnesium amalgam to give a pyridyl alcohol. Recently a quantitative study of the reaction of pyridine and 3-picoline with cyclohexanone and 2-methylcyclohexanone and of pyridine with 4-t-butylcyclohexanone has been carried out (119).

II. Properties

The properties of pyridine alcohols are the expected ones for alcohols containing a tertiary amine group. A number of physiological activities have been reported for various compounds and are mentioned here. Pharmacological properties of benzylpyridylcarbinols have been reported (120). Psychopharmacological activity has been ascribed to *trans*-2-(4-pyridyl)-α,α-diphenylcyclopropanemethanol (50). 3-Pyridinemethanol diminishes circulating cholesterol and fatty acids in the blood (121, 122). The toxicity and distribution of 1-(3-pyridyl)ethanol in mice has been investigated using ^{14}C tagging (123). 2-{2,6-Diethyl-α-[2-(methylamino)-ethoxy]benzyl}pyridine **(XIII-15)** and its

XIII-15

carbamate have diuretic properties in rats (124). The blood levels of 2,6-pyri-dinemethanol bis(N-methylcarbamate) after different dosages has been reported (125). Of six pyridylalkylcarbinols tested for rat choleretic activity, 1-(2-pyrid-yl)propanol had the most rapid action, and 1-(3-pyridyl)butanol had the most prolonged activity (126). Hypotensive and spasmolytic activity has been reported for a series of 4-pyridylcarbinols carrying various amino groups (9). Antihelminthic properties have been ascribed to 2-(β-methoxyethyl)pyridine (127) and its pathological effects noted (128).

The optical rotations of a group of alkyl 2-, 3-, and 4-phenylcarbinols have been recorded (49, 129). The dissociation constants (pK_a alcohol-pK_a amine) of a group of 2-, 3-, and 4-pyridine alcohols have been measured (130). N.m.r. spectral data aided in the determination of the structure of 3-hydroxy-2-pyri-dinemethanol (131). Intramolecular hydrogen bonding in 2-(2-hydroxyethyl)- and 2-(3-hydroxypropyl)pyridine has been studied (132).

For other physical properties one should consult the tables at the end of this chapter. Some quaternary salts of 2-pyridinemethanol have been prepared (133). The kinetics of hydrolysis of 2-, 3-, and 4-pyridylmethylphosphates have been determined (134).

III. Reactions

1. Oxidation

Oxidation of pyridine alcohols to aldehydes continues to be an important synthetic reaction and has been carried out recently by self-oxidation of the 1-oxides using phenylhydrazine (135) and by $PhCH=SO_2$ or sulfonate esters of the alcohols (136). The self-oxidation of 2-pyridinemethanol-1-oxide to the aldehyde by heating has been patented (137) (XIII-16). An example of carbinol

XIII-16

to acid oxidation where a ring hydroxyl group is protected by benzylation followed by potassium permanganate oxidation has been reported (138, 139). 3-Pyridinemethanol has been oxidized biologically to nicotinic acid (140). Air oxidation in a basic medium converted 3-pyridyl-2-thiazolylcarbinol to the ketone (8, 141) and manganese dioxide oxidized methyl-4-pyridylcarbinol to

4-pyridyl methyl ketone (15). Surprisingly, a thioether linkage in a pyridine-methanol could be oxidized to the sulfone with hydrogen peroxide without affecting the alcohol (71) (XIII-17) and, in several cases, the pyridine ring

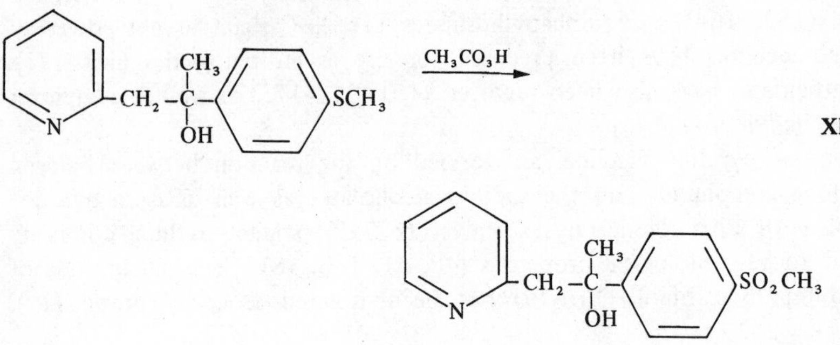

XIII-17

nitrogen of a pyridinemethanol has been converted to the 1-oxide using peracetic acid without oxidizing the alcohol (69, 142). 1-(α-Pyridyl)-3-butene-2-ol underwent the Oppenauer oxidation to the corresponding ketone (64).

2. Reduction

The reduction of pyridine alcohols to hydrocarbons takes place often with α-methylation, and this reaction (XIII-18) has been discussed by Reinecke and Kray (143–148). The more usual reduction is that of the pyridine ring to the

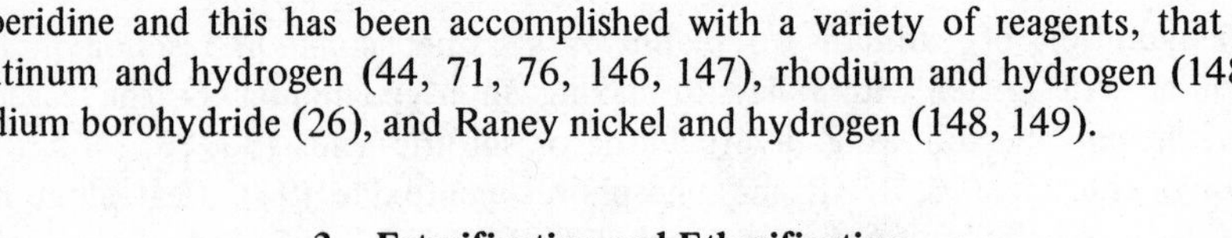

XIII-18

piperidine and this has been accomplished with a variety of reagents, that is, platinum and hydrogen (44, 71, 76, 146, 147), rhodium and hydrogen (148), sodium borohydride (26), and Raney nickel and hydrogen (148, 149).

3. Esterification and Etherification

Numerous esters of pyridine alcohols have been prepared by normal methods. Among esters synthesized recently are acetates (150, 151), camphorates (152), succinates (153), salicylates (154), tartrates (155), 2-(2,4,5-trichlorophenoxy)-propionates (156), phenylcarbonates (157), α-(p-chlorophenoxy)isobutyrate (158), 2,2-dimethyl-3-(2-methylpropenyl)cyclopropylcarboxylates (159), xanthates (160), various sulfonates (136), phosphates (20, 160–163), thionphosphates (162, 164), and diphenylborinates (165). Carbamate derivatives of pyridine alcohols have been prepared for use as antiphlogistics (166–171). Thiocarbamates have also been reported (151, 172–175) as well as complexes with triphenylboron (176).

Ethers of pyridine alcohols are formed by the reaction between halogen containing compounds and the sodium alcoholates as well as from pyridine alcohols with other alcohols using catalysts (177–179). Many pyridine ethers are claimed to have biological properties (40, 41, 180, 181). The glucuronide of 3-pyridylmethylcarbinol (XIII-19) has been isolated from rat urine (17).

$$HOOC-CH-\underset{\underset{OHH}{|}}{\overset{\overset{H}{|}}{C}}-\underset{\underset{OH}{|}}{\overset{\overset{OHH}{|}}{C}}-\underset{|}{\overset{\overset{H}{|}}{C}}-CH-O-\underset{\underset{CH_3}{|}}{\overset{\overset{H}{|}}{C}}-$$

XIII-19

Bis(2-pyridylmethyl)ether has been rearranged by $NaNH_2$ to give 1,2-di-(2-pyridyl)ethanol (182).

4. Replacement of Hydroxyl Groups by Halogen

The carbinol group can be replaced by halogen directly using hydrogen halides or thionyl chloride (183), and esters of carbinols such as the acetates can be converted to halides by hydrogen halides (184).

5. Dehydration to Olefins

Dehydration of carbinols to olefins is a rather common reaction in the pyridine series, often with a view to making vinyl-type monomers. The reagents used include the following: hydrochloric or sulfuric acids (60, 76, 146, 185), thionyl chloride (76, 186), and phosphorus pentoxide (94). Heat alone has effected the dehydration in many cases (30, 31, 36, 37, 187) and also heat plus sodium hydroxide (188) or alumina (189). The acetate esters of pyridine alcohols can be dehydrated directly by heat (190). Pyridylcyclohexyl alcohols have been dehydrated and aromatized by heating in a mixture of acetic and sulfuric acid (83, 84, 86).

6. Synthesis of Polycyclic Systems

Reductive cyclization of various substituted 2-pyridine alcohols produced indolizidines and quinolizidines (66). 2-(γ-Hydroxypropyl)pyridine-1-oxide gave a 2,3-dihydro-4H-oxazino[2,3-a]pyridinium salt **(XIII-20)** (on treatment with a

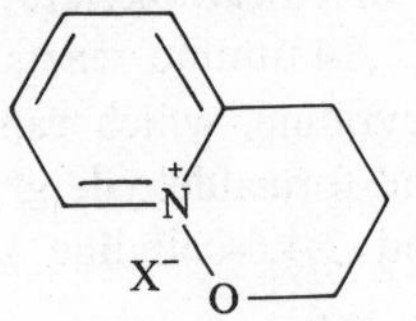

XIII-20

hydrogen halide) (191). 2-(2-Pyridyl)-1,3-propanediol was converted to the ditosylate and this was treated with zinc, acetamide, and NaI to prepare 2-cyclopropylpyridine (192).

7. Other Replacement Reactions

2- And 4-pyridylmethanols underwent replacement of the hydroxyl groups by an amino group when treated with aromatic amines (193, 194) **(XIII-21)**. On the

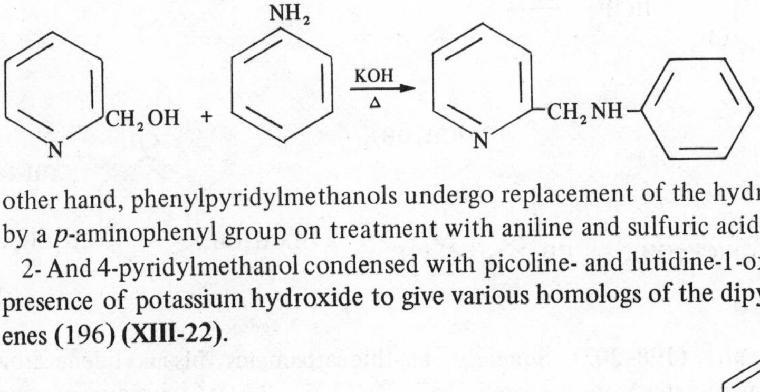

XIII-21

other hand, phenylpyridylmethanols undergo replacement of the hydroxyl group by a *p*-aminophenyl group on treatment with aniline and sulfuric acid (195).

2- And 4-pyridylmethanol condensed with picoline- and lutidine-1-oxides in the presence of potassium hydroxide to give various homologs of the dipyridylethyl- enes (196) **(XIII-22)**.

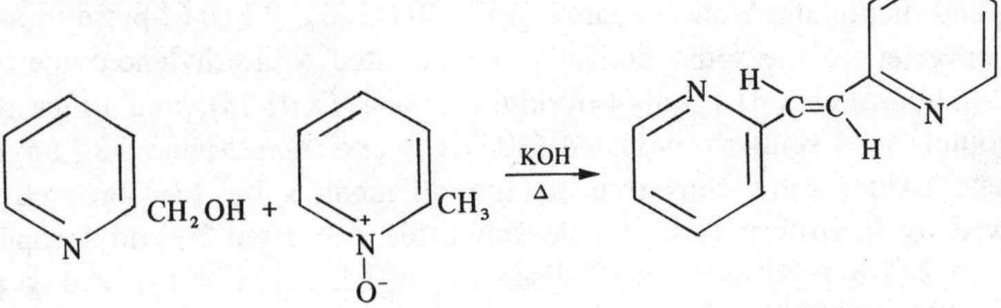

XIII-22

IV. Dihydric and Polyhydric Alcohols Containing One Pyridine Nucleus

2-Hydroxymethyl-6-methylpyridine-1-oxide was converted to 2,6-dihydroxy-methylpyridine by boiling under reflux in acetic anhydride (197, 208). Reduction of diethyl 2,6-pyridinedicarboxylate with sodium borohydride also gave the dialcohol (24, 25). 2,6-Lutidine reacts with formaldehyde to furnish some 2,6-(dihydroxymethyl)pyridine, which can be dehydrated to the divinyl-pyridine (30). 2,4-Lutidine and formaldehyde give some 2-(4-methyl-2-pyridyl)-propane-1,3-diol (**XIII-23**) and 2,4,6-collidine gives several pyridinediols (29)

$$
\underset{\text{N}}{\overset{\text{CH}_3}{\bigcirc}}\text{CH}_3 \;+\; \text{HCHO} \;\longrightarrow\; \underset{\text{N}}{\overset{\text{CH}_3}{\bigcirc}}\text{CH(CH}_2\text{OH)}_2 \;+\; \underset{\text{N}}{\overset{\text{CH}_2\text{OH}}{\bigcirc}}\text{CH}_2\text{OH}
$$

XIII-23

(**XIII-24**). 2,6-Bis(hydroxymethyl)pyridine has been converted to carbonates

$$
\underset{\text{CH}_3}{\overset{\text{CH}_3}{\bigcirc}}\text{CH}_3 \;+\; \text{HCHO} \;\longrightarrow\;
$$

$$
\underset{\text{CH}_3}{\bigcirc}\text{CH(CH}_2\text{OH)}_2 \;+\; \text{CH}_3\underset{\text{N}}{\bigcirc}\text{CH}_3 \;+\; \text{HOCH}_2\text{CH}_2\underset{\text{N}}{\overset{\text{CH}_3}{\bigcirc}}\text{CH}_2\text{CH}_2\text{OH}
$$

XIII-24

and carbamates (198–202). Similarly, bis-thiocarbamates, bis-alkylidenecarbi-zates, and nicotinates were prepared (157, 203–205). 1,3-Di-(4-pyridyl)propane was converted to the sodio derivative and treated with ethylene oxide to give 1,3-bis(β-hydroxyethyl)-1,3-di-(4-pyridyl)propane (**XIII-25**), and other related compounds were similarly prepared (206). *Cis*- and *trans*-2-phenyl-3-(2-pyridyl)-ethylene oxides were converted to the ethanediols by trichloroacetic acid followed by hydrolysis (10). Acetic anhydride converted 2-pyridylpropanol-1-oxide to 3-(2-pyridyl)propane-1,3-diacetate, which could be cyclized to pyrro-coline (207). 3-Methyl-2-(3-pyridyl)indene has been prepared by cyclization of

XIII-25

2-phenyl-3-(3-pyridyl)butane-2,3-diol (209). (2,3-Pyrido)cycloparaffins containing two alcoholic groups have been prepared from the cycloparaffin and formaldehyde (38) (**XIII-26**). When 2-pyridineglyoxalic acid was heated with

XIII-26

aldehydes there was obtained a series of 2-pyridyglycols (**XIII-27**) (210).

XIII-27

Phenyllithium and diethyl 4-phenylpyridine-2,5-dicarboxylate gave the 4-phenyl-2,5-bis(α-hydroxybenzhydryl)pyridine (211). Treatment of a variety of pyridine-1-oxides with BuLi at −65° followed by addition of an aldehyde or ketone gives the 2,6-di-α-hydroxyalkylpyridine-1-oxides (82).

V. Dihydric Alcohols Containing Two Pyridine Nuclei

The Gomberg-Bachmann pinacol synthesis was successful in the case of 2-benzoylpyridine and gave α,α-di-2-pyridylhydrobenzoin (**XIII-28**) (212). Electrolytic reduction of 3-acetylpyridine furnished 2,3-di-(3-pyridyl)-2,3-butanediol (213) as did the photochemical reduction in isopropanol (214). 4,6-Dimethyl-2-pyridinecarboxaldehyde with potassium cyanide gave 1,2-bis(4,6-dimethyl-2-pyridyl)-1,2-ethanediol (100). 2-Pyridylglyoxalic acid condensed with 2-pyridinecarboxaldehyde to give bis(2-pyridyl)ethanediol (210). Air oxidation of

XIII-28

2-hydroxymethylpyridinium methiodide produced 2,2'-(*N*-methylpyridinium)-glycol diiodide (215). Heating 1,3-di-(2-pyridyl)propane-1,1'-dioxide with acetic anhydride gave some 1,3-di-(2-pyridyl)-1,3-propanediol diacetate, which could be hydrolyzed to the glycol and 1-acetyl-3-(2-pyridyl)pyrrocoline (**XIII-29**) (216, 217). A series of 2-, 3-, and 4-bis-pyridylglycols were converted to their

XIII-29

di-1-oxides and oxidized by lead tetraacetate to the corresponding aldehyde-1-oxides (218). 1,2-(Di-3-pyridyl)-1,2-diphenylethane-1,2-diol rearranged to the pinacolone in sulfuric acid with exclusive migration of the phenyl group (214).

VI. Side-Chain Hydroxyacid Derivatives

Ethyl nicotinate and an arylacetonitrile condensed in the presence of a basic catalyst to give α-aryl-β-hydroxy-(3-pyridyl)acrylonitriles (219). Pyridinecarbox-aldehyde-1-oxides when treated with hydrogen cyanide afforded the cyan-hydrins which could also be made from the aldehyde cyanhydrins and hydrogen peroxide (218). A number of 5-substituted-2-picolinic acids containing alcohol groups in C-5 side chains were prepared by selenium dioxide oxidation of the corresponding picolines (220) (**XIII-30**). Several pyridyl hydroxyalkylsulfuric acids reacted with potassium cyanide and the products were hydrolyzed to the

XIII-30

hydroxyacids (221). 3- And 4-pyridyl-β-hydroxyacrylic esters were obtained from pyridylacetates and ethyl formate (222). γ-(3-Pyridyl)-γ-hydroxybutric acid was prepared from the corresponding ketoacid by reduction with sodium borohydride or biologically (223). 2,2′-Dipyridylglyoxal has been rearranged to 2,2′-pyridylic acid (224). The same rearrangement has been studied with 1-phenyl-2-(3-pyridyl)glyoxal and 1-(4-methoxyphenyl)-2-(pyridyl)glyoxal using ^{14}C tagging (225).

VII. Derivatives Containing Both Nuclear and Side-Chain Hydroxyl

1. Preparation and Reactions

When an alkaline solution of formaldehyde is heated with 3-hydroxy-6-methyl-pyridine, 3-hydroxy-6-methyl-2-pyridinemethanol is obtained (197, 226). Ethyl 4-ethoxymethyl-2-hydroxy-6-methylnicotinate was prepared by diazotization of the corresponding amine (227). 4-Ethoxy and -methoxy-2,6-bis(hydroxymethyl)pyridine dicarbamates have been synthesized from the dialcohols (228).

Reductive cleavage using zinc and sodium hydroxide has converted the 4-methoxymethyl group to methyl in the case of 5-hydroxymethyl-3-methoxy-4-methoxymethyl-2-methylpyridine (229). Piericidin A has been shown to be a derivative of a complicated nuclear trihydroxypyridine with an alcohol group in the side chain (XIII-31) (230). Pyridoxol (XIII-32) is undoubtedly the most

XIII-31

important compound in this series and a number of modifications of its structure have been studied. For example, 3,5-bis(hydroxymethyl)-6-methyl-2(1*H*)-pyridone (231), a positional isomer, and 3,5-dihydroxy-4-hydroxymethyl-6-methylpyridine (5-norpyridoxol) (XIII-33) (232) have been prepared.

XIII-32   XIII-33

5-Hydroxy-6-methyl-4-trifluoromethyl-3-pyridinemethanol has been made (233) and 6-methylpyridine-3,4-dimethanol has been found to be a by-product of pyridoxal syntheses (234). 2,4-Dimethyl-3-hydroxy-5-pyridinemethanol has been formed by replacement of the 4-hydroxyl group of pyridoxol by halogen followed by reduction (235). Homologs of pyridoxol such as 3-(3-hydroxy-4-hydroxymethyl-2-methyl)pyridinepropanol illustrate extension of the alcohol position of pyridoxol (236). Another norpyridoxol (3-hydroxy-4,5-bis-hydroxymethylpyridine) was prepared and tested for biological activity (237, 238), as was 4,5-dihydroxymethyl-6-hydroxy-2-methylpyridine, an isomer of pyridoxol (239). 3-Deoxypyridoxine has been isolated from pyridoxine synthesis mother liquors and has been shown to significantly arrest the growth of lymphosarcoma implants in rats maintained on a pyridoxine deficient diet (240). 4-Deoxypyridoxine has been prepared and tested for its vitamin B_6 antagonist activity (241, 242).

2. Pyridoxol Synthesis

A great deal of the pyridoxol technology was reported in the original chapter on pyridine alcohols, first edition, Part IV. A number of patents and papers have appeared since then, particularly outside the United States (243–267). A few reports on the conversion of furans to pyridines in connection with pyridoxol syntheses have been published (261–263).

A newer method for the syntheses of pyridoxol and related compounds involves the condensation of 5-alkoxy-4-methyloxazole and similar oxazoles with any one of the several dienophiles (XIII-34). Typical dienophiles used

XIII-34

include the cyclic acetal of formaldehyde and butenediol (268–273), 2,5-di-methoxy-2,5-dihydrofuran (274), 3,6-dihydro-1,2-dioxin (275), and various fumaric and maleic derivatives (276–286).

Pyridoxol has been labeled with tritium at specific positions (287). Further reports on the preparation of pyridoxol-1-oxide have appeared (288, 289).

3. Pyridoxol Derivatives

Because of the important biological activity of pyridoxol many modified pyridoxols and derivatives have been prepared. 2,6-Dimethyl-3-hydroxy-4-pyridinemethanol and its acetates has been prepared (290), and many workers have synthesized and studied the so-called 4-deoxypyridoxol (2,4-dimethyl-3-hydroxy-5-pyridinemethanol) (219, 292). 3-Hydroxy-4-hydroxymethyl-2-methyl-5-pyridineacetic and propionic acids have been described (293). Benzylation of pyridoxol gave the 3-benzyl ethyl (294). 6-Arylazo derivatives of pyridoxine were synthesized and tested for antisarcoma activity (295). A number of pyridine alcohols containing 3-phenylazo and 3-amino groups were prepared by ring closure and through these it was possible to prepare pyridoxol (296).

Pyridoxol, with its three hydroxyl groups, can naturally form a number of esters and many of these have been prepared and tested for physiological activity. Nicotinic acid esters of pyridoxol have been made and tested (297–305). A number of fatty acid esters were prepared and tested because of their lipid solubility. Among these are the lauroyl (304–306), palmitoyl (307–310), octanoyl (311–313), and linoleoyl (313) esters. Miscellaneous esters obtained include adamantoyl (314), ethyl carbonate (315), benzyl carbonate (315), glyoxylate (316), glycerrhetinate (317), succinate (318, 319), 4-amino-butyrate (320), glucuronate (321), malate (322), and phosphate (322–327). The quaternary salt of pyridoxol and methyl iodide has been tested against degenerative processes in the cardiovascular system (328). The N-acetylcystein-ate has also been used in cosmetics (329).

One of the very common modifications of pyridoxol is the replacement of one of the three oxygens by sulfur. Thus the "3-thiopyridoxol" (XIII-35) and its 4-deoxy derivative have been prepared (330, 331). The so-called "4-thiopyridoxol" (XIII-36) where the hydroxy of the 4-alcohol is replaced by SH has also

XIII-35   XIII-36   XIII-37

been made (235, 332–334). The most usual sulfur analog synthesized seems tc be the 5-thio or some modification thereof (335–340); this includes the cases where two pyridoxol groupings are linked through a 5-thioether bond (341–343) and where one pyridoxol and one thiamine unit are so linked (344, 345). The disulfide linkage has also been introduced at the 5-position with various alkyl groups attached (345–348). The case of two pyridoxol groups bonded through a disulfide linkage has been recorded (349–356), as well as a pyridoxol and a thiamine so joined (357). Other related derivatives containing a 5-sulfite ester and isothiuronium salts have been described (346, 349, 358).

Since both acetone ketals of pyridoxol have been made (α^4, α^5-O-isopropylidene- and α^3, α^4-O-isopropylidenepyridoxol), several ethers and esters of these have been prepared, and the acetone removed in some cases (359–362) (XIII-38).

Pyridoxol with ^{14}C at the 5-hydroxymethyl group has been synthesized (364). Several deuterated pyridoxols have been prepared both to study the deuteration and in connection with n.m.r. studies (365–369). Excited state ionization (370) and photochemical decomposition (371) of pyridoxol have been studied. Glyoxalic acid and pyridoxine combined to give a product called pyridoxilate (372). The 5-glycoside of pyridoxine has been synthesized (373). A dimer of pyridoxine has been prepared (374). 3-α^5-O-Dibenzylpyridoxol has been used to synthesize pyridoxine modified in the 4-position (375). Proceedings of an important conference on pyridoxol in the metabolism of the nervous system have been published (376). Several other papers on the physiological reaction of pyridoxol and its derivatives have appeared (377–387).

4. Pyridoxal and Pyridoxamine

Replacement of 4-hydroxymethyl group of pyridoxol by the aldehyde group gives pyridoxal (XIII-39), and when the 5-hydroxymethyl group is so replaced we have isopyridoxal (XIII-40). The latter compound has been synthesized and studied (388–390). Pyridoxal has been synthesized and some of its physical

CHO

HO⟍/⟍CH₂OH

CH₃

N

XIII-39

CH₂OH

HO⟍/⟍CHO

CH₃

N

XIII-40

properties studied (293, 391–394). Two pyridoxal groups can be joined through a 5-thioether linkage (395). Pyridoxal reacts with alkali cyanide as might be expected of an aldehyde (396). Oxidation of one alcohol grouping of pyridoxal or the aldehyde function of pyridoxal gives pyridoxic acid **(XIII-41)** (397, 398).

CHO

HO⟍/⟍CH₂OH $\xrightarrow{[O]}$ HO⟍/⟍CH₂OH

CH₃   CH₃

N   N

CO_2H

XIII-41

Pyridoxal phosphate **(XIII-42)** is known as codecarboxylase and its syntheses and physiological action are discussed in a number of papers (399–437).

CHO

HO⟍/⟍$CH_2OPO_3H_2$

CH₃

N

XIII-42

Pyridoxal gives a hydrazone (438, 439). Areas in which considerable work has been done are the syntheses and physiological action of the Schiff bases formed from pyridoxal and various amino acids (440–466). Pyridoxal-1-oxide has been prepared (287, 467).

5-Deoxypyridoxal has been prepared from pyridoxol (468) and 3-hydroxy-5-hydroxymethyl-2-methyl-6-pyridinealdehyde has been prepared (469). The triplet state of pyridoxal has been investigated by e.s.r. spectroscopy (470) and the effect of pH on the relative distribution of pyridoxal structures was examined by fluorescence spectroscopy (471). Pyridoxal esters have been prepared (472, 473) as well as pyridoxal azine (474, 475). 2-Norpyridoxal **(XIII-43)**, 2-nor-6-methylpyridoxal **(XIII-44)**, and 6-methylpyridoxal **(XIII-45)** have been synthesized (476).

$$\text{XIII-43} \qquad \text{XIII-44} \qquad \text{XIII-45}$$

Pyridoxamine **(XIII-46)** is the name of the compound in which an alcohol group of pyridoxal is replaced by an amino group and its syntheses has been described (477, 478). Pyridoxamine can be converted to pyridoxal (479). A

$$\text{XIII-46}$$

sulfur analog of pyridoxamine (331), a folic acid derivative (480), an ethyl carbamate (481), and Schiff's bases (482) have been prepared.

Pyridoxamine is involved in pyruvate transaminase, and there have been recent studies of its action in this capacity (483–487). Isopyridoxamine has been prepared from isopyridoxol (390). Pyridoxamine-1-oxide has been described (287, 467). Metabolites of pyridoxamine have been synthesized (488), and the role of pyridoxamine in its reaction with carbon disulfide has been discussed from the toxicological viewpoint (489).

VIII. TABLES

TABLE XIII-1. 2-Pyridinemethanols 2-PyC(OH)RR'

Py	R	R'	Physical properties (derivatives)	Ref.
2-Py	H	H	B.p. 114-116°/16-17 mm, 98-100°/7 mm; n_D^{26} 1.5248; picrate, m.p. 160.5-162°; MeI, m.p. 153.5-154.0°; EtI, m.p. 138°; PrI, m.p. 102°; benzoate, b.p. 124-125°/1.2 mm; phenylacetate, b.p. 122°/0.4 mm acetate, MeI, m.p. 154-155°; acetate, picrate, m.p. 101-102°; acetate, HCl, m.p. 94°; "O benzoate, b.p. 124-125°/1.2 mm; phosphate, m.p. 182-182.5°; phenylcarbamate, m.p. 98°; 3,4,5-tri-methoxyphenylcarbamate, m.p. 104° pK$_a$; i.r.; n.m.r.; furfuryl carbamate-1-oxide, m.p. 140-142°; 3-picolyl carbamate-1-oxide, m.p. 102-104°; methyl carbamate-1-oxide, m.p. 150-152°; isobutyl carbamate-1-oxide; picrate, m.p. 127-129°; dimethyl carbamate-1-oxide, m.p. 103-105°; alkylcarbamate-1-oxide, m.p. 140-142°; 3-pyridyl carbamate-1-oxide, m.p. 142-143°; acetate MeI, m.p. 160°; acetate EtI, m.p. 115° and 242°; acetate, PrI, m.p. 126°; diphenylborinate, m.p. 152-153°	20, 96, 99; 106, 108, 118; 130, 133, 135; 150, 165, 166; 167, 169, 171; 172, 173; 490
6-Methyl-2-pyridyl	H	H	B.p. 85°/2 mm; m.p. 28-30°; acetate, b.p. 84-85°/1 mm; diphenylborinate, m.p. 176-177°; diethyl carbamate, b.p. 138-143°/4 mm; diethyl carbamate picrate, m.p. 84-85°; methyl carbamate 1-oxide, m.p. 136-138°; α-(p-chlorophenoxy)isobutyrate, HCl, m.p. 108-110°	115, 130; 165, 166; 171, 172; 495
4-Methyl-2-pyridyl	H	H	α-(p-Chlorophenoxy)isobutyrate, HCl, m.p. 174-176°	495
4,6-Dimethyl-2-pyridyl	H	H	M.p. 48-50°; phenylurethane, m.p. 122°; acetate, b.p. 80-81°/1 mm; α-(p-chlorophenoxy)isobutyrate, HCl, m.p. 169-170°	100, 115; 495
6-Ethyl-2-pyridyl	H	H	Ethyl carbonate 1-oxide, m.p. 144-145°	166, 167, 172

TABLE XIII-1. 2-Pyridinemethanols 2-PyC(OH)RR' (Continued)

Py	R	R'	Physical properties (derivatives)	Ref.
5-Ethyl-2-pyridyl	H	H	B.p. 110°/2.5 mm; acetate, b.p. 80-81°/0.1 mm	115
5-Ethyl-4-phenyl-2-pyridyl	H	H	M.p. 69-71°; acetate, b.p. 160-165°/4 mm; acetate, $n_D^{19.5}$ 1.5650; acetate picrate, m.p. 144.5-145.5°	101
6-Chloro-2-pyridyl	H	H	Methyl carbamate, m.p. 54-56°	166, 167
4-Chloro-2-pyridyl	H	H	Methyl carbamate, m.p. 129-131°; b.p. 136-137°/1 mm; picrate 137-139°	168
4-Methoxy-2-pyridyl	H	H	B.p. 115-120°/1.2 mm; picrate, m.p. 140-142°; acetate, b.p. 118-122°/2 mm; methyl carbamate, m.p. 85-87°; 1-oxide, m.p. 170-172°	103, 168
4-Ethoxy-2-pyridyl	H	H	Methyl carbamate, m.p. 66-68°; picrate, m.p. 132-134°	168
3-Hydroxy-2-pyridyl	H	H	3-Benzyl ether, m.p. 80-81°; u.v.; i.r.	138, 139
6-Cyano-2-pyridyl	H	H	Methyl carbamate, m.p. 90-91°; phenyl carbamate, m.p. 124-125°	166, 167
5-Acetyl-2-pyridyl	H	H	Methyl carbamate, m.p. 93°; ethyl carbamate, m.p. 89°	170, 501
6-(5-Nitrofuryl)ethylene-2-pyridyl	H	H	M.p. 170-171°; acetate; m.p. 120-121°; propionate, m.p. 107-108°	104
2-Pyridyl	H	Ph	M.p. 75-76°; acetate, b.p. 135-139°/1.5 mm; b.p. 163°/9 mm; acetate, m.p. 46.5-47°; acetate, b.p. 139-140°/1.6 mm; acetate, n_D^{20} 1.5602; picrate, m.p. 105.5-102°; u.v., n.m.r.; CH$_3$I, m.p. 171-172° (decomp.)	19, 39, 98, 107, 115
2-Py	H	2-Tolyl	M.p. 60°; b.p. 178-180°/9 mm; HCl, m.p. 179-180°	39, 40
2-Py	H	2-Ethylphenyl	M.p. 76-77°	39, 40
2-Py	H	2-isopropylphenyl	M.p. 77°	39, 40
2-Py	H	2-t-Butylphenyl	M.p. 111°	39, 40
2-Py	H	3-Tolyl	M.p. 102°; HCl, m.p. 180-181°; u.v.	39, 40
2-Py	H	4-Tolyl	M.p. 96°; b.p. 178-180°/9 mm; u.v.	39, 40

2-Py	H	2,3-Dimethylphenyl	M.p. 114–115°; b.p. 160–162°/2 mm; u.v.	39, 40
2-Py	H	2,4-Dimethylphenyl	M.p. 57°, HCl, m.p. 178–179°; u.v.	39, 40
2-Py	H	2,5-Dimethylphenyl	M.p. 85–86°; b.p. 170–172°/6 mm; u.v.	39, 40
2-Py	H	2,6-Dimethylphenyl	M.p. 63°; b.p. 175–176°/7 mm; HCl, m.p. 192–194°, u.v.	39, 40
2-Py	H	2,6-Diethylphenyl	M.p. 42°; b.p. 140–145°/2 mm; u.v.	39, 40
2-Py	H	2,6-Diisopropylphenyl	M.p. 102°; u.v.	39, 40
2-Py	H	3,4-Dimethylphenyl	M.p. 86–87°; b.p. 178–180°/8 mm; u.v.	39, 40
2-Py	H	3,5-Dimethylphenyl	M.p. 92°; u.v.	39, 40
2-Py	H	2,4,6-Trimethylphenyl	M.p. 65°; m.p. HCl, 188–189°; u.v.	39, 40
2-Py	H	2,3,5,6-Tetramethylphenyl	M.p. 104°; u.v.; HCl, m.p. 187–188°	39, 40
2-Py	H	p-Nitrophenyl	M.p. 111.5–112.5°, acetate, m.p. 72–73°; u.v.; CH₃I, m.p. 209–210° (decomp.)	98, 107
2-Py	H	3,4-Dichlorophenyl	CH₃I, m.p. 199–200° (decomp.)	98
2-Py	H	2,4-Dichlorophenyl	CH₃I, m.p. 195–196°	98
2-Py	H	Cyclopentyl	B.p. 95–97°/0.5 mm; acetate, b.p. 115–117°/0.5 mm	115
2-Py	Ph	Ph	Diphenylborinate, m.p. 216°; i.r.	165
2-Py	Ph	3,4,5-Trimethylphenyl	M.p. 137–137.5°	81
2-Py	H	3,6-di-Me-pyrazinyl	M.p. 108–109.5°; i.r., n.m.r.	54

2-(1-hydroxy-2-methyl-cyclohexyl)pyridine

B.p. 156–157°/25 mm; m.p. 42.3°   115

B.p. 112–114°/15 mm; i.r.; picrate, m.p. 131°   119

TABLE XIII-1. 2-Pyridinemethanols 2-PyC(OH)RR' (Continued)

Py	R	R'	Physical properties (derivatives)	Ref.
3-Me–2-Py	OH (cyclohexane)		B.p. 110–114°/4 mm; m.p. 74°; i.r.; picrate, m.p. 111.5°	119
3-Me–2-Py	OH, CH₃ (cyclohexane)		M.p. 94°; i.r.; picrate, m.p. 162.5°	119
4-Cl–5-Me–2-Py	OH (cyclohexane)		1-Oxide, m.p. 164°	82
4-MeO–2-Py	OH (cyclohexane)		1-Oxide, m.p. 127–128°	82
2-Py	OH, t-Bu (cyclohexane)		1-Oxide, m.p. 164°	82
4-Me–2-Py	OH (cyclohexane)		Cis, m.p. 58–60°; i.r.; trans, m.p. 98–100°, i.r.	119
5-Me–2-Py	OH (cyclohexane)		M.p. 36°; i.r.; picrate, m.p. 126.5°, 1-oxide, m.p. 115°	82, 119
Ph, OH, 2-Py / C=2-Py, Ph			B.p. 152°/12 mm; m.p. 57.5–59°; i.r.; picrate, m.p. 146–146.5°	86
			M.p. 164–170°; 1,1'-dioxide, m.p. 99–100°	90, 92

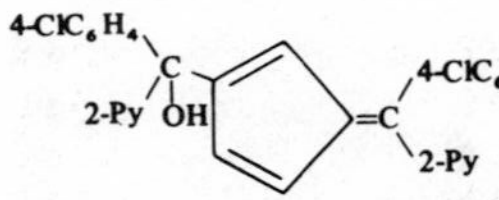

27

M.p. 90-108°   90

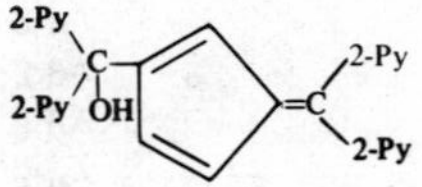

M.p. 146.5-147.5°   90, 91

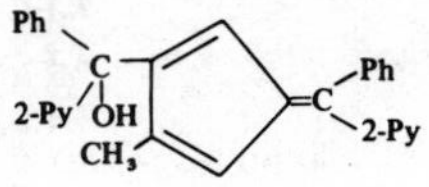

α-Isomer, m.p. 167-168°; β-isomer, m.p. 147-148°, 1,1'-dioxide, m.p. 220°   90

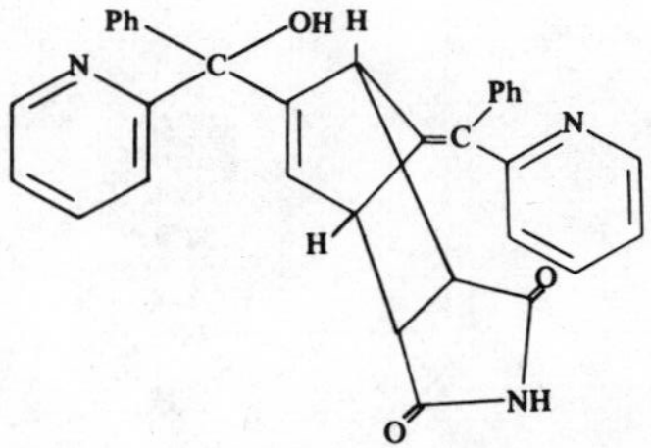

M.p. 190-198°; several isomers   92

TABLE XIII-2. 3-Pyridinemethanols 3-PyC(OH)RR'

Py	R	R'	Physical properties (derivatives)	Ref.
3-Py	H	H	B.p. 117.5–118°; picrate, m.p. 162.2–162.7°; phosphate, m.p. 194.5–195.5°; acetate, HCl, m.p. 96°; pK_a; CH_3I, m.p. 82.5°; C_2H_5I, m.p. 57°; acetate, MeI, m.p. 130°, methyl carbamate, m.p. 42°; b.p. 135–138°/1 mm; methyl carbamate, m.p. 132–134°; ethyl carbamate, m.p. 43°; phenyl carbamate, m.p. 136°; p-tolyl carbamate, m.p. 136°; p-chlorophenyl carbamate, m.p. 180–182°; p-anisyl carbamate, m.p. 148–150°; 3,5-dimethyloxyphenyl carbamate, m.p. 152–154°; 3,4,5-trimethoxyphenyl carbamate, m.p. 150°; 3-pyridyl carbamate, m.p. 150° cyclopropanecarboxylate, b.p. 90–95°/0.2–3 mm α-(p-chlorophenoxy) isobutyrate, HCl, m.p. 114°	20, 26, 27; 130, 150; 166, 167; 169, 173; 155, 490; 494; 495; 104
6-(5-Nitrofuryl)-vinyl-3-pyridyl	H	H	M.p. 150–151°; 1-oxide, m.p. 210–211°; acetate, m.p. 127–128°; 1-oxide, m.p. 178–179°	8
3-Py	H	2-Thiazoyl	M.p. 75–77.65°	43, 496
3-Py	Ph	Ph	M.p. 117–118°; 1-oxide, m.p. 140°	43, 496
3-Py	Ph	4-CH_3 S-Phenyl	M.p. 151–153°; HCl; m.p. 176–178°	43, 496
3-Py	Ph	Cyclobutyl	M.p. 125–126°	43, 496
3-Py	Ph	4-ClC_6H_4	M.p. 102°; HCl, m.p. 175–176°	43, 45, 496
3-Py	4-ClC_4H_6	4-ClC_6H_4	M.p. HCl, 215–218°; 1-oxide, m.p. 140°	45
3-Py	2-ClC_6H_4	2,4-$Cl_2C_6H_3$	M.p. 160–161°	43, 496
3-Py	4-ClC_6H_4	3,4-$Cl_2C_6H_3$	M.p. 149–150°	45
3-Py	4-Tolyl	p-Tolyl	M.p. 128–130°	43, 496
3-Py	3,4-$Cl_2C_6H_3$	3,4-$Cl_2C_6H_3$	M.p. 155–157°	43, 496
3-Py	2,4-$Cl_2C_6H_3$	2,4-$Cl_2C_6H_3$	HCl, m.p. 200–205°	43, 45, 496

3-Py	3-ClC_6H_4	3-ClC_6H_4	M.p. 186-190°	43, 496
3-Py	4-ClC_6H_4	p-Tolyl	M.p. 163-164°	43, 496
3-Py	Ph	p-Anisyl	M.p. 117.5-119°	45
3-Py	3-BrC_6H_4	3-BrC_6H_4	M.p. 169-171°	43, 45, 496
3-Py	Ph	Ph	M.p. 117-118°	45
3-Py	Ph	2-Thienyl	M.p. 117°; HCl, m.p. 172°	43, 45, 496
3-Py	Ph	Cyclopropyl	M.p. 125-126°	45
3-Py	4-ClC_6H_4	Cyclopropyl	M.p. 60°; HCl, m.p. 130-131°	43, 496
3-Py	3-$CF_3C_6H_4$	3-$CF_3C_6H_4$	M.p. 132-134°; HCl, m.p. 176-178°	43, 496
3-Py	3-ClC_6H_4	3-ClC_6H_4	M.p. 83-85°; HCl, m.p. 149°	43, 45, 496
3-Py	4-ClC_6H_4	Cyclopropyl	HCl, m.p. 149°	43, 45, 496
3-Py	Cyclopropyl	Cyclopropyl	B.p. 110-115°/3 mm, m.p. 97-99°; i.r.; picrate, m.p. 163-164°	43, 45, 496
3-Py	2-CH_3—cyclohexyl	H	B.p. 170-190°/20-25 mm, m.p. 81-82.5°; i.r.; picrate, m.p. 152-153°	83
3-Py	4-CH_3—cyclohexyl	H	M.p. 127-128°	83
5-CH_3-3-Py	Cyclohexyl	2-Thiazoyl	M.p. 144-144.5°; i.r.	84
4-Phenyl-3-pyridyl	Cyclohexyl	H	B.p. 136°/0.24 mm; m.p. 130.5-131.5°; i.r;	8
4-Phenyl-3-pyridyl	2-Methylcyclohexyl	H	picrate, m.p. 135-136°	86
4-Phenyl-3-pyridyl	H	Ph	M.p. 148-149°	12
4-Phenyl-3-pyridyl	H	Ph	M.p. 131-132°	12
5-F-3-Py	H	H	B.p. 83-85°/0.05-0.01 mm; HCl, m.p. 162°	23
3-Py	p-Anisyl	Isopropyl	M.p. 129-130°	491
3-Py	p-Anisyl	s-Butyl	B.p. 147-152°/0.08 mm	491

TABLE XIII-3. 4-Pyridinemethanols, 4-PyCOHRR'

Py	R	R'	Physical properties (derivatives)	Ref.
4-Py	H	H	B.p. 124–128°/6 mm; m.p. 51–55°; picrate, m.p. 156.5–158°; acetate, b.p. 99–107°/6 mm; acetate, n_D^{20} 1.5030°; phosphate, m.p. 201.5–202.5°; CH_3I, m.p. 97°; EtI, m.p. 84°; acetate, CH_3I, m.p. 142°; acetate, EtI, m.p. 113°; methyl carbamate-1-oxide, m.p. 116–118°, ^{18}O acetate, pK_a; phenylacetate, b.p. 134–136°/3 mm; isobutyrate, b.p. 122°/3.75 min; isobutyrate, n_D^{20} 1.4917, picrate, m.p. 127°; pivalate, n_D^{20} 1.4862, picrate, m.p. 142.3°; n.m.r.; *carbamates:* H, m.p. 122.4°, 130°; methyl, b.p. 130–135°/4 mm; picrate, m.p. 148–149°; 1-oxide, m.p. 116–118°; phenyl, m.p. 126°; p-tolyl, m.p. 143°; p-chlorophenyl, m.p. 169–170°; p-fluorophenyl, m.p. 148°; p-anisyl, m.p. 114°; 3,5-dimethoxyphenyl, m.p. 186°; 3,4,5-tri-methyloxyphenyl, m.p. 147°; 4-pyridyl, m.p. 182°; 3-pyridyl, n_D^{20} 1.5410, m.p. 8–9.5°;	20, 105, 108 109, 110, 130 150, 166, 169 187
4-Py	H		α-(p-Chlorophenoxy)isobutyrate, HCl, m.p. 165–166°	495
4-Py	H	Ph	Furoate, m.p. 70–72.5°, b.p. 155–160°/0.15 mm; nicotinate, m.p. 56–58°; di-HBr, m.p. 162–165° thenoate, b.p. 175–185°/0.15 mm; chrysanthemum monocarboxyate, oil	159
4-Py	H	2-Thiazolyl	M.p. 132–134°; HCl, m.p. 182–184°; MeI, m.p. 187–189°; diacetate, m.p. 127–128.5°; benzyl ether, m.p. 212–213°;	9

4-Py	Ph	2-Thiazolyl	M.p. 211.5–212.5°	9
4-Py	2-Me-cyclohexyl	H	B.p. 170–90°/18–25 mm; i.r.; picrate, m.p. 163–164°	83
4-Py	4-Me-cyclohexyl	H	B.p. 175–190°/20 mm; m.p. 92–94°; i.r.; picrate, m.p. 212.5–213°	83
4-Py	Cyclohexyl	2-Thiazolyl	M.p. 188–189°	8
3-CH₃,4-Py	Cyclohexyl	H	B.p. 205–209°/60; m.p. 95°; i.r.	84
4-Py	Cyclohexyl	H	M.p. 130.5–131°	80
3-CH₃,4-Py	2-Me-cyclohexyl	H	B.p. 55–60°/0.5 mm; m.p. 105–106°; i.r.; picrate, m.p. 163–164°	86
2,6-Dimethyl4-pyridyl	H	H	M.p. 98°; acetate, b.p. 81–82°/1 mm; phenyl urethane, m.p. 157°	100
4-Py	Piperidinyl	H	M.p. 134–136°	80
4-Py	3-Quinuclidinyl		M.p. 179–180°	80
4-Py	Dibenz(α,δ)-1,4,cyclo-heptadienyl		M.p. 175–176°	80

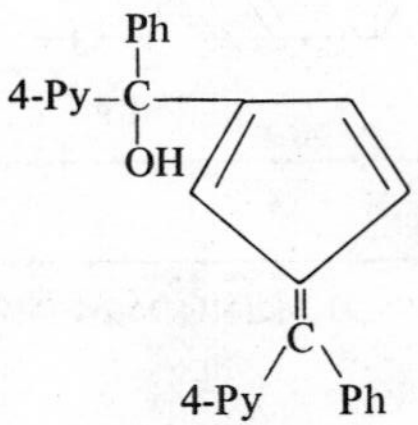

31

4-Py—C
Ph
OH
4-Py—C
Ph

M.p. 206–209°   90, 92

TABLE XIII-3. 4-Pyridinemethanols, 4-PyCOHRR' (Continued)

Py	R	R'	Physical properties (derivatives)	Ref.
			1,1'-Dioxide, m.p. 236-238°	90
4-Py-cyclopropyl	Ph	Ph	M.p. 160-162°; HCl, m.p. 236°	498

TABLE XIII-4. α-(2-Pyridyl)ethanols 2-PyC(OH)RCHR'R"

Py	R	R'	R"	Physical properties (derivatives)	Ref.
2-Py	H	H	H	Acetate, b.p. 111–116°/14 mm; n_D^{20} 1.4925; methylcarbamate-1-oxide, picrate, m.p. 113–114°; phosphotrithioate, 165° (decomp.); bisphosphotrithioate, m.p. 40–45°	129, 166; 172; 190; 164
2-Py	Ph	H	H	B.p. 135–137°/0.7 mm; i.r.; diphenyl-borinate, m.p. 233–235°	115; 165
2-Py	H	Ph	H	M.p. 160–161°	16
2-Py	H	Ph	NO	M.p. 99–100°; acetate, b.p. 140°/1 mm	115
2-Py	Ph	4-ClC$_6$H$_4$	H	M.p. 145–147°	53
2-Py	Ph	4-Py	H	M.p. 120–121°	53
2-Py	H	Ph	Br	M.p. 112–115°	10
2-Py	H	Ph	Cl	M.p. 120–122°	10
2-Py	Ph	Ph	H	HCl, m.p. 198–202°; acetate 125–132°; propionate 154–156°	46
2-Py	H	Br	Ph	Threo, m.p. 108–110°, HBr, m.p. 159–162°; erythro, m.p. 112–115°	10
2-Py	3-F$_3$CC$_6$H$_4$	Ph	H	HCl, m.p. 192–200°; propionate, m.p. 92–94°	46
2-Py	Ph	4-F$_3$CC$_6$H$_4$	H		46

33

TABLE XIII-5. α-(3-Pyridyl)ethanols 3-PyCOHRCHR'R'' (*Continued*)

Py	R	R'	R''	Physical properties (derivatives)	Ref.
2-Py	$3\text{-}F_3CC_6H_4$	$2\text{-}F_3CC_6H_4$	H		46
2-Py	H	Cl	Ph	*Threo*, m.p. 96–97°; HCl, m.p. 181–184°; *erythro*, m.p. 120–122°	10
2-Py	$3,6\text{-di-}CH_3\text{-pyrazinyl}$	H	H	M.p. 73–73.5°; i.r., n.m.r.	54
6-Me-2-Py	H	H	H	B.p. 96°/3 mm; pK_a	130

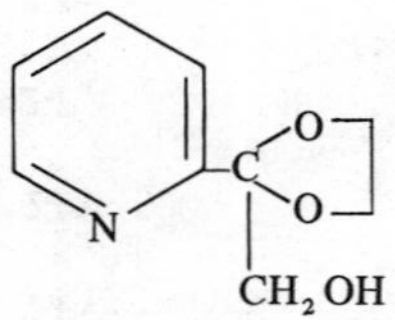

34

M.p. 120–121°

497

TABLE XIII-5. α-(3-Pyridyl)ethanols 3-PyCOHRCHR′R″

Py	R	R′	R″	Physical properties (derivatives)	Ref.
3-Py	H	H	H	B.p. 117–120°/2 mm; 106–107°/5–7 mm	3, 129
3-Py	H	H	Ph	B.p. 175–180°/0.3 mm, m.p. 105°	44
3-Py	H	$2,4\text{-}Cl_2C_6H_3$	Ph	M.p. 145–148°	43, 496
3-Py	H	$4\text{-}ClC_6H_4$	H	M.p. 132–133°	43, 496
3-Py	Ph	4-Pyridyl	H	M.p. 110°	53
3-Py	2-Thiazolyl	H	H	M.p. 116–117°	8
3-Py	Ph	H	1-Naphthyl	M.p. 223–227°	43, 496
3-Py	Ph	H	H	M.p. 134°	43, 45, 496
3-Py	Ph	$2,4\text{-}Cl_2C_6H_3$	H	M.p. 177–178°	43, 496
3-Py	$4\text{-}ClC_6H_4$	$3\text{-}BrC_6H_4$	H	M.p. 133.5–135°	43, 45, 496
3-Py	$4\text{-}BrC_6H_4$	$4\text{-}ClC_6H_4$	H	M.p. 170–172°	43, 496
3-Py	Ph	$4\text{-}ClC_6H_4$	H	M.p. 116–117°	43
3-Py	$3\text{-}ClC_6H_4$	$4\text{-}ClC_6H_4$	H	M.p. 132–133°	43, 45, 496
3-Py	$4\text{-}ClC_6H_4$	Ph	H	M.p. 130–131.5°	43, 45, 496

TABLE XIII-5. α-(3-Pyridyl)ethanols 3-PyCOHRCHR'R" *(Continued)*

Py	R	R'	R"	Physical properties (derivatives)	Ref.
3-Py	4-ClC$_6$H$_4$	2,4-Cl$_2$C$_6$H$_3$	H	M.p. 174–175°; HCl, m.p. 190–195° sulfate, m.p. 151–152°; nitrate, m.p. 172–174°; p-toluenesulfonate, m.p. 215–217°; methanesulfonate, m.p. 199–202°; 1-oxide, m.p. 156–160°	45
3-Py	3-ClC$_6$H$_4$	2,4-Cl$_2$C$_6$H$_3$		M.p. 145–148°	45
3-Py	Ph	R'R"H≡F$_3$		M.p. 127–128°	43, 496
3-Py	2-ClC$_6$H$_4$	2,4-Cl$_2$C$_6$H$_3$		M.p. 160–161°; HCl, m.p. 198–200°	45
3-Py	Ph	H	H	M.p. 76–77°	43, 496
3-Py	Ph	2-Py	H	M.p. 238–239°	43, 496
3-Py	Ph	4-Py	H	M.p. 132°	43, 45, 496
3-Py	4-ClC$_6$H$_4$	2-Py	H	M.p. 101°	43, 45, 496
5-Me-3-Py	4-ClC$_6$H$_4$	2-Py	H	M.p. 131°	43, 45, 496
3-Py	H	2-Py	NO	M.p. 156–158°	16
3-Py	H	Ph	NO		
2-Me-3-Py	H	H	H	B.p. 135°/0.2 mm	3

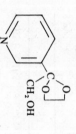

36

M.p. 71–72°   497

TABLE XIII-6. α-(4-Pyridyl)ethanols 4-PyC(OH)R CHR'R''

Py	R	R'	R''	Physical properties (derivatives)	Ref.
4-Py	H	H	H	Acetate, b.p. 124–131°/26 mm; n_D^{20} 1.4972	190
4-Py	Ph	H	H	dextro $[\alpha]_D^{31}$ 1.567° (c, 9.45, MeOH), 3.08° (c, 0.3982, CHCl$_3$), 2.114° (c, 0.62, benzene), 1.13° (c, 7.13, ethanol), 1.765° (c, 0.34, toluene) laevo, m.p. 145°; $[\alpha]_D^{31}$ −0.2° (c, 5.3, CHCl$_3$)	49
4-Py	H	Ph	NO	M.p. 167–170°	16
4-Py	Ph	4-ClC$_6$H$_4$	H	M.p. 177–178°	53
4-Py	2-Thiazoyl	H	H	M.p. 173–174°	9

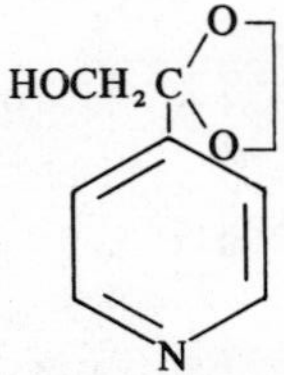

M.p. 75–76°

497

TABLE XIII-7. β-(2-Pyridyl)ethanols 2-PyCHRC(OH)R'R''

Py	R	R'	R''	Physical properties (derivatives)	Ref.
2-Py	H		H	1-Oxide, m.p. 93-95°, picrate, m.p. 96-98°; b.p. 98-130°/5 mm; pK_a; N-phenylcarbamate, m.p. 121-122°; methyl carbamate, n.p. 150-156°/0.5 mm; 1-oxide m.p. 150-152°; diphenyl-borinate, n.p. 157-158°; carbonate, b.p. 124-125°/0.7 mm; n_D^{20} 1.4855; picrate, m.p. 120-121°	16, 18, 31, 33-36 / 142 / 165 / 174
2-Py	Br	Ph	H	Threo, m.p. 90-91°; erythro, m.p. 85-88°	10
2-Py	Cl	Ph	H	Threo, m.p. 148-150°; erythro, m.p. 82-86°	
2-Py	H	Ph	H	M.p. 106-108°; or 110-111°; i.r.; picrate, m.p. 116°	10, 96
2-Py–CH₂ (1-hydroxycyclohexyl)	H		H	M.p. 40-42°; i.r.; perchlorate, m.p. 192-193°	96
2-Py–CH₂ (1-hydroxycyclopentyl)				B.p. 95°/0.7 mm; acetate, b.p. 115-117.5°/0.5 mm	115
3-Et-2-Py	H	H	H	B.p. 80-82°/0.07 mm	38
5-Et-2-Py	H	H	H	B.p. 118-120°/3 mm; n_D^{20} 1.5365	28, 34
4-Me-2-Py	H	H	H	B.p. 87-88°/0.5 mm; m.p. 39-40°, picrate, m.p. 119-120°; picrolonate, m.p. 175°; chloroplatinate, m.p. 164-165°	29, 63
4,6-Me₂-2-Py	H	H	H	M.p. 62-63°; b.p. 71°/0.05 mm; picrate, m.p. 129-129.5°; picrolonate, m.p. 171-172°	29

4-Me-2-Py	H	Ph	H	M.p. 92.1–92.3°	63
4,6-Me$_2$-2-Py	H	Ph	H	HgCl$_2$·HCl, m.p. 98–99°; H$_2$PtCl$_6$, m.p. 126.9°; HCl, m.p. 208–209°	63
6-Me-2-Py	H	H	H	B.p. 120°/3 mm; pK$_a$	130
2-Py	Ph	H	2-ClC$_6$H$_4$	M.p. 122–125°	71
2-Py	Ph	H	4-ClC$_6$H$_4$	M.p. 168–172°	71
2-Py	Ph	H	Ph	M.p. 144–146°	71
2-Py	Ph	H	3-Cyclohexenyl	M.p. 123–129°	71
2-Py	Ph	H	Cyclohexyl	M.p. 102–104°	71
2-Py	Ph	H	2-CH$_3$OC$_6$H$_4$	M.p. 126–129°	71
2-Py	Ph	H	3-CH$_3$OC$_6$H$_4$	M.p. 136–137°	71
2-Py	Ph	H	4-CH$_3$OC$_6$H$_4$	M.p. 160–162°	71
2-Py	Ph	(bicyclic structure, OH)		M.p. 115–118°	71
2-Py	Ph	H	4-(CH$_3$)$_2$NC$_6$H$_4$	M.p. 156–159°	71
2-Py	H	(tetrahydronaphthalenol structure, OH)		M.p. 107–112°	71

TABLE XIII-7. *β*-(2-Pyridyl)ethanols 2-PyCHRC(OH)R'R'' (*Continued*)

Py	R	R'	R''	Physical properties (derivatives)	Ref.
2-Py	H	H		M.p. 126–127°	71
2-Py	H	H		M.p. 130–131°	71
2-Py	H	H		M.p. 235–237°	71
2-Py	Ph	Ph	2-Thienyl	M.p. 153–159°	71

TABLE XIII-8. β-(4-Pyridyl)ethanols 4-PyCHRC(OH)R'R''

Py	R	R'	R''	Physical constant (derivatives)	Ref.
4-Py	H	H	H	M.p. 59–61°; acetate, b.p. 117–120°/15 mm; i.r.	15
4-Py	H	Cyclopropyl	Cyclopropyl	M.p. 51–53°; b.p. 135–141°/0.2 mm; HCl, m.p. 140–141°; i.r.	146
4-Py	H	Cyclopropyl	Ph	M.p. 98–99°; HCl, m.p. 189.5–190°; i.r.	146
4-Py	H	Ph	p-Tolyl	M.p. 159–160°	53
4-Py	H	p-Tolyl	p-Tolyl	M.p. 154–156°	53
3-Me-4-Py	H	p-Tolyl	Ph	M.p. 171°	53
2-Me-4-Py	H	H	H	M.p. 171–172°	53
2-Me-4-Py	H	Ph	H	B.p. 97–98°/0.5 mm; n_D^{23} 1.5370; picrate, m.p. 84–85°; picrolonate, m.p. 167°; chloroplatinate, m.p. 179°	29
2-Me-4-Py	Ph	H	H	M.p. 201–202°	53
2-Vinyl4-Py	H	H	H	B.p. 93–97°/0.5 mm; n_D^{20} 1.5530; picrate, m.p. 122°; picrate, m.p. 188–189°	29
4-Py	H	2-Thienyl	Ph	M.p. 153–154°; HCl, m.p. 238–241°	80
4-Py	H	$4\text{-}(CH_3)_2NC_6H_4$	$4\text{-}(CH_3)_2NC_6H_4$	M.p. 211–213°	80
4-Py	H	Ph	Piperidinyl	M.p. 125–128°; HCl, m.p. 168–178°	76
4-Py	H	$3\text{-}FC_6H_4$	H	M.p. 69–72°	7
4-Py	H	p-Anisyl	H	M.p. 138–139°	7
4-Py	H	2-Thienyl	H	M.p. 95–96.5°	7

TABLE XIII-9. 2-Pyridine Propanols

Compound	Physical properties (derivatives)	Ref.
2-PyCH$_2$CH$_2$CH$_2$OH	B.p. 106°/3 mm; pK$_a$; i.r.; 2-(2,4,5-trichlorophenoxy) propionate, m.p. 38–40.5°; 1-oxide, m.p. 52–54°; diphenyl-borinate, m.p. 119–120°	130, 156 165, 207
2-PyCH$_2$COHCH$_3$Ph	M.p. 30–38°; i.r.; picrolonate, m.p. 184–185°; picrate, m.p. 145–150°	96
2-PyCH$_2$CHOHCH$_3$	B.p. 80°/0.5 mm	115
3-Me-2-PyCH$_2$CH$_2$CH$_2$OH	B.p. 125.6°/0.2 mm; 2,4,6-trinitrobenzene sulfonate, m.p. 135–136°	66
2-PyCHOHCH$_2$CH$_3$	B.p. 70–72°/0.5 mm; acetate, b.p. 80–82°/1 mm	115
2-PyPhCHCOHCH$_3$(2-thienyl)	M.p. 93–99°	71
2-PyPhCHCOHCH$_3$(3-pyridyl)	M.p. 95–97°	71
2-PyPhCHCOHCH$_3$(4-pyridyl)	M.p. 118–119°	71
2-PyPhCHCOHCH$_3$(3-cyclohexenyl)	M.p. 146–152°	71
2-PyPhCHCOHCH$_3$(2,3-Cl$_2$C$_6$H$_3$)	M.p. 129–131°; m.p. 160–164°	71
2-PyPHCHCOHCH$_3$(2,4-Cl$_2$C$_6$H$_3$)	M.p. 142–144°; m.p. 152–153°; 1-oxide, m.p. 159–160°	71
2-PyPhCHCOHCH$_3$(2,5-Cl$_2$C$_6$H$_3$)	M.p. 125–126°; 1-oxide, m.p. 166–167°	71
2-PyPhCHCOHCH$_3$(3-BrC$_6$H$_4$)	M.p. 116–119°; m.p. 158–160°	71
2-PyPhCHCOHCH$_3$(2-ClC$_6$H$_4$)	M.p. 112–115°	71

2-PyPhCHCOHCH₃(3-ClC₆H₄)	M.p. 108–110°; m.p. 134–136°; 1-oxide, m.p. 168–170°	71
2-PyPhCHCOHCH₃(4-BrC₆H₄)	M.p. 122–124°; 1-oxide, m.p. 163–165°	71
2-PyPhCHCOHCH₃(4-ClC₆H₄)	M.p. 152–154°	71
2-Py(4-ClC₆H₄)CHCOHCH₃(4-ClC₆H₄)	M.p. 135–140°; m.p. 140–145°	71
2-PyPhCHCOHCH₃(2-FC₆H₄)	M.p. 103–105°	71
2-PyPhCHCOHCH₃(3-FC₆H₄)	M.p. 97.5–100°; m.p. 107–110°	71
2-PyPhCHCOHCH₃(4-FC₆H₄)	M.p. 116–118°	71
2-PyPhCHCOHCH₃Ph	M.p. 106–107°; m.p. 133–135°	71
2-PyPhCHCOHCH₃(3-HOC₆H₄)	M.p. 125.5–126.5°	71
2-PyPhCHCOHCH₃(4-HOC₆H₄)	M.p. 139–140°	71
2-Py(2-CH₃OC₆H₄)CHCOHCH₃(2,4-Cl₂C₆H₃)	M.p. 139–142°; m.p. 165–167°	71
2-Py(4-tolyl)CHCOHCH₃(3-ClC₆H₄)	M.p. 125–127°; m.p. 142–143.5°	71
2-Py(4-ClC₆H₄)CHCOHCH₃(4-tolyl)	M.p. 148–150°	71
2-Py(4-ClC₆H₄)CHCOHCH₃(4-anisyl)	M.p. 135–137°	71
2-Py(2-anisyl)CHCOHCH₃(3-ClC₆H₄)	M.p. 155–156°; m.p. 141.5–143°	71
2-PyPhCHCOHCH₃(2-tolyl)	M.p. 79–83°	71
2-PyPhCHCOHCH₃(3-tolyl)	M.p. 108–110.5°	71

TABLE XIII-9. 2-Pyridine Propanols (*Continued*)

Compound	Physical properties (derivatives)	Ref.
2-PyPhCHCHCOHCH₃ (4-tolyl)	M.p. 127-129°; m.p. 130-131°	71
2-PyPhCHCHCOHCH₃ (2-CH₃SC₆H₄)	M.p. 141-143°; m.p. 98-101°	71
2-PyPhCHCHCOHCH₃ (4-CH₃SC₆H₄)	M.p. 95-98°; m.p. 145-149°	71
2-PyPhCHCOHCH₂ (2-anisyl)	M.p. 109-110°; m.p. 140-142°	71
2-PyPhCHCOHCH₃ (3-anisyl)	M.p. 97-98°; m.p. 121-123°; 1-oxide, m.p. 156-157°	71
2-PyPhCHCOHCH₃ (4-anisyl)	M.p. 135-137°; m.p. 146-151°; 1-oxide, m.p. 143-145°	71
2-PyPhCHCOHCH₂ OCH₃ (Ph)	M.p. 130-133°; 1-oxide, m.p. 148°, 156-158°	71
2-PyPhCHCOHCH₃ (4-C₆H₄SO₂CH₃)	M.p. 148-152°	71
2-PyPhCHCOHCH₃ (4-CH₃CONHC₆H₄)	M.p. 164-167°	71
2-PyPhCHCOHCH₃ (2,4-Me₂C₆H₃)	M.p. 148-149°	71
2-PyPhCHCOHCH₃ (3,4-Me₂C₆H₃)	M.p. 108-109°	71
2-PyPhCHCOHCH₃ (2-C₂H₅OC₆H₄)	M.p. 112-115°	71
2-PyPhCHCOHCH₃ [2,5-(MeO)₂C₆H₃]	M.p. 128-134°	71
2-PyPhCHCOHCH₃ [4-(CH₃)₂NSOC₆H₄]	M.p. 166-168°	71
2-PyPhCHCOHCH₃ [4-(CH₃)₂NCH₂CH₂OC₆H₄]	M.p. 123-125°	71

44

Compound		Reference
2-PyPhCHCOHCH$_3$(3-OC$_5$H$_9$OC$_6$H$_4$)[a]	M.p. 135.5–136°	71
2-PyPhCHCOHCH$_3$(4-OC$_5$H$_9$OC$_6$H$_4$)[a]	M.p. 134.5–135.5°; m.p. 165.5–166°	71
2-PyPhCHCOHCH$_3$[4-(CH$_3$)$_2$N(CH$_2$)$_3$OC$_6$H$_4$][a]	M.p. 93.5	71
2-PyPhCHCOHCH$_3$(4-C$_6$H$_4$·C$_6$H$_5$)	M.p. 157–159°	71
2-PyPhCHCOHCF$_3$Ph	M.p. 110–112°; m.p. 145–146.5°	70
2-Py(4-tolyl)CHCOHCF$_3$Ph	M.p. 193–195°	70
2-PyPhCHCOHCF$_3$Ph	M.p. 190–190.5°; HCl, m.p. 151–153°	70
2-Py(4-ClC$_6$H$_4$)CHCOHCF$_3$Ph	M.p. 174–175°	70
2-PyPhCHCOHCF$_3$(2-ClC$_6$H$_4$)	M.p. 133–134.5°; m.p. 157–158°	70
2-PyPhCHCOHCF$_3$(3-ClC$_6$H$_4$)	M.p. 169–170°	70
2-PyPhCHCOHCF$_3$(4-ClC$_6$H$_4$)	M.p. 159–161°	70
2-PyPhCHCOHCF$_3$(4-tolyl)	M.p. 141–142°; m.p. 180–182°	70
2-PyPhCHCOHCF$_3$(4-anisyl)	M.p. 113–115°; m.p. 152–153°	70
2-PyPhCHCOHCF$_3$(4-C$_6$H$_4$C$_6$H$_5$)	M.p. 183–184°	70
2-PyCH$_2$COHCH$_3$(4-anisyl)	M.p. 52–53.5°	70
2-PyCH(OH)CHNO$_2$CH$_3$	n_D^{20} 1.5922; HCl, m.p. 142–146°	94

TABLE XIII-9. 2-Pyridine Propanols (*Continued*)

Compound	Physical properties (derivatives)	Ref.
	M.p. 84.5–86°; u.v.; acetate, b.p. 94°/0.3 mm; n_D^{21} 1.5260	102
2-PyCH(CH$_2$Ph)CH$_2$OH	B.p. 175°/10⁻⁴ mm	22
2-PyCH(4-ClC$_6$H$_4$CH$_2$)CH$_2$OH	B.p. 165–170°/10⁻⁵ mm	22
2-PyCH[3,4-(MeO)$_2$C$_6$H$_3$CH$_2$]CH$_2$OH	B.p. 210°/10⁻⁴ mm	22
2-PyC(OH)(CH$_3$)CH$_2$-4-Py	M.p. 56–58°; b.p. 152–153°/0.6 mm	53
	M.p. 85–86°; b.p. 175–180°/0.8 mm; HCl, m.p. 152–153°; acetate, b.p. 136–138°/0.2 mm; acetate, HCl, m.p. 181–182.5°	65
	B.p. 194–200°/0.6 mm; HCl, m.p. 165–166.5°	65

2-Py–CH₂

2-Py–CH₂ [structure, OH]	Cis, m.p. 149-150°; trans, m.p. 247-249°; u.v.	65
2-PyCHPhCOHCH₃ (3-CH₃OC₆H₄)	α, m.p. 121-123°; β, m.p. 97-98°, 1-oxide, m.p. 156-157°	69
2-PyCHPhCOHCH₃Ph	α, m.p. 133-135°; 1-oxide, m.p. 152-154°; β, m.p. 107-108°; 1-oxide, m.p. 148°	69

^{a}Tetrahydropyranyl.

TABLE XIII-10. 3-Pyridine Propanols

Compound	Physical properties (derivatives)	Ref.
3-PyCH$_2$CH$_2$CH$_2$OH	B.p. 126°/3 mm; pK_a	130
3-PyCH$_2$COHCH$_3$ (4-anisyl)	M.p. 90–91°	71
3-PyCHOHCHOH(NO$_2$)CH$_3$	n_D^{21} 1.4520; b.p. 180–182°/0.3 mm; HCl, m.p. 154–156°	94
	B.p. 76–78°/0.25 mm; picrate, m.p. 144.5–145.5°	38
3-PyPhCOHC$_2$H$_5$	M.p. 98–99°	43
3-Py(4-ClC$_6$H$_4$)COHC≡CH	M.p. 111–112°	43
3-PyCOHCH$_2$Ph(4-ClC$_6$H$_4$CH$_2$)	M.p. 121–123°; HCl, m.p. 180–183°	43, 45
3-PyCOH(4-ClC$_6$H$_4$CH$_2$)(2,4-Cl$_2$C$_6$H$_3$CH$_2$)	M.p. 154–156°; HCl, m.p. 198–200°	43
3-PyCOH(CH$_2$Ph)(2,4-Cl$_2$C$_6$H$_3$CH$_2$)	M.p. 133–134°	43
3-PyCOH(CH$_3$)CH$_2$(4-ClC$_6$H$_4$)	B.p. 176–184°/0.1 mm; m.p. 84–86°	53
6-Me-3-PyCOH(CH$_3$)CH$_2$(4-ClC$_6$H$_4$)	M.p. 177–180°/0.8 mm	53
3-PyCOH(4-ClC$_6$H$_4$CH$_2$)$_2$	M.p. 125–126°; HCl, m.p. 215–218°	43
3-PyCOH(CH$_3$)CH$_2$Ph	M.p. 158°	44
3-PyCOH(CH$_3$)CH$_2$(4-ClC$_6$H$_4$)		44

48

TABLE XIII-11. 4-Pyridine Propanols

Compound	Physical properties (derivatives)	Ref.
4-PyCH$_2$CH$_2$CH$_2$OH	B.p. 121°/3 mm; pK_a	26, 37, 130
4-PyPhCHCOHCH$_3$ (4-ClC$_6$H$_4$)	M.p. 150–158°	71
4-PyPhCHCOHCH$_3$ (4-CH$_3$OC$_6$H$_4$)	M.p. 128–130°	71
4-PyCHOHCH(NO$_2$)CH$_3$	M.p. 165–166°	94
4-PyCH$_2$COHCH$_2$Ph(4-Py)	CH$_3$I, m.p. 189–191°	76
4-PyCH$_2$COH(cyclohexyl)CH$_2$-4-Py	M.p. 176–178°	76
4-PyCH$_2$COHCH$_3$Ph	M.p. 83–85°	80
4-PyCH$_2$COHCH$_3$ (2-ClC$_6$H$_4$)	M.p. 139–141°	80
4-PyCH$_2$COHCH$_3$ (4-Py)	M.p. 135–138°	80
4-Py-CH$_2$COHPhCH$_2$ (α-piperidyl)	M.p. 89–91	80
4-Py-CH$_2$C(OH)PhCH$_2$-α-morpholyl	M.p. 104–105°	80
4-Py-CH$_2$C(OH)PhCH$_2$-4-piperidyl	M.p. 79–81°	80
4-PyCH$_2$COHCH$_3$ (4-ClC$_6$H$_4$)	M.p. 121–124°	80
4-PyCH$_2$COHCF$_3$Ph	M.p. 132°; HCl, m.p. 204–205°	80
4-PyCH(CH$_2$Ph)CH$_2$OH	M.p. 43°	22
4-PyCH(CH$_2$C$_6$H$_4$NH$_2$-4)CH$_2$OH	M.p. 154°	22
4-PyCH(CH$_2$C$_6$H$_4$OCH$_3$-4)CH$_2$OH	B.p. 220–225°/10^{-4} mm	22
4-PyCH(CH$_3$)COHPh$_2$	M.p. 137–138°	53
4-PyCH$_2$COHCH$_3$ (4-Py)	B.p. dec	53
4-PyCOH(CH$_3$)CH$_2$-4-Py	M.p. 194–196°	53
4-PyCOH(CH$_3$)CH$_2$-4-ClPh	M.p. 146–148°	53
(4-PyCH$_2$)$_2$COHPh	M.p. 92°, 126–127°, 117–118°, 167.5–169°; HCl, m.p. 227–229°	73, 74 79
(4-PyCH$_2$)$_2$COH(p-tolyl)	M.p. 100–110°, 160–161°	74
(4-PyCH$_2$)$_2$COH(m-FC$_6$H$_4$)	M.p. 108–112°; 189–191°	75
(4-PyCH$_2$)$_2$COH(p-FC$_6$H$_4$)	M.p. 122.5°; m.p. 155.5–157°	75
(4-PyCH$_2$)$_2$COH(p-ClC$_6$H$_4$)	M.p. 87–102°; m.p. 181–183°	75, 77
(4-PyCH$_2$)$_2$COH[3,4,5,(CH$_3$O)$_3$C$_6$H$_2$]	M.p. 151–154°; m.p. 168–170°	77
(4-PyCH$_2$)$_2$COH(4-CH$_3$SC$_6$H$_4$)	M.p. 164–167°	77
(4-PyCH$_2$)$_2$COH(cyclohexyl)	M.p. 176–178°	78
(4-PyCH$_2$)$_2$COH(cyclopropyl)	M.p. 141–143°	78
(4-PyCH$_2$)$_2$COH(cycloheptyl)	M.p. 151–153°	78
(4-PyCH$_2$)$_2$COH(4-Py)	M.p. 217–219°	78
(4-PyCH$_2$)$_2$COH(2-Py)	M.p. 155–180°	78
(4-PyCH$_2$)$_2$COH(2-furyl)	M.p. 142–144°	78
(4-PyCH$_2$)$_2$COH(2-thienyl)	M.p. 96–99°	78
4-PyCOH(CH$_3$)$_2$	M.p. 135–136°	4

TABLE XIII-12. Higher 2-Pyridine Alcohols

Compound	Physical properties (derivatives)	Ref.
2-PyPhCHCHCOH(2-thienyl)CH(CH$_3$)$_2$	M.p. 115–116°; m.p. 137–139°	71
2-PyPhCHCOH(CH$_3$)CH(CH$_3$)$_5$	M.p. 133–139°	71
2-PyPhCHCOH(CH$_3$)CH$_2$Ph	M.p. 89.5–92°	71
2-PyPhCHCOH(CH$_3$)CH$_2$Ph	M.p. 174–176°	71
2-PyPhCHCOH(CH$_3$)CH$_2$CH$_2$(3-indolyl)	B.p. 161–163°/5 mm; b.p. 112–114°/0.2 mm; n_D^{20} 1.5222; HCl,	71
2-PyCH$_2$–1-cyclohexanol-2CH$_2$OCH$_3$	m.p. 154–155°; picrate, m.p. 140–141°; MeI, m.p. 156–158°;	59
	1-benzyloxy, b.p. 163°/0.18 mm; n_D^{20} 1.5545	
2-PyCH$_2$CHOHCH=CH$_2$	M.p. 36°; Phenylurethane, m.p. 104–105°; HBr, m.p. 97°	64
2-PyCH$_2$CHOHCH$_3$	B.p. 125–130°/18 mm; Phenylurethane, m.p. 235°	64
2-Py-cyclopropylCOHPh$_2$	trans, m.p. 126–128°; HCl, m.p. 188–189°; CH$_2$Br, m.p. 216–218°	50
2-PyCOHPh(n-octyl)	M.p. 38–40°; (+)- m.p. 38°; [α]$_D^{35}$ 3.96° (EtOH); (–)- m.p. 39°	129
	[α]$_D^{35}$ –2.64°	
5-Et-2-Py-cyclopropylCOHPh$_2$	trans, m.p. 134–135°; HCl, m.p. 190–191°.	50
2-PyCHC(2-Py)(CH$_3$)CH$_2$OH	B.p. 142–145°/0.01 mm; n_D^{21} 1.5720; dipicrate, m.p. 154–155°	21
2-PyCH$_2$CH$_2$C(2-Py)(CH$_3$)CH$_2$OH	B.p. 144–145°/0.03 mm; n_D^{20} 1.7500; dipicrate, m.p. 175–176°	21
2PyCHCH$_3$COHCF$_3$Ph	M.p. 138–140°	70
2-PyCH(CH$_2$CH$_2$CH$_3$)COHCF$_3$Ph	M.p. 129–130°	70
2-Py-CHPhCOH(C$_2$F$_5$)Ph	M.p. 166–167°	70
2-PyCHPhCOH(C$_2$H$_5$)Ph	M.p. 151–153°	70
2-PyCCH$_3$PhCOHCF$_3$Ph	M.p. 137–137.5°	70
2-PyCHPhCOHCH$_3$(2-F$_3$CC$_6$H$_4$)	M.p. 144–146°; m.p. 156–159°	70
2-PyCHPhCOHCH$_3$(3-F$_3$CC$_6$H$_4$)	M.p. 156–157°	71

Compound	Properties	Ref.
2-PyCHPhCOHC₂H₅(4-ClC₆H₄)	M.p. 126–128°; m.p. 158°	71
2-PyCHPhCOHCH(CH₃)₂(4-BrC₆H₄)	M.p. 112–116°	71
2-PyCHPhCOHCH(CH₃)₂(4-ClC₆H₄)	M.p. 118–122°	71
2-PyCHPhCOHCH(CH₃)₂(4-FC₆H₄)	M.p. 128–130°	71
2-PyCHPhCOHCH(CH₃)₂Ph	M.p. 141–142°	71
2-PyCHPh—COHC₃H₇Ph	M.p. 150–153°	71
2-PyCHPhCOHCH(CH₃)₂(4-tolyl)	M.p. 115–116°; m.p. 137.5–139.5°	71
2-PyCHPh—COHCH(CH₃)₂(4-anisyl)	M.p. 105–106°	71
2-PyCHCH₃COH(CH₃)₂	B.p. 75–76°/0.7 mm	115
2-PyCH(CH₂OH)C(CH₃)₂	B.p. 96–98°/0.8 mm	115
2-PyCH₂COHPhCHCH₃CH₂NCH₃OCH₃	B.p. 180°/0.2 mm; i.r.; dipicrate, m.p. 155°	55
2-PyCH₂COHPhCHCH₃CH₂NEtOEt	B.p. 192°/0.2 mm; i.r.; α-picrate, m.p. 147–148.5°; β-picrate, m.p. 136–138.5°; propionate, b.p. αβ 195°/0.3 mm, β, 205°/0.2 mm	55
5-(2-Picolinyl)-5-hydroxydibenzo-[a,d]cyclohepta[1,4]-diene	M.p. 113–115°	61
2-PyPhCOHCH₂C≡CH	M.p. 50°; i.r.; n.m.r.	13
2-PyCH₂CH₂CH₂OH	B.p. 98–100°/0.05 mm; CH₃I, m.p. 98–99°	66
3-Methyl-2-PyCH₂CH₂CH₂OH	B.p. 152–153°/0.5 mm; TNBS[a], m.p. 135–136°	66
2-PyCH(CH₃)CH₂CH₂OH	B.p. 112–113°/0.09 mm; TNBS[a], m.p. 137–138°	66
3-Methyl-2-PyCH(CH₃)CH₂CH₂OH	B.p. 128–130°/0.09 mm; TNBS[a], m.p. 138–139°	66
2-PyC(CH₃)₂CH₂CH₂OH	B.p. 123–125°/0.1 mm; TNBS[a], m.p. 146–148°	66
2-PyCHOCH₂CH₂CH₃	Acetate, b.p. 115–125°/9 mm; n²⁰_D 1.4891	66
2-PyCH₂CHOHCH₂CH₃	M.p. 91–91.5°	190
2-PyCH₂CH₂CHOHCH₂NO₂	M.p. α, 85.5°, β 56–58°, γ 66–72°	221
2-PyCH₂CH₂CHOHCH₂NO₂	M.p. 64–66°; b.p. 169–173°/1 mm	221
2-PyCH₂CH₂CHOHPh		221

TABLE XIII-12. Higher 2-Pyridine Alcohols (*Continued*)

Compound	Physical properties (derivatives)	Ref.
2-PyCH$_2$CHOH(CH$_2$CH$_2$-2-Py)	B.p. 170-175°/1 mm; picrate, m.p. 212°; di-HI, m.p. 209-210°	221
2-PyCH$_2$CH$_2$CHOHCH$_2$CHCH$_3$NO$_2$	M.p. 78-78.5°	221
2-Py-cyclopropyl-COH(CH$_2$=CH-CH$_2$)$_2$	M.p. 126-128°; HCl, m.p. 188-189°, CH$_3$Br, m.p. 216-218°	52
2-PyCHOHC(CH$_3$)$_3$	B.p. 69-72°/0.5 mm; acetate, b.p. 94-95°/0.9 mm	115
2-PyPhCOHCH(CH$_3$)$_2$	M.p. 90-92°, d-, [α]$_D^{30}$ 106° (c, 2.2, CHCl$_3$), 32.08° (c, 0.281, acetone), 83° (c, 1.06, benzene), 81.7° (c, 1.04, methyl acetate), 30.3° (c, 0.732, dioxane), 35° (c, 0.795, CH$_3$OH); l-, [α]$_D^{31}$ -4.4° (c, 10.15, MeOH)	49
	(c, 0.834, EtOH), 37.7° (c, 0.71,	
2-PyPhCOHCH$_2$CH$_2$CH$_3$	M.p. 61-62°	49
2-PyPhCOHC$_4$H$_9$	B.p. 180°/22 mm; n$_D^{33}$ 1.5630; d^{34} 1.2629	49
2-PyPhCOHCH$_2$CH(CH$_3$)$_2$	M.p. 74°	49
2-PyPhCOHCH$_2$CH$_2$CH(CH$_3$)$_2$	B.p. 220°/12 mm; n$_D^{33}$ 1.5540; d^{34} 1.2380	49

^{a}Trinitrobenzenesulfonate.

TABLE XIII-13. Higher 3-Pyridine Alcohols

Compound	Physical properties (derivatives)	Ref.
3-PyCOHPhCH(CH$_3$)$_2$	M.p. 128–129°	43, 45
3-PyCOH(CH$_3$)C$_2$H$_5$	B.p. 114–116°/5–7 mm	129
3-PyCOH(2-thienyl)CH$_2$CH$_2$CH$_3$	M.p. 104°	43
3-PyCOH(4-ClPh)CH(CH$_3$)$_2$	M.p. 150–151°	42
6-Me–3-Py–C≡C—COHCH$_3$CH$_2$CH$_2$CH$_3$	M.p. 30–32°; b.p. 146–150°/2 mm	87–89
6-Me–3-Py–C≡C—COHCH$_3$C(CH$_3$)$_3$	M.p. 100–101°	87–89
6-Me–3-Py–C≡C—COH(CH$_3$)Ph	M.p. 104–105°	87–89
6-Me–3-Py–C≡C—C(OH)C$_5$H$_{10}$	M.p. 116–118°; acetate, b.p. 165–168°/2 mm; m.p. 52–53°	87, 89
6-Me–3-Py–C≡C—COH(CH$_3$)$_2$	M.p. 101–102°; acetate, b.p. 124–125°/1 mm; benzoate, m.p. 60–62°	87, 89
6-Me–3-Py–C≡C—COH(CH$_3$)C$_2$H$_5$	M.p. 98–99°; picrate, m.p. 150°	87–89
6-Me–3-Py–C≡C—CHOHCH$_2$CH$_2$CH$_3$	Phenylurethane, m.p. 112–114°	87, 88
3-PyPhCHOHCH$_2$CH$_2$CH$_2$CH$_3$	Picrate, m.p. 103–104°	87
6-Me–3-Py—CHOHCH$_2$CH$_2$CH$_3$	M.p. 137°; i.r., n.m.r.	13
6-Me–3-PyCH$_2$CH$_2$COHCH$_3$C(CH$_3$)$_3$	B.p. 159°/3 mm; m.p. 82°; picrate, m.p. 149°	88
6-Me–3-PyCH$_2$CH$_2$COH(CH$_3$)(CH$_2$)$_3$CH$_3$	B.p. 166–167°/1 mm; n_D^{22} 1.5070, picrate, m.p. 106–106.5°	88
6-Me–3-PyCH$_2$CH$_2$CHOHCH$_2$CH$_2$CH$_3$	B.p. 150°/1 mm; n_D^{22} 1.5075; picrate, m.p. 110–111°	88
6-Me–3-PyCH$_2$CH$_2$COH(C$_4$H$_8$)	B.p. 175–181°/1 mm; picrate, m.p. 129–129.5°	88
6-Me–3-PyCH$_2$CH$_2$COH(C$_5$H$_{10}$)	B.p. 184°/2 mm; m.p. 56–58°; picrate, m.p. 137–138°	88
6-Me–3-PyCH$_2$CH$_2$COH(CH$_3$)$_2$	B.p. 151°/6 mm; n_D^{20} 1.5150; picrate, m.p. 149.5–150°	88
6-Me–3-PyCH$_2$CH$_2$COH(CH$_3$)C$_2$H$_5$	B.p. 150–155°/5 mm; n_D^{20} 1.5219; picrate, m.p. 115°	88
3-PyCHOHC(3-Py)(CH$_3$)$_2$		
3-PyCOHC$_2$H$_5$CH$_2$Ph	B.p. 149–151°/0.35 mm	18
3-PyCOH(4-anisyl)CH(CH$_3$)$_2$	M.p. 129–130°	44
3-PyCOH[CH$_2$CH(CH$_3$)$_2$]$_2$	B.p. 134°/0.1 mm	85

53

TABLE XIII-14. Higher 4-Pyridine Alcohols

Compound	Physical properties (derivatives)	Ref.
4-Py-cyclopropylCOH(Ph)₂	trans, m.p. 169–170°; HCl, m.p. 219–220°; MeBr, m.p. 200–202°, n.m.r.; cis, m.p. 157–158°; HCl, m.p. 196°	50, 51, 52, 146
4-Py-cyclopropylCOH(CH₂CH₂CH₃)₂	trans, m.p. 89–90°; HCl, m.p. 154–154.5°; cis, m.p. 75–77°; HCl, m.p. 173°	50
4-Py-cyclopropylCOH(o-tolyl)₂	trans, HCl, m.p. 179°	50, 52
4-Py-cyclopropylCOH(cyclohexyl)₂	trans, m.p. 151–152°; HCl, m.p. 197°	50
4-PyCHC₂H₅COH(Ph)₂	M.p. 136–137°	53
4-PyCH₂COH(CH₃)₂	B.p. 99–101°/0.7 mm; maleate, m.p. 94–98°	80
4-PyCH₂C(OH)(C₂H₅)₂	M.p. 63–64°	80
4-PyCH₂COH(4-Py)CH=CH-Ph	M.p. 188.5–189.15°	80
4-PyCH₂COH(CH₂-2-PyCH₂CH(Ph)₂	M.p. 174–176°	80
4-PyCH₂C(OH)CH₃CH[CH(CH₃)₂]C₂H₅	M.p. 102–103.5°	80
4-PyCH₂COHCH₃(CH₂OPh)	M.p. 67–80°	80
4-PyPhCOHCH₂CH₂CH₂CH₃	M.p. 110°, 119–120°; i.r.; n.m.r.	80
4-PyPhCOHCH₂C≡CH	M.p. 96°; i.r.; n.m.r.	13
4-PyCHOH—CH₂CH=C(Ph)	HCl, m.p. 216–217°; i.r.; n.m.r.	146
4-PyCH₂CH₂COH(Ph)₂	M.p. 159–160°; HCl, m.p. 182–183°; i.r.	146
4-PyCH(CH₃)CH₂COH(Ph)₂	M.p. 141–142°; i.r.	146
4-PyC≡C—COH(Ph)₂	M.p. 187–188°; MeSO₄, m.p. 164–165°; i.r.	146
4-PyCH=CH—COH(Ph)₂	trans, m.p. 178.5–180°; MeSO₄, m.p. 178°; cis, m.p. 147–149°; HCl, m.p. 168–168.5°; i.r.	146
4-PycyclopropylCOH(CH₂CH₂CH₃)₂	trans, m.p. 89–90°; HCl, m.p. 154–154.4°; cis, m.p. 75–77°; HCl, m.p. 173°	52
4-PyPhCOHCH(CH₃)₂	M.p. 136°	49
4-PyPhCOHCH₂CH₂CHCH(CH₃)₂	M.p. 117–118°	49
4-PyCOHPh(n-octyl)	M.p. 118–120°; (+)-m.p. 114° [α]$_D^{35}$ 4.2° (EtOH); (−)-m.p. 113–114°; [α]$_D^{35}$ −3.50° (EtOH)	129

54

TABLE XIII-15. Ethers of Pyridine Alcohols

Compound	Physical properties (derivatives)	Ref.
$2\text{-PyCH}_2\text{OCH}_3$	1-Oxide, b.p. 127–129°/3 mm; HCl, m.p. 102–104°; u.v.; n.m.r.	135
$2\text{-PyCH}_2\text{OCH}_2\text{-2-Py}$	MeI, m.p. 192–194°; u.v.	181
$2\text{-PyCH}_2\text{OCH}_2\text{CH}_2\text{-2-Py}$	B.p. 141–143°/1 mm; MeI, m.p. 161–162°; u.v.	181
$2\text{-PyCH}_2\text{CH}_2\text{OCH}_3$	B.p. 96°/17 mm; 1-oxide, b.p. 128°/1 mm; u.v.	67, 68, 177, 178, 191
$2\text{-PyPhCHOCH}_2\text{CH}_2\text{N(CH}_3)_2$	B.p. 155–158°/3 mm; HCl, m.p. 102–104°; u.v.	39, 40, 155
$2\text{-Py}(o\text{-tolyl})\text{CHOCH}_2\text{CH}_2\text{N(CH}_3)_2$	HCl, m.p. 114–115°; u.v.	39, 40, 179
$2\text{-Py}(2\text{-ethyl } \text{C}_6\text{H}_4)\text{CHOCH}_2\text{CH}_2\text{N(CH}_3)_2$	B.p. 160–165°/3 mm; HCl, m.p. 180–182°; u.v.	39, 40, 179
$2\text{-Py}(2\text{-isoPrC}_6\text{H}_4)\text{CHOCH}_2\text{CH}_2\text{N(CH}_3)_2$	B.p. 165–170°/2 mm; HCl, m.p. 173–174°; u.v.	39, 40, 179
$2\text{-Py}(2\text{-}t\text{-BuC}_6\text{H}_4)\text{CHOCH}_2\text{CH}_2\text{N(CH}_3)_2$	M.p. 50–51°; b.p. 155–160°/1 mm; HCl, m.p. 186–187°, u.v.	39, 40, 179
$2\text{-Py}(m\text{-tolyl})\text{CHOCH}_2\text{CH}_2\text{N(CH}_3)_2$	B.p. 142–146°/0.03 mm; HCl, m.p. 103–105°; u.v.	39, 40, 179
$2\text{-Py}(p\text{-tolyl})\text{CHOCH}_2\text{CH}_2\text{N(CH}_3)_2$	M.p. 45–46°; 133–136°/0.01 mm; HCl, m.p. 122–123°; u.v.	39, 40, 179
$2\text{-Py}(2,3\text{-Me}_2\text{C}_6\text{H}_3)\text{CHOCH}_2\text{CH}_2\text{N(CH}_3)_2$	B.p. 158–160°/3 mm; HCl, m.p. 164–167°; u.v.	39, 40, 179
$2\text{-Py}(2,4\text{-Me}_2\text{C}_6\text{H}_3)\text{CHOCH}_2\text{CH}_2\text{N(CH}_3)_2$	B.p. 170–175°/4 mm; HCl, m.p. 113–115°; u.v.	39, 40, 179
$2\text{-Py}(2,5\text{-Me}_2\text{C}_6\text{H}_3)\text{CHOCH}_2\text{CH}_2\text{N(CH}_3)_2$	B.p. 150–155°/1 mm; HCl, m.p. 140–142°; u.v.	39, 40, 179
$2\text{-Py}(2,6\text{-Me}_2\text{C}_6\text{H}_3)\text{CHOCH}_2\text{CH}_2\text{N(CH}_3)_2$	B.p. 166–167°/3 mm; HCl, m.p. 146–149°; u.v.	39, 40, 179
$2\text{-Py}(2,6\text{-Et}_2\text{C}_6\text{H}_3)\text{CHOCH}_2\text{CH}_2\text{N(CH}_3)_2$	B.p. 172–174°/2 mm; HCl, m.p. 156–157°; u.v.	39, 40, 179
$2\text{-Py}(3,4\text{-Me}_2\text{C}_6\text{H}_3)\text{CHOCH}_2\text{CH}_2\text{N(CH}_3)_2$	B.p. 160–164°/1 mm; HCl, m.p. 126–128°; u.v.	39, 40, 179
$2\text{-Py}(3,5\text{-Me}_2\text{C}_6\text{H}_3)\text{CHOCHOCH}_2\text{CH}_2\text{N(CH}_3)_2$	B.p. 150–153°/1 mm; HCl, m.p. 139–141°; u.v.	39, 40, 179
$2\text{-Py}(2,4,6\text{-Me}_3\text{C}_6\text{H}_2)\text{CHOCH}_2\text{CH}_2\text{N(CH}_3)_2$	B.p. 167–174°/2 mm; HCl, m.p. 173–174°; u.v.	39, 40, 179
$2\text{-Py}(2,3,5,6\text{-Me}_4\text{C}_6\text{H})\text{CHOHOHCH}_2\text{CH}_2\text{N(CH}_3)_2$	B.p. 175–180°/1 mm; HCl, m.p. 197°; u.v.	39, 40, 179

TABLE XIII-16. Pyridine Glycols

Compound	Physical properties (derivatives)	Ref.
2-PyCHOHCHOHPy-2	M.p. 156°; bis-MeI, m.p. 241-242°; 1,1'-dioxide, m.p. 213°	210, 215, 218
2-PyCHOHCH$_2$CHOHPy-2	Dipicrate, m.p. 180°; MeI, m.p. 173-180°; diacetate, b.p. 141-142°/0.005 mm; n_D^{21} 1.5540; dipicrate, m.p. 205-206°	218
2-PyCHOHCH$_2$OH	M.p. 98°	210
2-PyCHOHCHOHCCl$_3$	M.p. 120°	10, 210
2-PyCHOHCHOHPh	—	210
2-PyCHOHCH$_2$OH	M.p. 143°; threo, m.p. 150-151°; erythro, m.p. 135-137°	207
2-PyPhCOHCHOHPh-2	Diacetate, b.p. 115-118°/0.7 mm; n_D^{28} 1.4892	212
2-PyCH(CH$_2$CH$_2$OH)CH(4-Py)CH$_2$CH$_2$OH	M.p. 141-142°	206
5-Et-2-PyCH(CH$_2$CH$_2$OH)CH(4-Py)CH$_2$CH$_2$OH		206
2-PyCH(CH$_2$OH)$_2$	Ditosylate, m.p. 106-107°; i.r.	192
4-Me-2-PyCH(CH$_2$OH)$_2$	M.p. 84-85°; picrate, m.p. 108°; picrolonate, m.p. 169-170°; chloroplatinate, m.p. 157-158°	29
4,6-Me$_2$-2-PyCH(CH$_2$OH)$_2$	M.p. 95-95.5°; b.p. 120-121°/0.05 mm; picrate, m.p. 128-129°; picrolonate, m.p. 154-154.5°	29
4,6-Me$_2$-2-PyCHOHCHOH(4,6-Me$_2$-2-Py)	M.p. 234°	100

Compound	Properties	Ref.
6-Me-2-PyCHOHCHOHCHOH(6-Me-2-Py)	M.p. 140°; hydrate, m.p. 104°; 1-oxide, m.p. 217°; m.p. 242°	218
3-PyCHOHCHOHPy-3	M.p. 245°; 1,1'-dioxide, m.p. 264°	218
3-PyCOH(CH₃)COH(CH₃)-3-Py	M.p. 218-225°	213
3-PyCH(CH₂CH₂OH)CH₂CH₂CH₂OH)4Py		206
4-PyCHOHCHOHPy-4	Bis-MeI, m.p. 217-219°; racemate, m.p. 178°; 1,1'-dioxide, m.p. 200°; meso, m.p. 214°; 1,1'-dioxide, m.p. 247°	215, 218
4-Py-C(CH₃)(CH₂OH)₂		37
4-PyCH(CH₂CH₂OH)CH(CH₂CH₂OH)4-Py	M.p. 150°	206
4-PyCH(CH₂CH₂OH)CH(CH₂CH₂OH)4Py		206
2,6-Me₂-4-PyCH(CH₂CH₂OH)	M.p. 150-155°; picrate, m.p. 133-134°	29
2,6-Me₂-4-PyCH(CH₂CH₂OH)CH(CH₂CH₂OH)4Py		207
5-Me–4Ph–2-Py (structure with OH, OH)	M.p. 316-317°	211
(pyridine structure: N ring with CH₂ and C(CH₂OH)₂)	M.p. 101-102.5°; b.p. 120-124°/0.2 mm; picrate, m.p. 142°	38

TABLE XIII-16. Pyridine Glycols (Continued)

Compound	Physical properties (derivatives)	Ref.
(structure: pyridine with CH$_2$ and C(CH$_2$OH)$_2$ substituents)	M.p. 100.5–101.5°; b.p. 98–105°/0.05 mm	38
(structure: Ph-pyridine with CH$_2$ and C(CH$_2$OH)$_2$ substituents)	M.p. 122.5–126.5°	38
(structure: Ph-pyridine with CH$_2$CH$_2$ and C(CH$_2$OH)$_2$ substituents)	M.p. 154–155°	38
(structure: pyridine with HOCH$_2$ and CH$_2$OH substituents)	M.p. 45°; m.p. 113°, m.p. 113–115°; monoacetate, HCl, m.p. 135°; diacetate, m.p. 132°; MeI, m.p. 164°; diacetate, m.p. 149–150°; bisphenoxyformate, m.p. 50–51°; picrate, m.p. 108–109°; substituted bis-carbazates, substituted carbonates, substituted carbamates, bis-thioncarbamates, bis-nicotinate, m.p. 126–127°; HCl, m.p. 175°, mono α-(p-chlorophenoxy)isobutyrate, HCl, m.p. 126–129°; di-α-(p-chlorophenoxy) isobutyrate, HCl, m.p. 120–123°	6, 24 12, 30 34, 157 197, 198, 200–205, 208, 492, 495, 502, 503

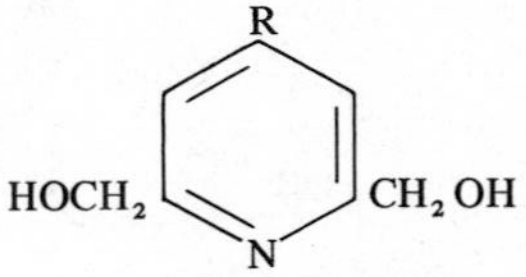

R = CH₃	M.p. 70°; picrate, m.p. 98–99°	29

R = CH_3 — M.p. 70°; picrate, m.p. 98–99° — 29
R = 4-OCH_3 — M.p. 164° — 24, 228
R = 4-SCH_3 — M.p. 173° — 24, 228
R = 4-Cl — M.p. 123°; m.p. 142° — 24, 25, 228
R = 4-SC_2H_5 — M.p. 118° — 24, 25, 228
R = 4-PhS — M.p. 127°; m.p. 142° — 24, 25, 228
R = 4-$N(CH_3)_2$ — M.p. 174° — 24, 25, 203
R = 4-OC_2H_5 — M.p. 125° — 24, 25
R = 4-$SO_2C_2H_5$ — N-Methylcarbamate, m.p. 173° — 24

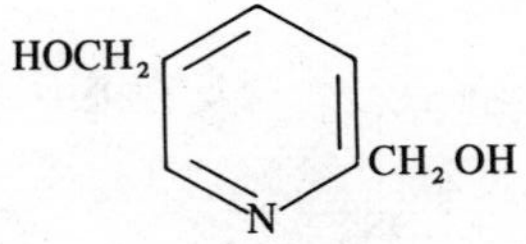

Bis-N-methylcarbamate, m.p. 129° — 24, 29, 200

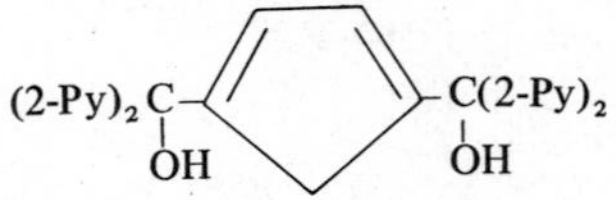

M.p. 133–135° — 93

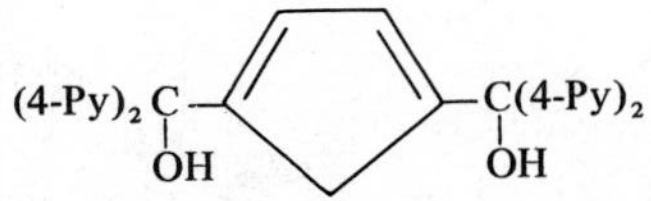

M.p. — 93

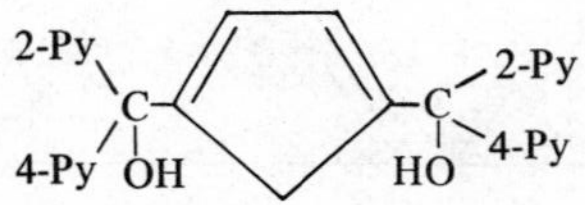

93

TABLE XIII-16. Pyridine Glycols (*Continued*)

Compound	Physical properties (derivatives)	Ref.
	B.p. 145°/0.05 mm; picrolonate, m.p. 146–147°	93
	M.p. 166–167°; HCl, m.p. 130–131°; triacetate, b.p. 165–170°/ 0.01 mm; m.p. 50–51°	226
	M.p. 150–151°	211
	1-Oxide, m.p. 158°	82

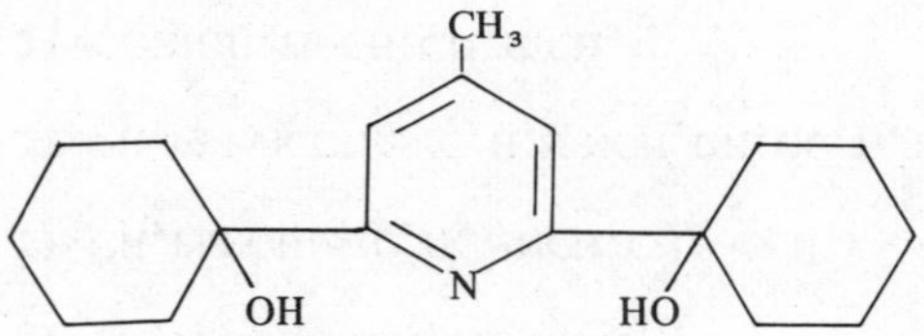

1-Oxide, m.p. 198–199° 82

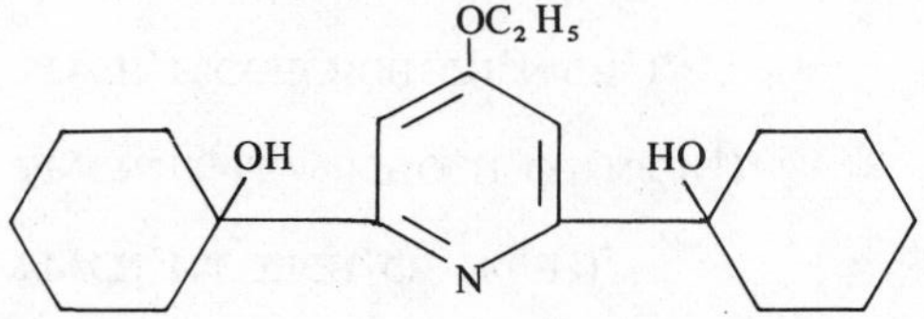

1-Oxide, m.p. 166–167° 82

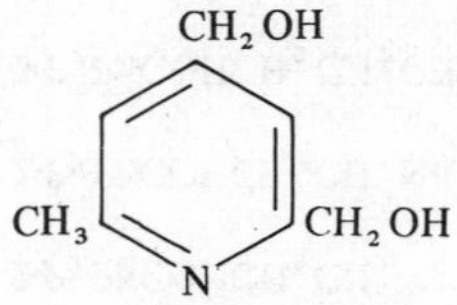

HCl, m.p. 206–208°; u.v.; n.m.r. 240

TABLE XIII-17. Pyridine Aminocarbinols

Compound	Physical properties (derivatives)	Ref.
2-PyCHOHCH$_2$Ph	M.p. 135–138°; HCl, m.p. 213–215°; diacetate, m.p. 156–158°, dibenzoate, 209–210°	16
2-PyCHOHCHNH$_2$Ph		
2-PyPhCOHCH$_2$N(CH$_3$)$_2$	M.p. 101°; maleate, m.p. 150–151°; i.r.; n.m.r.	13
2-PyPhCOHCH$_2$N(C$_2$H$_5$)$_2$	M.p. 60°; maleate, m.p. 100°; i.r.; n.m.r.	13
2-PyPhCOHCH$_2$CH$_2$N(CH$_3$)$_2$	M.p. 90°; maleate, m.p. 166°; i.r.; n.m.r.	13
2-PyPhCOHCH$_2$CH$_2$N(CH$_2$CH$_2$)$_5$	M.p. 93°; HCl, m.p. 224° decomp.; i.r.; n.m.r.	13
2-PyPhCOHCH$_2$CH$_2$CH$_2$N(CH$_3$)$_2$	M.p. 114°; i.r.; n.m.r.	13
2-PyPhCOHCH$_2$CH$_2$CH$_2$N(CH$_2$CH$_2$)$_2$O	M.p. 176–177°	221
2-PyCH$_2$CH$_2$CHOHCH$_2$NH$_2$	B.p. 190–200°/4.5 mm; dipicrate, m.p. 183–184°	56, 58, 60, 185
2-PyCH$_2$PhCOHCH$_2$CH$_2$N(CH$_3$)$_2$		
2-PyCH$_2$(2-thienyl)CHOHCH$_2$CH$_2$N(CH$_3$)$_2$	M.p. 85°	60
2-PyCH$_2$PhCOHCHCH$_2$CH$_2$N(CH$_3$)$_4$	B.p. 150–160°/0.5 mm; α-propionate, m.p. 90–91°	56, 58, 60
2-PyCH$_2$PhCOHCHCH$_2$CH$_2$N(CH$_3$)$_2$	M.p. 63–65°	57
2-PyCH$_2$PhCOH(p-(C$_2$H$_5$)$_2$NCH$_2$CH$_2$OC$_6$H$_4$)	M.p. 70–71°	57
2-PyCH$_2$(3-Py)COH(p-(C$_2$H$_5$)$_2$NCH$_2$CH$_2$OC$_6$H$_4$)	B.p. 160–167°/0.5 mm; propionate, m.p. 63–66°	48
2-PyCOH(CH$_2$Ph)CH$_2$CH$_2$N(CH$_3$)$_2$		

2-PyCOH(Ph)CHCH$_3$CH$_2$N(CH$_3$)$_2$ — B.p. 150–160°/0.6 mm; propionate, m.p. 90–91° — 48

2-Py, OH, CH$_2$N(CH$_3$)$_2$ — Propionate, m.p. 118–120° — 48

63

2-Py, OH, CH$_2$N(CH$_3$)$_2$ — Propionate, m.p. 110°; maleate, m.p. 146–150° — 48

3-PyCHOHCH(Ph)NH$_2$ — M.p. 125–128°; di-HCl, m.p. 220–222°; diacetate, m.p. 189–192°; dibenzoate, m.p. 265–267° — 16

3-PyPhCOHCH$_2$CH$_2$N(CH$_3$)$_2$ — M.p. 147°; maleate, m.p. 139°; i.r.; n.m.r. — 16

3-PyPhCOHCH$_2$CH$_2$N(C$_2$H$_5$)$_2$ — M.p. 63°; maleate, m.p. 123°; i.r.; n.m.r. — 13

3-PyPhCOHCH$_2$CH$_2$N(CH$_2$)$_5$ — M.p. 90°; maleate, m.p. 154°; i.r.; n.m.r. — 13

3-PyPhCOHCH$_2$CH$_2$N(CH$_2$CH$_2$)$_2$O — M.p. 93°; maleate, m.p. 158°; i.r.; n.m.r. — 13

3-Py[C$_6$H$_4$OCH$_2$CH$_2$N(C$_2$H$_5$)$_2$-4]COHCH$_2$(4-ClC$_6$H$_4$) — M.p. 123–125° — 57

3-Py[C$_6$H$_4$OCH$_2$CH$_2$N(C$_2$H$_5$)$_2$-4]COHCH$_2$(4-CH$_3$OC$_6$H$_4$) — M.p. 124–126° — 57

2-Cl-5-PyCHOHCH$_2$NHCH(CH$_3$)$_2$ — M.p. 95–97°; 1-oxide, m.p. 204–207°; maleate, m.p. 136.5–137.5° — 14

TABLE XIII-17. Pyridine Aminocarbinols (*Continued*)

Compound	Physical properties (derivatives)	Ref.
4-PyCHOHCH$_2$NH$_2$	di-HCl, m.p. 204° (decomp.)	11
4-PyCHOHCH$_2$NHCH$_3$	M.p. 106°; b.p. 155°/3 mm; di-HCl, m.p. 190° (decomp.)	11
4-PyCHOHCH$_2$NHC$_2$H$_5$	M.p. 116°; di-HCl, m.p. 204°	11
4-PyCHOHCH$_2$NHCH$_2$CH$_3$	M.p. 63°; b.p. 163°/0.2 mm; di-HCl, m.p. 183°	11
4-PyCHOHCH$_2$NHCH(CH$_3$)$_2$	di-HCl, m.p. 186° (decomp.)	11
2-Cl-4-PyCHOHCH$_2$NHCH(CH$_3$)$_2$	M.p. 110-112°; HCl, m.p. 156-158°	14
4-PyCHOHCH$_3$NHCH$_2$CH$_2$CH$_3$	di-HCl, m.p. 171°	11
4-PyCHOHCH$_2$NHCHCH$_3$CH$_2$CH$_3$	di-HCl, m.p. 156°	11
4-PyCHOHCH$_2$N(CH$_3$)$_2$	HCl, m.p. 164°	11
4-PyCHOHCH$_2$N(C$_2$H$_5$)$_2$	HCl, m.p. 110°	11
4-PyCHOHCH$_2$N(CH$_2$CH$_2$CH$_3$)$_2$	HCl, m.p. 150°	11
4-PyCHOHCH$_2$N(CH(CH$_3$)$_2$)	HCl, m.p. 199°	11
4-PyCHOHCH$_2$N(CH$_2$CH$_2$CH$_2$CH$_3$)$_2$	HCl, m.p. 53°	11
4-PyCHOHCH$_2$NHC$_6$H$_{11}$	B.p. 185°/2 mm; HCl, m.p. 114°	11
4-PyCHOHCH$_2$NHCH$_2$Ph	M.p. 101°; HCl, m.p. 196°	11

64

Compound	Properties	Ref.
4-PyCHOHCH$_2$N(CH$_2$CH$_2$)$_2$O	HCl, m.p. 184° (decomp.)	11
4-PyCHOHCH$_2$N(CH$_2$)$_4$	HCl, m.p. 183° (decomp.)	11
4-PyCHOHCH$_2$N(CH$_2$)$_5$	HCl, m.p. 173° (decomp.)	11
4-PyCOHCH$_2$CH$_2$N(CH$_3$)$_2$	M.p. 155°; maleate, m.p. 153°; i.r.; n.m.r.	13
4-PyPhCOHCH$_2$CH$_2$N(CH$_2$CH$_2$)$_2$O	M.p. 66°; i.r.; n.m.r.	13
4-PyPhCOHCH$_2$CH$_2$N(CH$_3$)$_2$	M.p. 89°; maleate, m.p. 216°	13
4-PyPhCOHCH$_2$CH$_2$N(C$_2$H$_5$)$_2$	M.p. 114°; maleate, m.p. 210°	13
4-PyPhCOHCH$_2$CH$_2$N(CH$_3$)$_2$	M.p. 61°	13
4-PyCHOHCH(Ph)NH$_2$	M.p. 158-161°; di-HCl, m.p. 200-205°; diacetate di-HCl, m.p. 205-208°; dibenzoate di-HCl, m.p. 249-250°	16
4-PyPhCOHCH(CH$_3$)N(CH$_2$)$_5$	M.p. 123.5-124.5°; HCl, m.p. 198-202°; di-CH$_3$I, m.p. 161-163°	47
4-PyPhCOHCH(CH$_3$)N(CH$_2$CH$_2$)$_2$O	M.p. 140-141°; di-HCl, m.p. 164-168°; di-CH$_3$I, m.p. 63° (decomp.)	47
4-PyPhCOHCH(CH$_3$)N(CH$_3$)$_2$	M.p. 129-130.5°; di-HCl, m.p. 113-119°; di-CH$_3$I, m.p. 173-175°	47

TABLE XIII-18. Side Chain Hydroxypyridine Acids and Derivatives

Compound	Physical properties (derivatives)	Ref.
2-PyCH$_2$CH$_2$CHOHCOOH	M.p. 93–94°; HCl, m.p. 162°; amide, m.p. 161.5–162°; nitrile, m.p. 81–82°	221
2-PyCH$_2$CH$_2$CH$_2$CHOHCOOH	M.p. 143°; amide, m.p. 135–136°; nitrile, m.p. 62°	221
3-PyCHOHCH$_2$CH$_2$COOH	R$_f$ values	223
	M.p. 171–172°	220
	M.p. 153–154°	220
	M.p. 166–168°	220
(2-Py)$_2$COHCOOH		224
3-Py(4-CH$_3$OC$_6$H$_4$)COHCOOH	[14]C labeled	225
	M.p. 111–113°; MeI, m.p. 192–193°	499
	M.p. 92–94°; MeI, m.p. 100°	499
	M.p. 126–128°; MeI, m.p. 116–117°	499

66

TABLE XIII-18. Side Chain Hydroxypyridine Acids and Derivatives (*Continued*)

Compound	Physical properties (derivatives)	Ref.
pyridine-CHOH–CH(C$_6$H$_5$)–CH$_2$N(morpholine)	M.p. 142–144°; MeI, m.p. 113–117°	499
pyridine-CHOH–CH(C$_6$H$_5$)–CH$_2$N(pyrrolidine)	M.p. 135–136°; MeI, m.p. 181–182°	499
pyridine-CHOH–CH(C$_6$H$_5$)–CH$_2$N(pyrrolidine)	M.p. 107–109°; MeI, m.p. 60° (decomp.)	499
pyridine-CHOH–CH(C$_6$H$_5$)–CH$_2$N(CH$_2$)$_2$	M.p. 102–104°; MeI, m.p. 178–180° (decomp.)	499
pyridine-CHOH–CH(C$_6$H$_5$)–CH$_2$N(CH$_3$)$_2$	M.p. 106–108°, MeI, m.p. 167–170°	499

TABLE XIII-19. Polyhydroxy Compounds Related to Pyridoxol

Compound	Physical properties (derivatives)	Ref.
	Tritium labeled: 3-benzyl ether, m.p. 116–117°; 1-oxide HCl, m.p. 145–146°; dimer, di-HCl, m.p. 194–197°; n.m.r.; u.v.	239, 249, 253, 255, 259, 260, 271, 272, 275, 278, 280, 286, 287, 294
	4-Methyl ether, m.p. 204–206°; 3-acetyl 68–68.5°; HCl, m.p. 148°	287–289, 264–266, 267, 285, 374
	3-Methyl ether, m.p. 141–142°; 5-benzoate, m.p. 200–201°; HCl, m.p. 180–181°	
	3-p-Toluene sulfonate, m.p. 186–187°	
	5-p-NO₂ benzoate, m.p. 197–198.5°	
	5-Palmitate, m.p. 104–105°; camphorsulfonate, m.p. 115–120°	
	5-Nicotinate, m.p. 174–175°	
	5-Adamantoyl, HCl, m.p. 173–175°; u.v.	
	4-Adamantoyl, HCl, m.p. 182–183°; u.v.	
	3-Benzoate, m.p. 141–142°	

69

TABLE XIII-19. Polyhydroxy Compounds Related to Pyridoxol (*Continued*)

Compound	Physical properties (derivatives)	Ref.
	Phosphate, n.m.r.	366
	6-Arylazo derivatives	295
	4,5-Di-Me-ether, m.p. 63–65°	273, 276, 279
	3,4-Di-Me-ether, b.p. 120°/0.5 mm; HCl, m.p. 156–158°	
	3-Methyl ether-4-benzoate, m.p. 100°, n.m.r.	363
	5-Benzoate, m.p. 138–140°	363
	3-Methanesulfonate-5-benzoate, m.p. 138°	363
	4-*p*-nitrobenzoate, m.p. 162–166°	
	4,5-Dibenzoate, m.p. 143–144°	363
	3,4-Dibenzoate, HCl, m.p. 152–154°; n.m.r.	363
	3,5-Dinicotinate, m.p. 112–113°	298
	3,4-Dipalmitate, m.p. 88–89°; HCl, m.p. 154–155°	308, 309
	3,4-Di-(ethyl carbonate) m.p. 161°	315
	3,4-Di-(benzyl carbonate) m.p. 148°	315
	3-Acetate-5-palmitate, m.p. 65–67°; 5-benzoate, m.p. 108–109°	410
	3,5-Dipalmitate, m.p. 68–69°	410

TABLE XIII-19. Polyhydroxy Compounds Related to Pyridoxol (*Continued*)

Compound	Physical properties (derivatives)	Ref.
	3,4-Isopropylidene, m.p. 111–112°; HCl, m.p. 205–211° (decomp.); benzoate, m.p. 85–87°; p-NO_2-benzoate, m.p. 185–187°; n.m.r.	298, 299, 301, 314 332, 362, 363, 367, 410, 412
	Nicotinate, m.p. 105–106°	
	Tri-palmitate, m.p. 50–51°, m.p. 73–74°	310
	Succinate and malate	319, 322
	Adamantoyl HCl, m.p. 173–174.5°	
	5-Phosphate, m.p. 211–212° dec.; u.v.	326
	4,5-Monophosphate	325, 327
	4,5-Isopropylidene, m.p. 184–185°; HCl, m.p. 170–175°	337, 359
	3-Benzoate, m.p. 107–109°; R_f values; HCl, m.p. 142–144°	359, 363
	3-p-Toluenesulfonate, m.p. 145–146°	359
	3-Methanesulfonate, m.p. 72–73°	359
	5-Benzyl ether, m.p. 117–118°; HCl, m.p. 152–153°	375
	3,5-Dibenzyl ether, m.p. 68–69°	375
	3-Nicotinate, m.p. 107–108.5°	298

73

3-Adamantoyl; HCl, m.p. 183–184° 314

3-Palmetoyl, m.p. 55–56° 363

3-Acetyl, oil; picrate, m.p. 163° 363

4-α-O-Glucoside; HCl, m.p. 154–157° 373

3-Methyl-5-benzyl ether; HCl, m.p. 140–141° 375

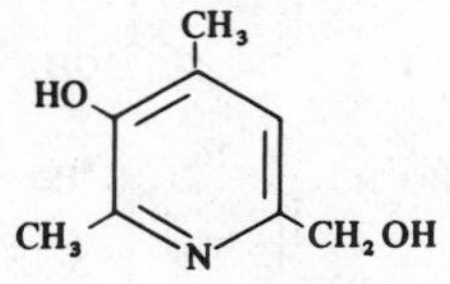

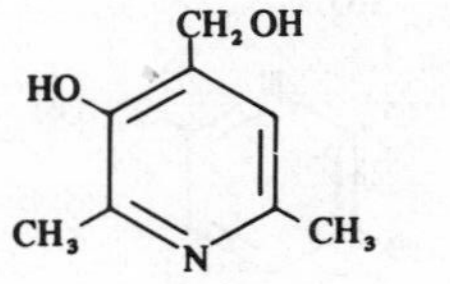

M.p. 179–181°; HCl, m.p. 139–142° 380, 412, 468

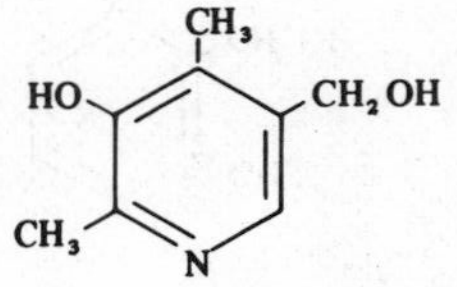

5-Acetate HCl, m.p. 209–210; m.p. 274° 291, 292, 331

Benzoate; HCl, m.p. 225–226°; picrate, m.p. 210–211°; HCl, m.p. 235° 235

3-Methyl ether, b.p. 120°/0.5 mm; HCl, m.p. 137–139° 226

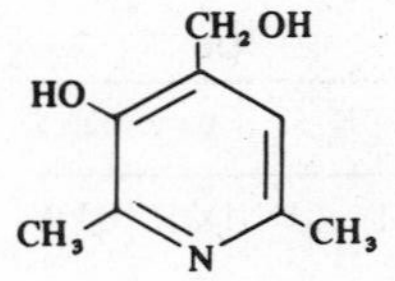

M.p. 160.5–161.5°; HCl, m.p. 159–160°; b.p. 128–130°/1.2 mm 290

TABLE XIII-19. Polyhydroxy Compounds Related to Pyridoxol (*Continued*)

Compound	Physical properties (derivatives)	Ref.
(structure)	3-Methyl ether, m.p. 169–171°; 2,3,4-trimethyl ether, m.p. 48–49°; 3,4-dimethyl ether, m.p. 176–177°; 3-methyl ether-4,6-diacetate, m.p. 143–146°	2
(structure)	N.m.r	131
(structure)	N.m.r.	131
(structure)	M.p. 152–157°; picrate, m.p. 174.5–175°	197
(structure)	M.p. 153.5°; picrate, m.p. 190–191°; acetate; b.p. 112–117°/1.2 mm; HCl, m.p. 157–158°	197, 226
(structure)	M.p. over 270°; diethyl ether, HCl, m.p. 124–126°; i.r.; 3,4-isopropylidene, m.p. 67–68°	413, 427

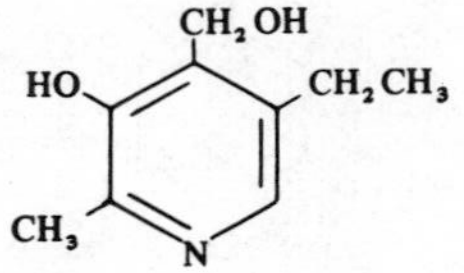

75

Structure 1: HO, CH_3, N, CH_2OH, CH_2CH_3 — M.p. 109°; 3,4-isopropylidene, m.p. 37–38°   380

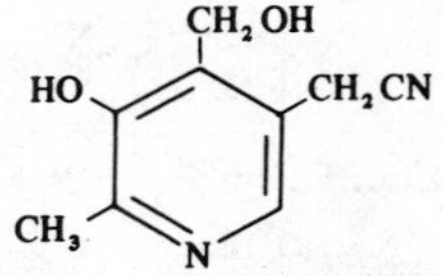

Structure 2: HO, CH_3, N, CH_2OH, CH_2CN — 3,4-Isopropylidene, m.p. 89–90°   380

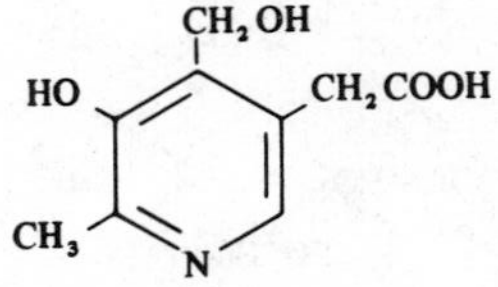

Structure 3: HO, CH_3, N, CH_2OH, CH_2COOH — HCl, m.p. 150–155°; methyl ester, m.p. 140–142°; 3,4-isopropylidene, m.p. 194°; amide, HCl, m.p. 164°; methyl ester, HCl, m.p. 168–170°   380

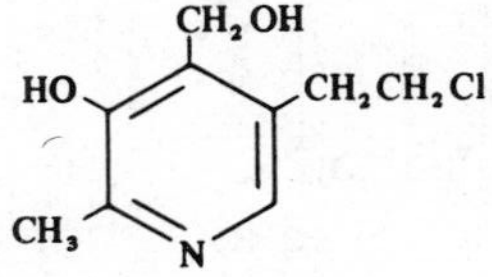

Structure 4: HO, CH_3, N, CH_2OH, CH_2CH_2Cl — 3,4-Isopropylidene, HCl, m.p. 181–183°   380

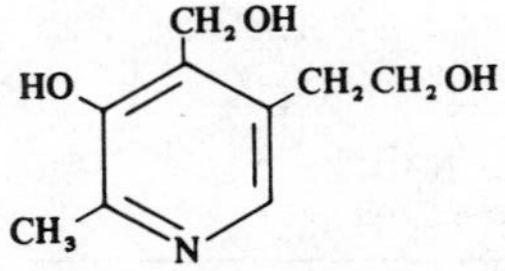

Structure 5: HO, CH_3, N, CH_2OH, CH_2CH_2OH — HCl, m.p. 109–111°; 3,4-isopropylidene, m.p. 167–169°   380

5-Phosphate   427

TABLE XIII-19. Polyhydroxy Compounds Related to Pyridoxol (Continued)

Compound	Physical properties (derivatives)	Ref.
Pyridine ring: HO, CH3, N; substituents CH2OH and CH2CH2CH2OH	HCl, m.p. 142-143°; u.v., i.r.; 3,4-isopropylidene, m.p. 84°	236, 380
Pyridine ring: HO, CH3, N; substituents CH2OH and CH2CH2COOH	HCl, m.p. 208-209°; 3,4-isopropylidene, m.p. 160-161°	380
Pyridine ring: HO, CH3, N; substituents CH2OH and CH=C(COOH)2	3,4-Isopropylidene, m.p. 200-201°; u.v.; i.r.	236
Pyridine ring: HO, CH3, N; substituents CH2OH and CH=CHCOOH	HCl, m.p. 255-260°, 3,4-isopropylidene, m.p. 220-221°; u.v.; i.r.	236
Pyridine ring: HO, CH3, N; substituents CH2OH and CH2CH2COOH	HCl, m.p. 214-215°; 3,4-isopropylidene, m.p. 188-190°; u.v.; i.r.	236

Structure	Properties	Ref.
CH₃, HO, N, CH₂OH, COCH₃ pyridine	Oxime, m.p. 195–196°; 3,4-isopropylidene, m.p. 107–108°; oxime, m.p. 194°	380
CH₃, HO, N, CH₂OH, CH=CHNO₂ pyridine	M.p. 197–198°; 3,4-isopropylidene, m.p. 172–173°	380
CH₃, HO, N, CH₂OH, CH₂CH₂NH₂ pyridine	di-HCl, m.p. 204–205.5°	380
CH₃, HO, N, CH₂OH, CHOHCH₃ pyridine	M.p. 160°; 3,4-isopropylidene, m.p. 125–128°	380
CH₃, HO, N, CH₂OH, COH(CH₃)₂ pyridine	HCl, m.p. 123–124°; isopropylidene, m.p. 125–126°	380

TABLE XIII-19. Polyhydroxy Compounds Related to Pyridoxol (Continued)

Compound	Physical properties (derivatives)	Ref.
[structure: pyridine ring with CH_3, HO, CH_2OH, $CH_2CHNHCOCH_3$, COOH]	Ethyl ester, m.p. 248–249° dec.; 3,4-isopropyli-dene, ethyl ester, m.p. 117–118°; 3,4-isopro-pylidene, methyl ester, m.p. 194–195°	380
[structure: pyridine ring with CH_3, CH_3O, CH_2OH, OH, CH_3]	methyl ester, m.p. 194–195°	380
[structure: pyridine ring with CH_3O, CH_3O, N, $CH_2CH=CCH=CHCH_2$, CH_3; $C=CHCH_2$, CH_3, CH_3; $C=CHCHCHOHC=CHCH_3$, CH_3, CH_3]	N.m.r. and octahydro, n.m.r.	230
[structure: pyridine ring with R, HO, CH_2OH, CH_2OH, N; R = H, HCl; = C_2H_5, HCl; = $CH(CH_3)_2$]	M.p. 133–135° M.p. 186–191° M.p. 193–195°	411 377, 411 377
[structure: pyridine ring with CH_3, CH_3, HO, CH_2OH, CH_2OH, OH, N]		
[structure: pyridine ring with CH_3, HO, CH_2OH, CH_2OH, N]	M.p. 250° (decomp.); 2-methyl ether, m.p. 75–76°; i.r.	239
		366

Nor-pyridoxine, m.p. 130-135°; HCl, m.p. 125-127°, u.v.; i.r.   237, 401, 414, 476

M.p. 181-181.5°; diacetate, m.p. 146-148°; 2-benzyl ether, m.p. 86.5-87°; i.r.   231

HCl, m.p. 204-205°   232

HCl, m.p. 210-212°; diacetate, m.p. 165-167°   234, 377

5-Benzoate, m.p. 190.5°   332

M.p. 169-171° (decomp.); 3,4-isopropylidene, HCl, m.p. 192-193°   332, 349, 412

TABLE XIII-19. Polyhydroxy Compounds Related to Pyridoxol (*Continued*)

Compound	Physical properties (derivatives)	Ref.
	M.p. 188–192°	377
	4-Methyl ether; HCl, m.p. 180–182°; u.v.	234
	5-Benzoate; HCl, m.p. 178–179°	235
	HCl, m.p. 192.5–194.5°	233
	HCl, m.p. 177–178°; dibenzyl ether, m.p. 72°; n.m.r	375

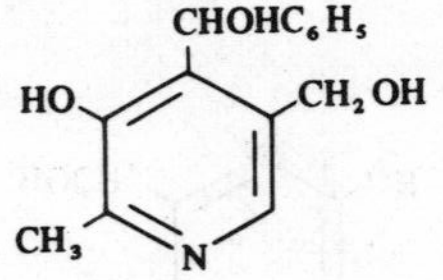

M.p. 190–191°; dibenzyl ether, 102–104°; n.m.r.　375

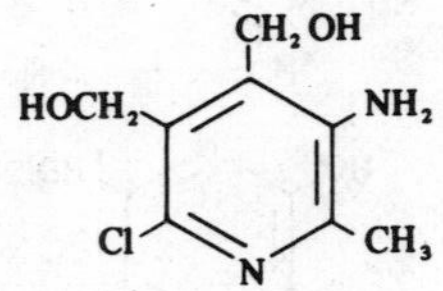

M.p. 217–219°　377

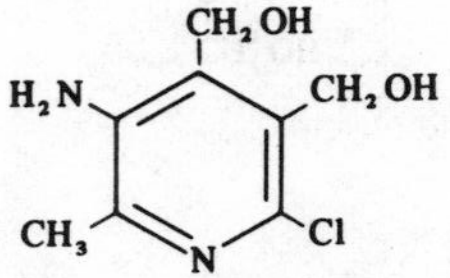

HCl, m.p. 175–180°　239

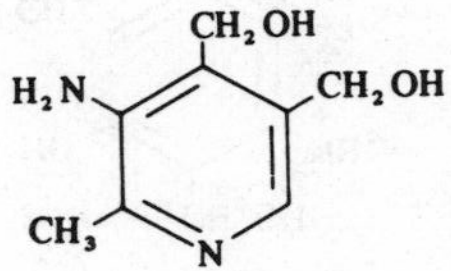

M.p. 150–160°; picrate, m.p. 214.5–216.5°;
HCl, m.p. 160–170°; acetyl, m.p. 155–156°;
triacetyl, m.p. 130–131°; i.r.　239
292
377

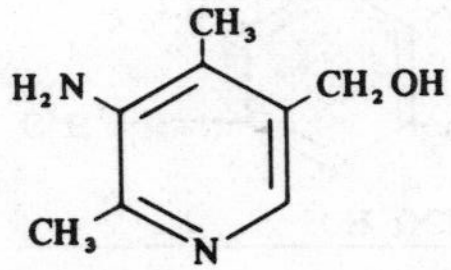

M.p. 145–147°; HI, m.p. 190–196°　292, 330, 336

TABLE XIII-19.   Polyhydroxy Compounds Related to Pyridoxol (*Continued*)

Compound	Physical properties (derivatives)	Ref.
$C_6H_5-N=N$ structure (pyridine ring with CH_3, CH_3, CH_2OH, CN, OH)	5-Tosylate, m.p. 195°	505
HO / CH_3 / CH_2OH / NH_2 pyridine structure	HCl, m.p. 189–190°; 5-benzoyl, m.p. 210–211°; dibenzoyl, m.p. 218–219°; methylsulfonate, m.p. 187–188°; dimethylsulfonate, m.p. 195–196°; 3,4-isopropylidene, m.p. 159–160°; acetate, m.p. 130–131°; benzoate, m.p. 170–171°; dibenzoate, m.p. 174–175°; methylsulfonate, m.p. 217–218°; di-(methylsulfonate), m.p. 188–189°	232
NH_2 / CH_3 / CH_2OH / CN / OH pyridine structure	4-Methyl ether-2-tosylate, m.p. 126°	296, 505
$HOCH_2$ / H_2N / CH_3 / CN / OH pyridine structure	2-Methyl ether-6-tosylate	296

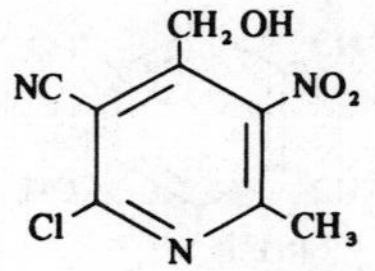

4-Methyl ether, m.p. 73–74°

248, 503, 504, 506–508

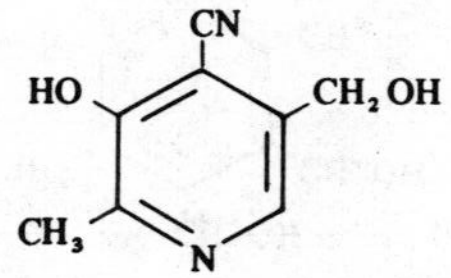

M.p. HCl, 190°

277, 282

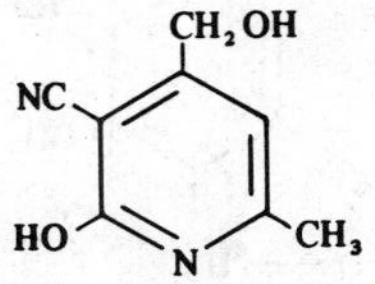

4-Methyl ether, m.p. 237–238°; 5-nitro, m.p. 210°

247

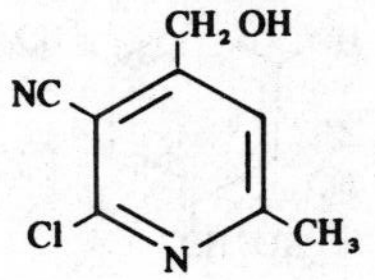

4-Methyl ether-5-nitro, m.p. 74–75°; 4-ethyl ether, m.p. 224°

247, 257

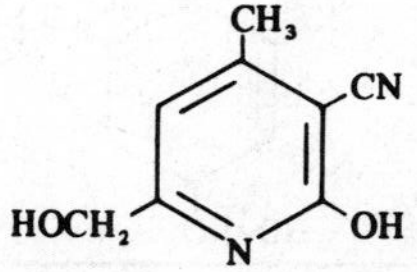

2-Methyl ether, m.p. 152–152.5°

258

TABLE XIII-19. Polyhydroxy Compounds Related to Pyridoxol (*Continued*)

Compound	Physical properties (derivatives)	Ref.
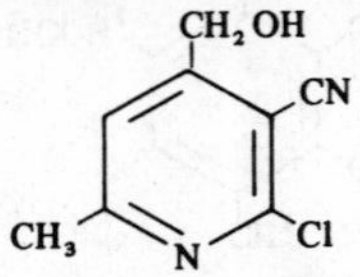	4-Ethyl ether, m.p. 146°	245
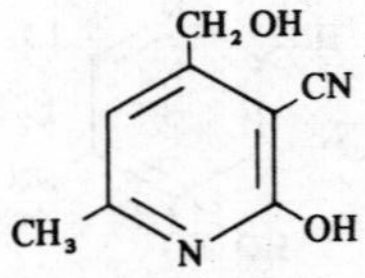	4-Methyl ether, m.p. 238–239°	258
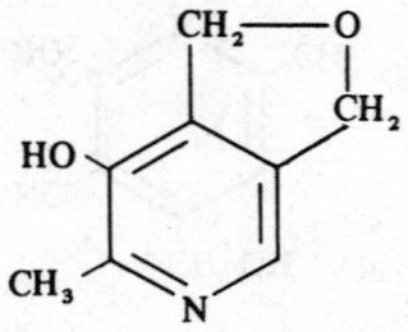	HCl, m.p. 235–240°	270, 274, 280
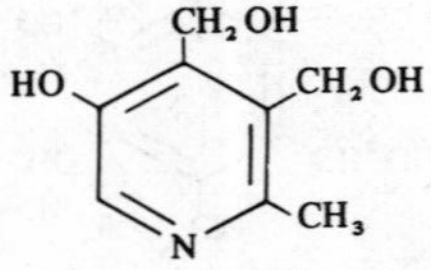	M.p. 197–199°	476
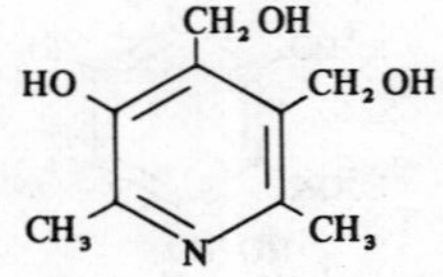	M.p. 174–179°	476

HCl, m.p. 177–178°; n.m.r.; dibenzyl ether, m.p. 72–74°   375

M.p. 190–191°; dibenzyl ether, 102–104°   375

M.p. 64–65°; HCl, m.p. 161–162°; n.m.r.   375

TABLE XIII-20. Sulfur Derivatives Related to Pyridoxol

Compound	Physical properties (derivatives)	Ref.
(pyridine: 2-CH₃, 3-HS, 4-CH₃, 5-CH₂OH)	HCl, m.p. 201–203°	330
(pyridine: 2-CH₃, 3-HS, 4-CH₂OH, 5-CH₂OH)	HCl, m.p. 189–191°	330
(pyridine: 2-CH₃, 3-HS, 4-CH₂SH, 5-CH₂OH)	HCl, m.p. 172°; 3,5-dibenzoate-4-acetate, m.p. 142–143°; 5-benzoate HCl, m.p. 182–184°; 3,4-diacetate, m.p. 153°; 4,5-diacetate, m.p. 116.5°; triacetate, m.p. 64.5°; 3,4-diacetate-5-methyl ether, m.p. 88.5°; 5-methyl ether HCl, m.p. 144–146°	235, 332, 334
(pyridine: 2-CH₃, 3-HO, 4-CH₂OH, 5-CH₂SH)	HCl, m.p. 119–120°; HCl·H₂O, m.p. 132–133°; 4-methyl ether HCl, m.p. 169–170°	336, 338

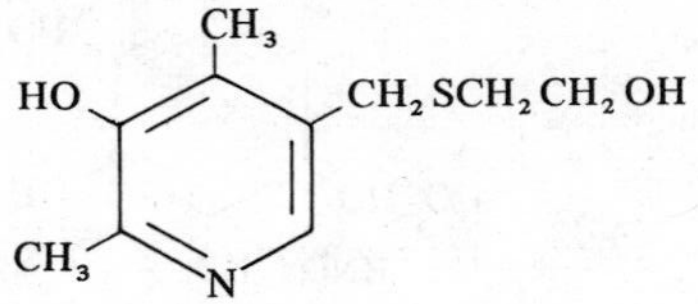	M.p. 140–142°; HCl, m.p. 180–182°	337
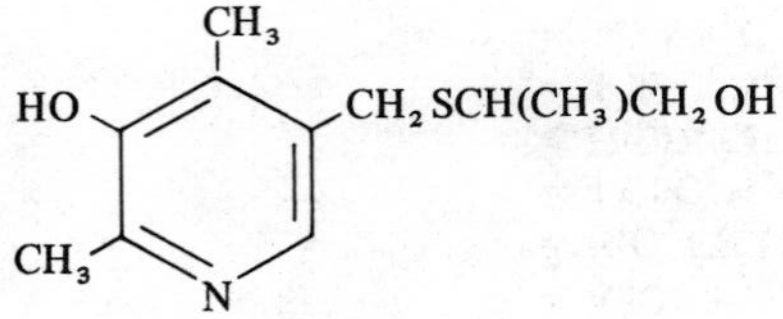	M.p. 143–146°	337
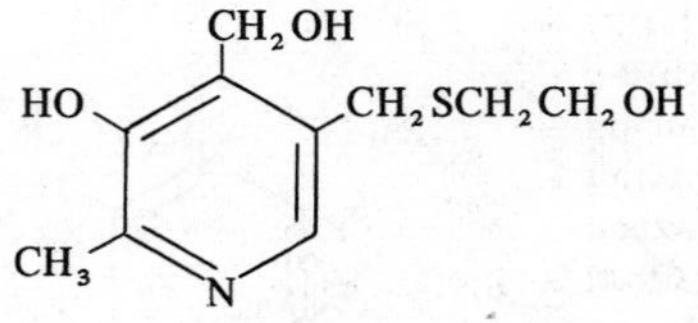	HCl, m.p. 146–148°	337
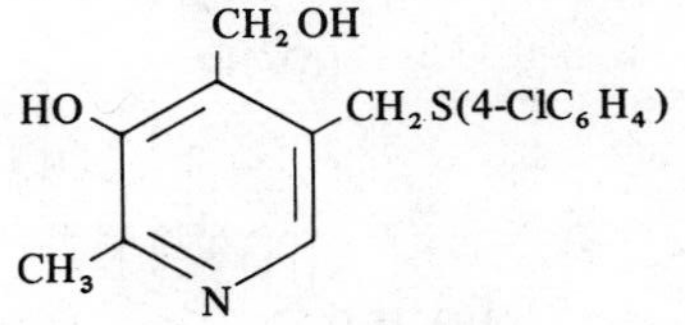	HCl, m.p. 153–155.5°	337
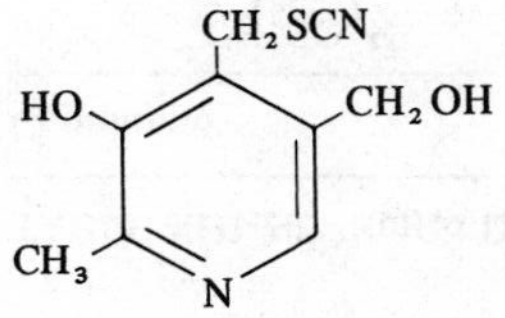	5-Benzoate, m.p. 184°; i.r.	235

TABLE XIII-20. Sulfur Derivatives Related to Pyridoxol (*Continued*)

Compound	Physical properties (derivatives)	Ref.
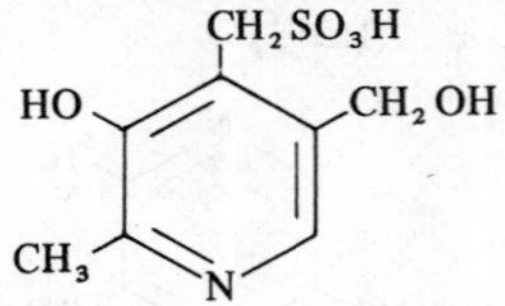 (structure: HO, CH₃, N-pyridine ring with CH₂SO₃H)	5-Benzoate, no m.p.; i.r.	235
(structure: HO, CH₃, N-pyridine ring with CH₂OH and CH₂SO₃H)		
(structure: HO, CH₃, N-pyridine ring with CH₂OH and S₂R)		
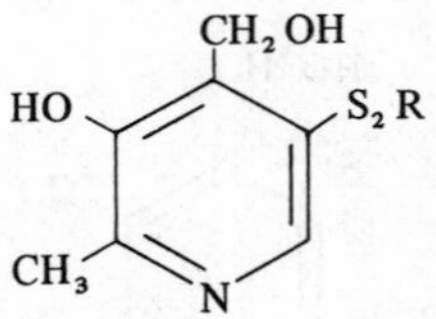 R = propyl	M.p. 116°	345
= benzyl	M.p. 139° (decomp.)	346
= butyl	M.p. 111–113°; diacetate, m.p. 37-38°	347
= phenyl	M.p. 105°	352
= *p*-nitrophenyl	M.p. 163°	
= lauryl	M.p. 109–109.5°	
= isoamyl	M.p. 114–115°	
= AcO(Ch₂)₄	M.p. 71–72°	
= EtO₂C(CH₂)₂	M.p. 71°	
= 4-Cl-C₆H₄OCH₂CH₂	M.p. 145–146°	346
= PhCH₂CH₂	M.p. 142°	
= 4-Br-C₆H₄CH₂	M.p. 156–157°	
= 2,5-Me₂C₆H₃	M.p. 135–136°	346
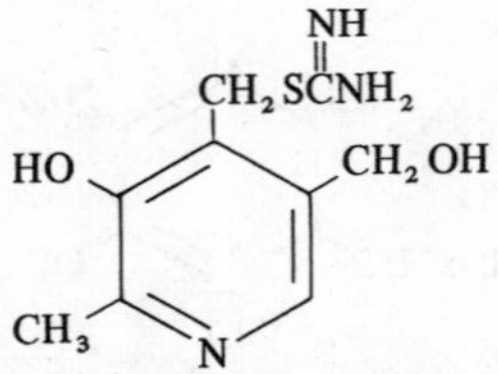 (structure: HO, CH₃, N-pyridine ring with CH₂OH and CH₂SCNH₂ / NH)	di-HBr, m.p. 224–226°	349

The structures shown:

First compound:
$$\text{HO, } CH_3 \text{ on pyridine ring, } CH_2SO_3H$$

Second compound:
$$\text{HO, } CH_3 \text{ on pyridine ring, } CH_2OH, CH_2SO_3H$$

Third compound:
$$\text{HO, } CH_3 \text{ on pyridine ring, } CH_2OH, S_2R$$

Fourth compound:
$$\text{HO, } CH_3 \text{ on pyridine ring, } CH_2OH, CH_2S\overset{NH}{\underset{}{C}}NH_2$$

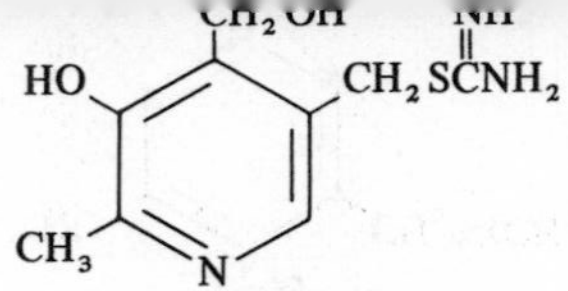

di-HCl, m.p. 165–167°; di-HBr, m.p. 170–172° 349

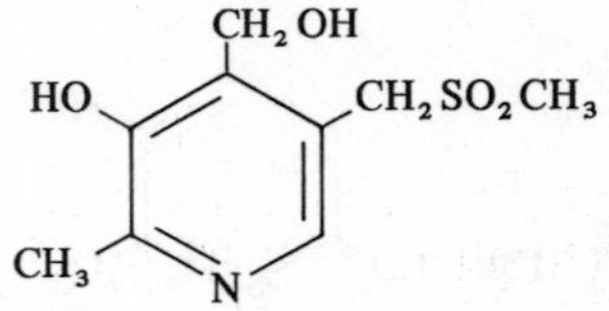

M.p. 219° 358

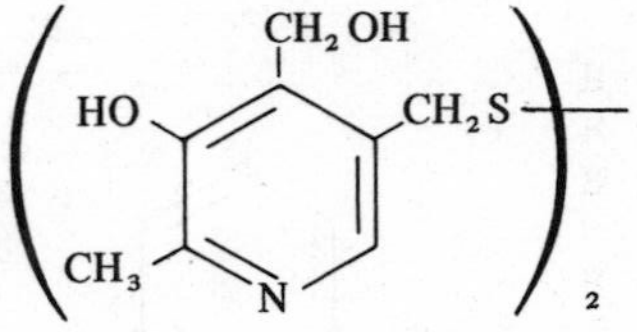

HCl, m.p. 184–186° (220°) 350, 351, 354–356

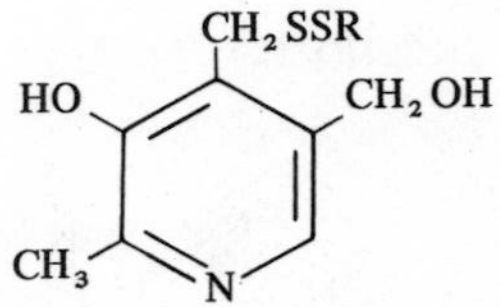

R = benzyl
= butyl
= lauryl

M.p. 123–126°
M.p. 112–114°
M.p. 105–107° 348

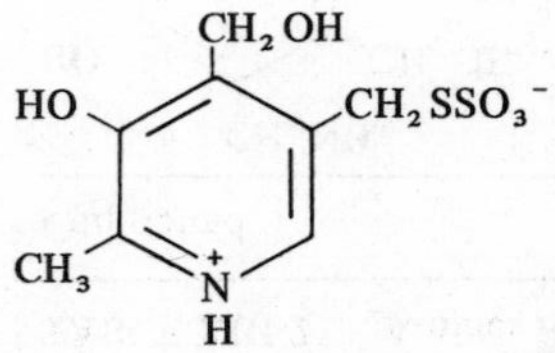

M.p. 182° 353

TABLE XIII-21. Amines Related to Pyridoxamine

Compound	Physical properties (derivatives)	Ref.
Pyridine ring: Me, HO, CH₂OH, CH₂NH₂ (CH_2NH_2)	di-HCl, m.p. 225–226° (decomp.); u.v.; n.m.r.	250, 285, 300, 331, 336, 448, 477, 478
Pyridine ring: Me, HO, CH₂OH, CH₂NHR (CH_2NHR) — Schiff's bases		482
Pyridine ring: Me, HO, CH₂OH, CH₂NHR: R = H; R = Ph; R = 2-thiazolyl; R = benzyl	di-HCl, m.p. 178°; 3,4-isopropylidene, m.p. 89–90° M.p. 182–183°; isopropylidene, m.p. 114–115° M.p. 180–182°; isopropylidene, m.p. 118–119° M.p. 119–120°; HCl, m.p. 217–218°; isopropylidene, m.p. 118–119°	332 390
Pyridine ring: Me, HO, CH₂OH, CH₂N(CH₂)₅ ($CH_2N(CH_2)_5$)	M.p. 128.5–130.5°	337
Pyridine ring: Me, HO, CH₂OH, CH₂N(CH₂)₄ ($CH_2N(CH_2)_4$)	HBr, m.p. 200–203°	337

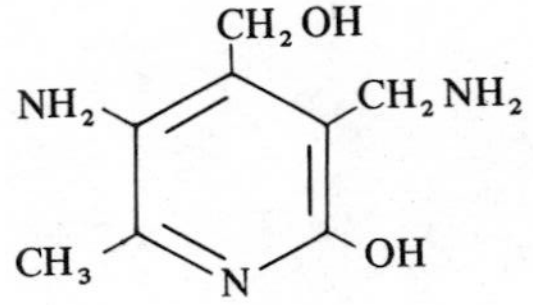

NH₂ ... CH₃ ... CH₂OH ... OH

4-Methyl ether-6-tosylate; picrate, m.p. 188°   296

4-Methyl ether, di-HCl, m.p. 228°   505

H₂N ... CH₃ ... N ... CH₂OH ... CH₂NH₂

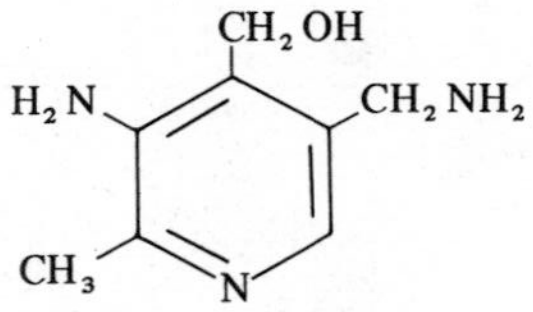

4-Ethyl ether di-HCl, m.p 127°; dipicrate, m.p. 225°   245, 296

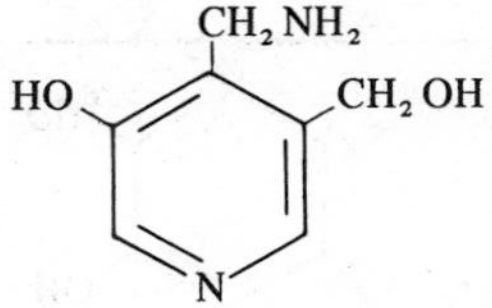

HO ... N ... CH₂NH₂ ... CH₂OH

di-HCl, m.p. 165–169°; u.v.
N-Acetyl, m.p. 184–185°
N-(N′,N′-Dimethylaminoacetyl), m.p. 203–205°   439
N-isopropyl di-HCl, m.p. 157–160°
N-2-(1-Phenylpropyl) HCl, m.p. 197–198°   401, 438

HO ... CH₃ ... N ... CH₂NHN=CH–R ... CH₂OH

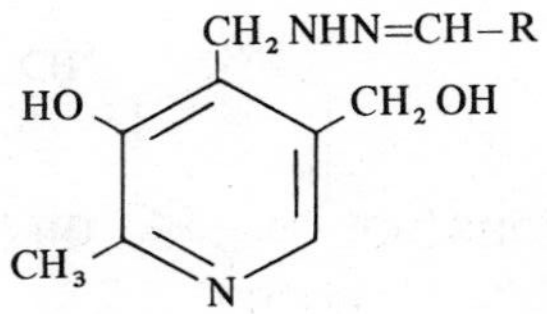

R = p-NO₂ Ph
R = 5-nitro-2-furyl
R = 5-nitro-2-thienyl
R = 1-bromo-2-(5-nitro-2-thienyl)-vinyl

M.p. 212–215° (decomp.)
M.p. 203–205° (decomp.)
M.p. 212–215° (decomp.)   438
M.p. 200° (decomp.)

TABLE XIII-21. Amines Related to Pyridoxamine (*Continued*)

Compound	Physical properties (derivatives)	Ref.
(structure: pyridine ring with CH_3, HO, CH_2OH, $CH_2NHC_6H_4-p-CO_2H$)	M.p. 250°; ethyl ester, m.p. 120–121°, derivatives with various amino acids	480
(structure: pyridine ring with CH_3, HO, CH_2NH_2, CH_2NH_2)	Tri-HCl, m.p. 290° (decomp.)	280

TABLE XIII-22. Aldehydes Related to Pyridoxal

Compound	Physical properties (derivatives)	Ref.
	HCl, m.p. 173–174°; oxime, m.p. 225° (decomp.); 5-phosphate, m.p. 139–142°; 3-benzyl ether, m.p. 165–166°; u.v.; n.m.r.; i.r.; isopropylidene, HCl, m.p. 209–211°; acetyl hydrazone, m.p. 200–202°; isopropylhydrazone, HCl, m.p. 219–221°; 1-phenyl-2-propylhydrazone, HCl, m.p. 201–204° Schiff's bases with amine 5-palmitate, m.p. 66–68°; 5-phosphate, and hydrazone, m.p. 252–253° (decomp.)	279, 294, 366, 391, 392, 394, 438, 439, 468, 441, 446, 410, 417, 418, 440, 441, 446
	3-p-Nitrobenzoyl, m.p. 209–210°; n.m.r.	472
	3-Benzoyl, m.p. 171–172°; n.m.r.	472
	3-Palmitoyl, m.p. 118–120°; n.m.r.	472
	Hemiacetal 3, α^4, 3,4-dipalmitoyl, m.p. 76–77°	472
	Hemiacetal 3, α^4, 3,4-di-p-nitrobenzoyl, m.p. 170–171°; n.m.r.	472
	Hemiacetal 3, α^4, 3,4-dibenzoyl, HCl, m.p. 129–130°; n.m.r.	472

TABLE XIII-22. Aldehydes Related to Pyridoxal (*Continued*)

Compound	Physical properties (derivatives)	Ref.
![structure: 3-hydroxy-pyridine with CH₂OH and CHO]	M.p. 185–186°; methyl acetal HCl, m.p. 155–157°; oxime, m.p. 198–199°; diethylmercaptal, m.p. 99–100°; 3,4-isopropylidene, m.p. 61–62°; oxime, m.p. 205–206°; hydrazones	232, 388, 389, 390, 469
![structure: (pyridine with CHO and CH₂–S)₂]	M.p. 169°; oxime, m.p. 253°	395
![structure: pyridine with CHO and CH₂CH₂OH]	Oxime, m.p. 201–203°; HCl, m.p. 144–147° (decomp.)	338, 348, 351

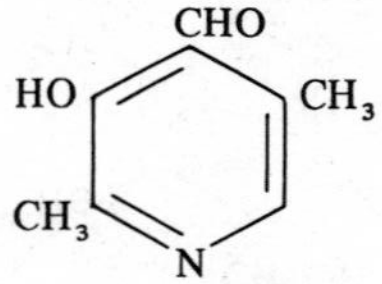

M.p. 111.5–113°   349

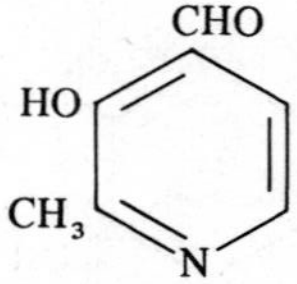

                                                                  383

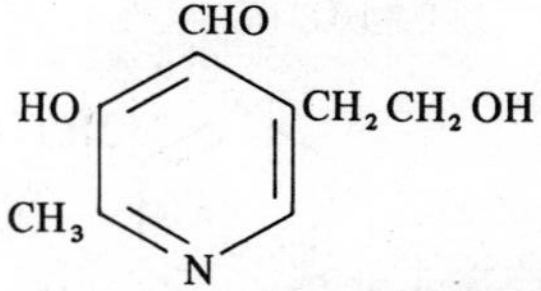

5-Phosphate   427

95

TABLE XIII-23. Pyridine Carboxylic Acids with Nuclear and Side Chain Hydroxyl Groups

Compound	Physical properties (derivatives)	Ref.
	Ethyl ester, m.p. 104–105°	27
	M.p. 256° (decomp.); 3-acetyl, m.p. 169–170°; lactone, m.p. 263–265°; nitrile, di-HCl, m.p. 225–226°; amide, m.p. 195–196°; dimethylamide, m.p. 141–142°; lactone 3-benzyl ether, m.p. 135–136°	294, 367, 397, 477
	Isopropylidene, m.p. 219–220°; lactone, m.p. 275–278°; amide, m.p. 273–275°; u.v.	293, 389
	M.p. 217–218°; nitrile, m.p. 90–91°; isopropylidene, m.p. 186–187°	293

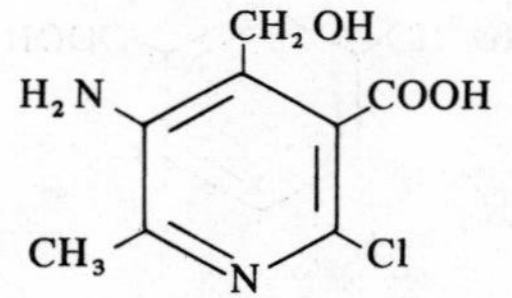

Lactone, m.p. 280–282°   239

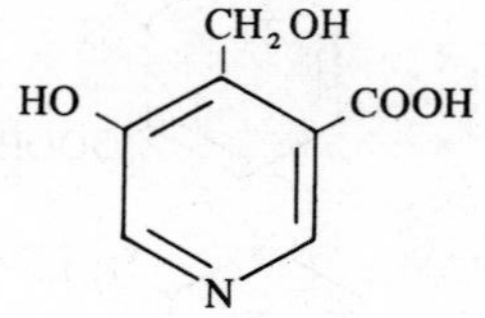

Lactone, decomp. 250°; i.r.   237

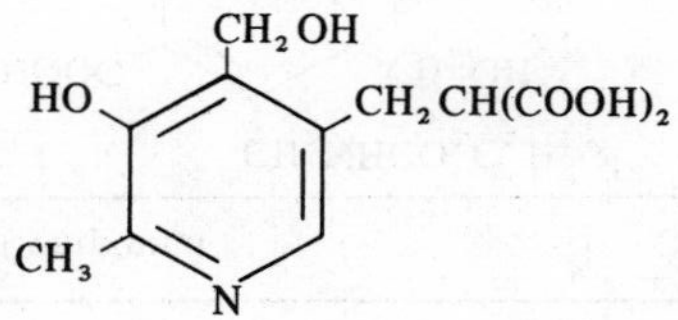

Diethyl ester, m.p. 134–135°; isopropylidene, HCl, m.p. 148–149°   293

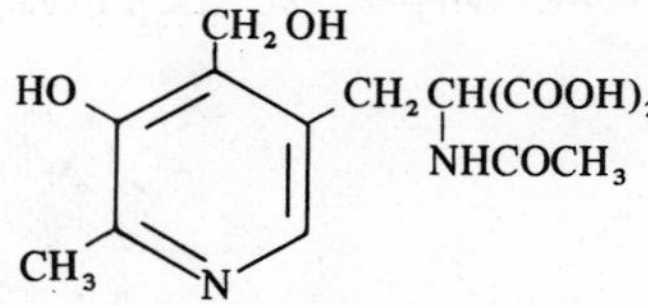

Diethyl ester, m.p. 187–188°; isopropylidene, m.p. 122–123°   293

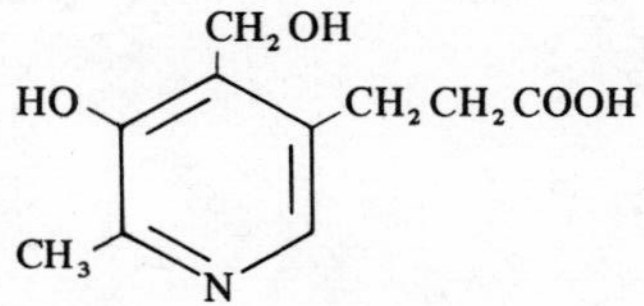

HCl, m.p. 213–215°   293

TABLE XIII-23. Pyridine Carboxylic Acids with Nuclear and Side Chain Hydroxyl Groups (*Continued*)

Compound	Physical properties (derivatives)	Ref.
	Ethyl ester, m.p. 70–71°	481
	Ethyl ester, methyl carbamate, m.p. 74°; amide, methyl carbamate, m.p. 116–117°	170
	3-pyridyl carbamate, m.p. 135–136°	170

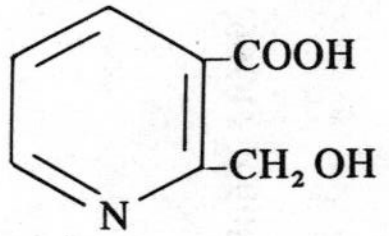

Phenylurethane, m.p. 91–93°

493

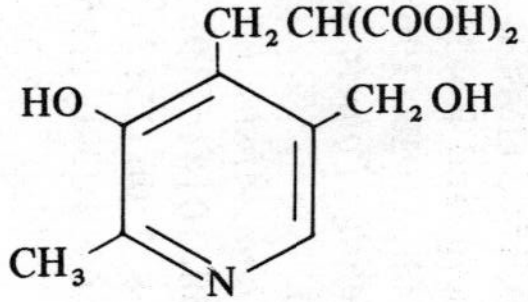

Dibenzyl ether, m.p. 173–174°; dibenzyl ether, diethyl ester, 109–110°

375

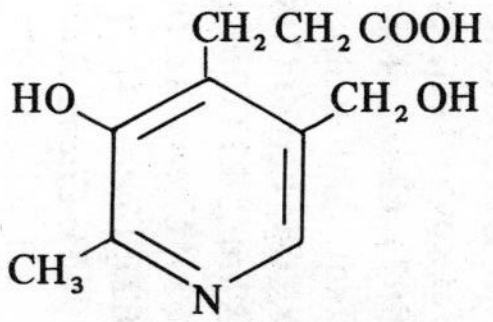

Dibenzyl ether, m.p. 175–176°

375

References

1. Chumakov, Martynova, and Marchenko, U.S.S.R. Patent 196,845 (1967); *Chem. Abstr.*, **68**, 105005n (1968).
2. Suzuki, Takahashi, and Tamura, *Agr. Biol. Chem.* (Tokyo), **30**, 13 (1966).
3. Imperial Chemical Industries, Ltd., British Patent 1,143,994 (1969); *Chem. Abstr.*, **71**, 3282v (1969).
4. Drillat, Torres, and Bordier, *C. R. Acad. Sci., Paris, Ser. C*, **266**, 1381 (1968).
5. Levi and Ivamov, *Farmatsiya* (Sofia), **15** 85 (1965).
6. Merck and Co., French Patent 1,394,362 (1965); *Chem. Abstr.*, **63**, 8326e (1965).
7. Hoffmann-La Roche and Co., Belgian Patent 668,701 (1966); *Chem. Abstr.*, **65**, 5446c (1966).
8. Ermolaeva and Shchukina, *Zh. Obshch. Khim.*, **33**, 2716 (1963).
9. Ermolaeva and Shchukina, *Zh. Obshch. Khim.*, **32**, 2664 (1962).
10. Berti, Bottari, Macchia, and Nuti, *Ann. Chim.* (Rome), **54**, 1253 (1964).
11. Friz, *Farmaco, Ed. Sci.*, **18**, 972 (1963).
12. Lyle and Nelson, *J. Org. Chem.*, **28**, 169 (1963).
13. Gautier, Miocque, Fauran, D'Engenieres, and LeCloarec, *Bull. Soc. Chim. Fr.*, 3162 (1965).
14. Allen and Hanburys, Ltd., German Patent 1,811,833 (1969); *Chem. Abstr.*, **71**, 91325q (1969).
15. Nielsen, Moore, Mazur, and Berry, *J. Org. Chem.*, **29**, 2898 (1964).
16. Kuczynski, *Diss. Pharm. Pharmacol.*, **20**, 163 (1968).
17. Neuhoff and Kuehler, *Naturwissenschaften*, **52**, 475 (1965).
18. Kraulis, Traikove, Li, Lantos, and Birmingham, *Can. J. Biochem.*, **46**, 465 (1968).
19. Shafiee and Hite, *J. Med. Chem.*, **12**, 269 (1969).
20. Murakami, Takagi, and Nishi, *Bull. Chem. Soc. Jap.*, **39**, 1197 (1966).
21. Bodalski and Michalski, *Rocz. Chem.*, **41**, 939 (1967).
22. Chatterji, Mukerji, Gautam, and Anand, *Indian J. Chem.*, **6**, 235 (1968).
23. Aktiebolag Astra, South African Patent 67,06685 (1968); *Chem. Abstr.*, **70**, 87582h (1969).
24. Shimamoto, Ishikawa, Ishikawa, and Inoue, Japanese Patent 8620, 1967; *Chem. Abstr.*, **68**, 264 (1968).
25. Inoue, French Patent 1,396,624 (1965); *Chem. Abstr.*, **63**, 5610a (1965).
26. Brown and Rapport, *J. Org. Chem.*, **28**, 3261 (1963).
27. Yamuda and Kikugawa, *Chem. Ind.* (London), 2169 (1966).
28. Farberov, Ustavshchikov, Kut'in, and Bukhareva, *Metody Polucheniya Khim. Reaktivovo Preparatov. Gos. Kom. Sov. Min. SSSR Khim.*, 108 (1964); *Chem. Abstr.*, **65**, 7586 (1966).
29. Bodalski, Michalski, and Studniarski, *Rocz. Chem.*, **40**, 1505 (1966).
30. Kuindzhi, Zepalova, Gluzman, Val'kova, Tsin, Zaitseva, and Rokk, *Metody Poluch. Khim. Reaktivov Prep.*, **15**, 93 (1967); *Chem. Abstr.*, **68**, 11019 (1968).
31. Kuindzhi, Zepalova, Gluzman, Tsin, and Rokk, *Metody Poluch. Khim. Reaktivov Prep.*, **15**, 34 (1967); *Chem. Abstr.*, **68**, 11020 (1968).
32. Gluzman, Rokk, and Tsin, *Koks Khim.*, 43 (1966); *Chem. Abstr.*, **65**, 16999c (1966).
33. Aries, U. S. Patent 3,042,682 (1962); *Chem. Abstr.*, **58**, 10181c (1963).
34. Aries, British Patent 901,654 (1962); *Chem. Abstr.*, **58**, 510b (1963).
35. Brajtburg, Polish Patent 52306 (1967); *Chem. Abstr.*, **69**, 10370j (1968).

36. Rok and Gluzman, *Sb. Nauchn. Tr., Ukr. Nauchn.-Issled. Uglekhim. Inst.*, 144 (1965); *Chem. Abstr.*, **65**, 16814a (1966).
37. Carey, British Patent 1,117,384 (1968); *Chem. Abstr.*, **69**, 43804g (1968).
38. Hahn and Epsztajn, *Rocz. Chem.*, **37**, 395 (1963).
39. Roukema, *Sci, Commun., Brocades-Stheeman Pharmocia*, **10**, 1 (1960-1961); *Chem. Abstr.*, **58**, 4513 (1963).
40. Nanta, U.S. Patent 3,354,168 (1967); *Chem. Abstr.*, **68**, 104988y (1968).
41. CIBA, Ltd., British Patent 907,070 (1962); *Chem. Abstr.*, **59**, 2777h (1963).
42. CIBA, Ltd., British Patent 901,700 (1962); *Chem. Abstr.*, **58**, 9031f (1963).
43. Van Heyningen, U.S. Patent 3,396,224 (1968); *Chem. Abstr.*, **69**, 96485k (1968).
44. Pesson and Antoine, *C. R. Acad. Sci., Paris, Ser. C*, **256**, 193 (1963).
45. Eli Lilly and Company, British Patent, 1,175,693 (1969); *Chem. Abstr.*, **72**, 78897d (1970).
46. Aldrich Chemical Company, U.S. Patent 3,413,298 (1968); *Chem. Abstr.*, **70**, 68188t (1969).
47. Breganowska and Kuczynski, *Acta Pol. Pharm.*, **25**, 1 (1968); *Chem. Abstr.*, **69**, 77077n (1968).
48. DeStevens, Halamandaris, Strachan, Donoghue, Dorfman, and Huebner, *J. Med. Chem.*, **6**, 357 (1963).
49. Thakar and Pathok, *J. Indian Chem. Soc.*, **41**, 555 (1964).
50. Gray and Kraus, *J. Org. Chem.*, **31**, 399 (1966).
51. Neisler Laboratories, Inc., Netherlands Patent 6,406,527 (1965); *Chem. Abstr.*, **65**, 695a (1966).
52. Neisler Laboratories, Inc., Belgian Patent 649,145 (1964); *Chem. Abstr.*, **64**, 8151c (1966).
53. Wright, Dunnigan, and Biermacher, *J. Med. Chem.*, **7**, 113 (1964).
54. Hertz, Kabacinski and Spoerri, *J. Heterocycl. Chem.*, **6**, 239 (1969).
55. Major, Fitzi, and Hess, *J. Med. Chem.*, **8**, 127 (1965).
56. CIBA, Ltd., British Patent 885,801 (1961); *Chem. Abstr.*, **59**, 1601d (1963).
57. Bencze, U.S. Patent 3,007,934 (1960); *Chem. Abstr.*, **56**, 4736h (1962).
58. Huebner, U.S. Patent 3,036,082 (1962); *Chem. Abstr.*, **57**, 12443c (1962).
59. Rubtsov, Yakhontov, and Krasnokutskaya, *Zh. Obshch. Khim.*, **34**, 2610 (1964).
60. DeStevens and Huebner, U.S. Patent 3,120,519 (1964); *Chem. Abstr.*, **60**, 9253r (1964).
61. Villiani, U.S. Patent 3,232,950 (1966); *Chem. Abstr.*, **64**, 12650h (1966).
62. Huebner, U.S. Patent 3,085,094 (1963); *Chem. Abstr.*, **59**, 10,000a (1963).
63. Cole, McGinnis, and Teague, *J. Org. Chem.*, **25**, 1507 (1960).
64. Wischmann, Logan, and Stuart, *J. Org. Chem.*, **26**, 2795 (1961).
65. Sam, Plampin, and Alwani, *J. Org. Chem.*, **27**, 4543 (1962).
66. Reinecke and Kray, *J. Org. Chem.*, **29**, 1736 (1964).
67. Eilhauer, Steinke, and Kurtschinski, East German Patent 45082 (1966); *Chem. Abstr.*, **64**, 17552b (1966).
68. Midland Tar Distillers, Ltd., British Patent 918,179 (1963); *Chem. Abstr.*, **58**, 13923a (1963).
69. Dice and Westland, U.S. Patent 3,128,281 (1966); *Chem. Abstr.*, **60**, 15842h (1964).
70. Dice, Scheinman, and Berrodin, *J. Med. Chem.*, **9**, 176 (1966).
71. Burckhalter, Dixon, Black, Westland, Werbel, DeWald, Dice, Rodney, and Kaump, *J. Med. Chem.*, **10**, 565 (1967).
72. Cislak, U.S. Patent 2,891,959 (1959); *Chem. Abstr.*, **53**, 20096a (1959).
73. Fryer, Brust, Earley, and Sternback, *J. Org. Chem.*, **31**, 2415 (1966).

74. Hoffmann-LaRoche and Company, Netherland Patent 6,402,889 (1964); *Chem. Abstr.*, **62**, 6462h (1965).
75. Hoffmann-LaRoche and Company, Belgian Patent 645,241 (1964); *Chem. Abstr.*, **63**, 13223 (1965).
76. Derieg, Brust, and Fryer, *J. Heterocycl. Chem.*, **3**, 165 (1966).
77. Brust, Fryer, and Sternbach, U.S. Patent 3,309,375 (1967); *Chem. Abstr.*, **68**, 12862z (1968).
78. Brust, Fryer, and Sternbach, U.S. Patent 3,400,131 (1968); *Chem. Abstr.*, **69**, 106562z (1968).
79. Gurien and Rachlin, U.S. Patent 3,268,541 (1966); *Chem. Abstr.*, **65**, 15346h (1966).
80. Hoffmann-LaRoche and Company, Netherlands Patent 6,511,532 (1966); *Chem. Abstr.*, **65**, 3847a (1966).
81. Bennington, Morin, and Clark, *J. Org. Chem.*, **25**, 1913 (1960).
82. Abramovitch, Saha, Smith, and Coutts, *J. Amer. Chem. Soc.*, **89**, 1537 (1967). Abramovitch, Smith, Knaus, and Saha, *J. Org. Chem.*, **37**, 1690 (1972).
83. Abramovitch and Saha, *J. Chem. Soc.*, 2175 (1969).
84. Abramovitch and Saha, *Can. J. Chem.*, **44**, 1765 (1966).
85. Eli Lilly and Company, South African Patent 67,07609 (1968); *Chem. Abstr.*, **70**, 87589r (1969).
86. Abramovitch, Giam, and Poulton, *J. Chem. Soc.*, C, 128 (1970).
87. Terent'ev, Kost, Shchegolev, and Terent'ev, *Dokl. Akad. Nauk. S.S.S.R.*, **141**, 110 (1961); *Chem. Abstr.*, **56**, 11564 (1962).
88. Kotlyarevskii, Vereshchagin, Yashina, Vasil'ev, and Faershtein, *Izv. Sibirsk. Otd. Akad. Nauk S.S.S.R.*, 80 (1962); *Chem. Abstr.*, **59**, 1584a (1963).
89. Kost, Terent'ev, and Shchegolev, *Zh. Obshch. Khim.*, **32**, 2606 (1962).
90. McNeil Laboratories, Inc., Belgian Patent 660,853 (1965); *Chem. Abstr.*, **64**, 5050a (1966).
91. Stevenson and Welkner, U.S. Patent 3,264,312 (1966); *Chem. Abstr.*, **65**, 13666f (1966).
92. Roszkowski, Poos, and Mohrbacher, *Science*, **144**, 412 (1964).
93. McNeil Laboratories, Inc., U.S. Patent 3,426,033 (1969); *Chem. Abstr.*, **70**, 68171g (1969).
94. Biniecki and Emiljan, *Acta Polon. Pharm.*, **24**, 345 (1967); *Chem. Abstr.*, **68**, 11021 (1968).
95. Boehme and Koo, *J. Org. Chem.*, **26**, 3589 (1961).
96. Betts and Brown, *J. Chem. Soc.*, C, 1730 (1967).
97. Bowen, *Biogr. Mem. Fellows Roy. Soc.*, **13**, 107 (1967); *Chem. Abstr.*, **68**, 4457 (1968).
98. Howe and Ratts, *Tetrahedron Lett.*, 4743 (1967).
99. Kazarinova, Rybak, and Balykina, *Metody Poluch. Khim. Reaktivov Prep.*, **14**, 98 (1966); *Chem. Abstr.*, **67**, 3075 (1967).
100. Syper, Skrowaczewska, and Bytnar, *Rocz. Chem.*, **41**, 1027 (1967).
101. Prostakov, Gaworonskaya, Mikhailova, and Kirillova, *Zh. Obshch. Khim.*, **33** 2573 (1963).
102. Robison, *J. Amer. Chem. Soc.*, **80**, 6254 (1958).
103. Instylut Farmaceutyczny, Polish Patent 51,702 (1966); *Chem. Abstr.*, **68**, 21839z (1968).
104. Dainippon Pharaceutical Company, British Patent 1,053,730 (1967); *Chem. Abstr.*, **66**, 115605f (1967).
105. Oae, Kitao, and Kitaoka, *Ann. Rept. Radiation Center Osaka Prefect.*, **2**, 133 (1961); *Chem. Abstr.*, **56**, 10088a (1962).

106. Oae, Kitao, and Kitaoka, *J. Amer. Chem. Soc.*, **84**, 3359 (1962).
107. Traynelis and Pacini, *J. Amer. Chem. Soc.*, **86**, 4917 (1964).
108. Cohen and Fager, *J. Amer. Chem. Soc.*, **87**, 5701 (1965).
109. Traynelis and Gallagher, *J. Amer. Chem. Soc.*, **87**, 5710 (1965).
110. Koenig, *J. Amer. Chem. Soc.*, **88**, 4045 (1966).
111. Varma and Lal, *J. Indian Chem. Soc.*, **43**, 613 (1966).
112. Cohen and Deets, *J. Amer. Chem. Soc.*, **89**, 3939 (1967).
113. Oae, Tamagaki, Negoro, Ogino, and Kazuka, *Tetrahedron Lett.*, 917 (1968).
114. Kozuka, Tamagoki, Negoro, and Oae, *Tetrahedron Lett.*, 923 (1968).
115. Bodalski and Katritzky, *J. Chem. Soc.*, *B*, 831 (1968).
116. Traynelis, "Rearrangements of *O*-Acetylated Heterocyclic *N*-Oxides," in "Mechanism of Molecular Migration," Vol. 2, B. S. Thyagarajan, Ed., Interscience, New York, 1969, pp. 1–42.
117. Ochiai, "Aromatic Amine Oxides," Elsevier, Amsterdam, 1967, Chap. 7.
118. Vozza, *J. Org. Chem.*, **27**, 3856 (1962).
119. Abramovitch and Vinutha, *J. Chem. Soc.*, *C*, 2104 (1969).
120. Wilimowski, *Arch. Immunol. Therap. Exptl.*, **10**, 739 (1962); *Chem. Abstr.*, **59**, 5650f (1963).
121. Zoelner and Gudenzi, *Progr. Biochem. Pharmacol.*, **2**, 406 (1967).
122. Nye and McCaw, *J. Pharmacol. Exp. Ther.*, **167**, 374 (1969).
123. Bederka, Hansson, Bowman, and McKennis, *Biochem. Pharmacol.*, **16**, 1 (1967).
124. Herstel, Van Hell, Mulder, and Nauta, *Arzneim.-Forsch.* (Weinheim), 827 (1968).
125. Schettino and La Rotonda, *Bull. Soc. Ital. Biol. Sper.*, **44**, 2210 (1968).
126. Buu-Hoi, Delbarre, Jacquignon, Rose, Sabathier, and Sinh, *Med. Exptl.*, **7**, 166 (1962); *Chem. Abstr.*, **58**, 7274c (1963).
127. Guilhom, *Bull. Acad. Vet. Fr.*, **34**, 361 (1962); *Chem. Abstr.*, **57**, 3978 (1962).
128. Thorpe, *J. Comp. Pathol. Therap.*, **72**, 29 (1962); *Chem. Abstr.*, **56**, 13504f (1962).
129. Sharkh and Thaker, *J. Indian Chem. Soc.*, **45**, 378 (1968).
130. Tissier and Tissier, *Bull. Soc. Chim. Fr.*, 3155 (1967).
131. Williams, *J. Chromatog.*, **24**, 203 (1966).
132. Kuhn, Wires, Ruoff, and Kwart, *J. Amer. Chem. Soc.*, **91**, 4790 (1969).
133. Golding and Katritzky, *Can. J. Chem.*, **43**, 1250 (1965).
134. Murakami and Takagi, *J. Amer. Chem. Soc.*, **91**, 5130 (1969).
135. Chilton and Butler, *J. Org. Chem.*, **32**, 1270 (1967).
136. Hackley and Daniber, *J. Org. Chem.*, **32**, 2624 (1967).
137. Mathes and Sauermilch, U.S. Patent 3,008,963, 1961; *Chem. Abstr.*, **57**, 5895h (1962).
138. Sheehan, *J. Org. Chem.*, **31**, 636 (1966).
139. Ondetti and Sheehan, U.S. Patent 3,291,805 (1966); *Chem. Abstr.*, **67**, 21835z (1967).
140. Raaflaub, *Experientia*, **22**, 258 (1966).
141. Ermolaeva and Kalmanson, *Dokl. Akad. Nauk S.S.S.R.*, **158**, 436 (1964).
142. Boekelheide and Feely, *J. Amer. Chem. Soc.*, **80**, 2217 (1958).
143. Reinecke and Kray, *J. Amer. Chem. Soc.*, **86**, 5355 (1964).
144. Hofling and Reckling, British Patent 974,113 (1964); *Chem. Abstr.*, **62**, 7734d (1965).
145. VEB Leuna-Werke, French Patent 1,296,376 (1962); *Chem. Abstr.*, **58**, 3402g (1963).
146. Gray, Kraus, Heitmeier, and Shiley, *J. Org. Chem.*, **33**, 3007 (1968).
147. Societa Farmaceutici Italia, British Patent 1,110,637 (1968); *Chem. Abstr.*, **69**, 77119c (1968).

148. Freifelder, Robinson, and Stone, *J. Org. Chem.*, **27**, 284 (1962); **30**, 1319 (1965).
149. Murakami and Takagi, *Bull. Chem. Soc. Jap.*, **42**, 3478 (1969).
150. Augustinsson and Hasselquist, *Acta Chem. Scand.*, **18**, 1006 (1964).
151. Taylor, *J. Chem. Soc.*, 4881 (1962).
152. Aries, French Patent 1,532,234 (1968); *Chem. Abstr.*, **71**, 61233r (1969).
153. Baizer, U.S. Patent 3,312,713 (1967); *Chem. Abstr.*, **68**, 21844x (1968).
154. Aktiebolag Bofors., Netherlands Patent 6,608,905 (1966); *Chem. Abstr.*, **68**, 29614t (1968).
155. Shukla and Choudry, *Def. Sci. J.*, **18**, 107 (1968).
156. Koch, U.S. Patent 3,312,711 (1967); *Chem. Abstr.*, **67**, 82113j (1967).
157. Banyu Pharmaceutical Company, British Patent 1,087, 020 (1967); *Chem. Abstr.*, **68**, 29615u (1968).
158. Laboratoire Perrier, French Patent M3489 (1965); *Chem. Abstr.*, **63**, 18043a (1965).
159. Rorig, U.S. Patent 3,098,857 (1963); *Chem. Abstr.*, **60**, 508d (1964).
160. Murakami and Takagi, *Bull. Chem. Soc. Jap.*, **40**, 2724 (1967).
161. Kampe, *Chem. Ber.*, **98**, 1031 (1965).
162. Farbenfabriken Bayer A.-G., Netherland Patent 6,409,877 (1965); *Chem. Abstr.*, **63**, 8325d (1965).
163. All-Union Sci. Res. Institute, U.S.S.R. Patent 172,801 (1965); *Chem. Abstr.*, **64**, 712f (1966).
164. Gupalo and Zemlyanskii, *Zh. Obshch. Khim.*, **38**, 2550 (1968).
165. Zimmerman, *Texas J. Sci.*, **15**, 192 (1963); *Chem. Abstr.*, **60**, 5446e (1964).
166. Ishikawa, Shimamoto, and Ishikawa, Japan Patent 67 25,666 (1967); *Chem. Abstr.*, **69**, 67238b (1968).
167. Ishikawa, Shimamoto, and Ishikawa, Japan Patent 67 25,665 (1967); *Chem. Abstr.*, **69**, 67236z (1968).
168. Matsumoto and Okazawa, Japan Patent 68 02,714 (1968); *Chem. Abstr.*, **69**, 86823e (1968).
169. Billiotte and Debay, *Chim. Therap.*, 164 (1966); *Chem. Abstr.*, **65**, 2213d (1966).
170. Matsumoto, Nakagawa, and Horiuchi, Japan Patent 68 00,518 (1968); *Chem. Abstr.*, **69**, 52012z (1968).
171. Juby, Babel, Bocian, Cladel, Godfrey, Hall, Hudyma, Luke, Matishella, Minor, Montzka, Partyka, Standridge, and Cheney, *J. Med. Chem.*, **10**, 491 (1967).
172. Ishikawa, Shimamoto, and Ishikawa, Netherland Patent 6,515,966 (1966); *Chem. Abstr.*, **65**, 15340a (1966).
173. Société Amilloise de Produits Chimiques, French Patent M4,898 (1967); *Chem. Abstr.*, **69**, 106565c (1968).
174. Profft and Schmuck, *Arch. Pharm.* (Weinheim), **296**, 209 (1963).
175. Bristol-Myers Co., U.S. Patent 3,290,319 (1966); *Chem. Abstr.*, **66**, 115609k (1967).
176. Roth and Sarraj, *Arch. Pharm.* (Weinheim), **300**, 44 (1967).
177. Imperial Chemical, Ind., British Patent 946,880 (1964); *Chem. Abstr.*, **60**, 10655f (1964).
178. Midland Tar Distillers, British Patent 927,613 (1963); *Chem. Abstr.*, **59**, 11443b (1963).
179. N. V. Koninklijke Pharm. Fabrieken, French Patent M1801 (1963); *Chem. Abstr.*, **59**, 12769f (1963).
180. N. V. Koninklijke Pharm. Fabrieken, Netherland Patent 6,511,354 (1966); *Chem. Abstr.*, **65**, 10569f (1966).
181. Buchmann and Franz, *Z. Chem.*, **6**, 107 (1966).
182. Franz and Buchmann, *Pharmazie*, **24**, 301 (1969).

183. Sullivan, Kester, and Norton, *J. Med. Chem.*, **11**, 1172 (1968).
184. Kasuga and Taguchi, *Chem. Pharm. Bull.* (Tokyo), **13**, 233 (1965); *Chem. Abstr.*, **63**, 6961 (1965).
185. Casy and Pocha, *Tetrahedron*, **23**, 633 (1967).
186. Wrobel and Dabrowski, *Rocz. Chem.*, **40**, 321 (1966).
187. Goldman, *J. Org. Chem.*, **28**, 1921 (1963).
188. Ruetgerswerke, A.-G., British Patent 956,398 (1964); *Chem. Abstr.*, **61**, 5618c (1964).
189. Abramovitch and Poulton, *J. Chem. Soc., B,* 267 (1967).
190. Chumakov and Shapovalova, *Zh. Org. Khim.*, **1**, 940 (1965).
191. Imperial Chemical, Ind., British Patent 974,168 (1964); *Chem. Abstr.*, **62**, 4015b (1965).
192. Mariella and Brown, *J. Org. Chem.*, **34**, 3191 (1969).
193. Miyano, *Chem. Pharm. Bull.* (Tokyo), **13**, 1135 (1965).
194. Miyano, Uno, and Abe, *Chem. Pharm. Bull.* (Tokyo), **15**, 515 (1967).
195. Nordisk, Droge, and Kemklieforretning, Netherland Patent 289,700 (1965); *Chem. Abstr.*, **63**, 16314f (1965).
196. Miyano and Abe, *Chem. Pharm. Bull.* (Tokyo), **15**, 511 (1967).
197. Kato, Kitaguwa, Shibata, and Nakai, *Yakugaku Zasshi,* **82**, 1647 (1962); *Chem. Abstr.*, **59**, 559b (1963).
198. Juby, Babel, Bocian, Cladel, Godfrey, Hall, Hudyma, Luke, Matishella, Minor, Montzka, Partyka, Standridge, and Cheney, *J. Med Chem.*, **10**, 491 (1967).
199. Banyu Pharmaceutical Company, Japanese Patent 66 22186 (1966); *Chem. Abstr.*, **66**, 75911u (1967).
200. Shimamoto, Ishikawa, Ishikawa, and Inoue, Japanese Patent 67 6353 (1967); *Chem. Abstr.*, **67**, 90692n (1967).
201. Matsumoto, Japanese Patent 68 02,355 (1968); *Chem. Abstr.*, **69**, 67233w (1968).
202. Banyu Pharmaceutical Company, Japanese Patent 66 22,185, 1966; *Chem. Abstr.*, **66**, 75907x (1967).
203. Shimamoto, Ishikawa, Ishikawa, and Inoue, Japanese Patent 67 9784 (1967); *Chem. Abstr.*, **68**, 49460t (1968).
204. Shimamoto, Ishikawa, Ishikawa, and Inoue, Japanese Patent 67 6354 (1967); *Chem. Abstr.*, **67**, 90682j (1967).
205. Banyu Pharmaceutical Company, Japanese Patent 66 22,392 (1966); *Chem. Abstr.*, **66**, 65395q (1967).
206. Cislak, McGill, and Campbell, U.S. Patent 3,317,550 (1967); *Chem. Abstr.*, **67**, 21842z (1967).
207. Boekelheide and Feely, *J. Org. Chem.*, **22**, 589 (1957).
208. Banyu Pharmaceutical Company, Japanese Patent 68 14,222 (1968); *Chem. Abstr.*, **70**, 19944c (1969).
209. Allen, U.S. Patent 3,118,895, 1964; *Chem. Abstr.*, **60**, 10656e (1964).
210. Dr. F. Raschig GMBH, German Patent 1,122,067 (1962); *Chem. Abstr.*, **56**, 15490f (1962).
211. Prostakov, Mathew, and Pkhal'gumani, *Zh. Org. Khim.*, **1**, 1128 (1965).
212. Ebetino and Amstulz, *J. Org. Chem.*, **28**, 3249 (1963).
213. Yost, U.S. Patent 3,200,053 (1965); *Chem. Abstr.*, **64**, 6624c (1966).
214. Bencze, Burckhardt, and Yost, *J. Org. Chem.*, **27**, 2867 (1962).
215. Roth, Sarraj, and Jaeger, *Arch. Pharm.* (Weinheim), **299**, 605 (1966).
216. Michalski, Wojaczynski, and Zajac, *Bull. Acad. Polon. Sci., Ser. Sci. Chem.*, **8**, 285 (1960); *Chem. Abstr.*, **56**, 7263 (1962).

217. Michalski, Wojaczynski, and Zajac, *Bull. Acad. Polon. Sci., Ser. Sci. Chim.*, 9, 401 (1961); *Chem. Abstr.*, 60, 6710f (1964).
218. Mathes and Sauermilch, *Ann. Chem.*, 618, 152 (1958).
219. Bencze and Walker, U.S. Patent 3,337,568 (1967); *Chem. Abstr.*, 68, 87179r (1968).
220. Kost, Terente'v, and Moslentseva, *Khim.-Farm. Zh.*, 1, 12 (1967); *Chem. Abstr.*, 68, 6640 (1968).
221. Profft and Stumpf, *Arch. Pharm.* (Weinheim), 296, 79 (1963).
222. Banashek and Shchukina, *Zh. Obshch. Khim.*, 31, 1479 (1961).
223. McKennis, Schwartz, Turnbull, Tamaki, and Bowman, *J. Biol. Chem.*, 239, 3981 (1964).
224. Black and Srivastava, *Aust. J. Chem.*, 22, 1439 (1969).
225. Novelli and Barris, *Tetrahedron Lett.*, 3671 (1969).
226. Smirnov, Lezina, Bystrov, and Dywmaev, *Izv. Akad. Nauk SSSR, Ser. Khim.*, 1836 (1965).
227. Seiyaku Company, Japanese Patent 63 4086 (1963); *Chem. Abstr.*, 59, 11446c (1963).
228. Shimamoto, Ishikawa, Ishikawa, and Inoue, Japanese Patent 14,550 (1967); *Chem. Abstr.*, 68, 104998b (1968).
229. Budesinsky, Prikryl, Vanecek, and Svatek, *Collect. Czech. Chem. Commun.*, 33, 2266 (1968).
230. Takahashi, Suzuki, and Tamura, *Agr. Biol. Chem.* (Tokyo), 30, 1 (1966); *Chem. Abstr.*, 64, 12636b (1966).
231. Jaques and Hubbard, *J. Med. Chem.*, 11 178 (1968).
232. Korytnyk and Paul, *J. Heterocycl. Chem.*, 2, 144 (1965).
233. Greene and Montgomery, *J. Med. Chem.*, 6, 294 (1963).
234. Harris, Harris, Peterson, and Rogers, *J. Med. Chem.*, 10, 261 (1967).
235. Singh and Korytnyk, *J. Med. Chem.*, 8, 116 (1965).
236. Korytnyk, *J. Med. Chem.*, 8, 112 (1965).
237. Van Der Wal, DeBoer, and Huisman, *Rec. Trav. Chim. Pays-Bas*, 80, 203 (1961).
238. Pol and Obbink, *Rec. Trav. Chim. Pays-Bas*, 80, 217 (1961).
239. Van Der Wal, DeBoer, and Huisman, *Rec. Trav. Chim. Pays-Bas*, 80, 221 (1961).
240. Harris, Harris, Peterson, and Rogers, *J. Med. Chem.*, 10, 261 (1967).
241. Kolodynska and Weiniaswski, *Acta Pol. Pharm.*, 26, 271 (1969); *Chem. Abstr.*, 72, 422 (1970).
242. Rosen, Milholland, and Nichol, *Symposium on Pyridoxol Enzymes*, 3rd ed., Maruzen Company, Ltd., Tokyo, 1967, p. 177.
243. Pollak, U.S. Patent 3,024,244 (1962); *Chem. Abstr.*, 57, 4639a (1962).
244. Pollak, U.S. Patent 3,024,245 (1962); *Chem. Abstr.*, 57, 4638i (1962).
245. Kon'kova and Petrova, *Trudy Vsesoyuz. Nauch.-Issledovatel. Vitamin. Inst.*, 6, 10 (1959); *Chem. Abstr.*, 55, 12399g (1961).
246. Suzuki, *Yakugaku Zasshi*, 81, 792 (1961); *Chem. Abstr.*, 55, 27476e (1961).
247. Cuiban and St. Cilianu, *Rev. Chim.* (Bucharest), 10, 74 (1959).
248. Seiyaku Company, Japanese Patent 26855 (1964); *Chem. Abstr.*, 62, 10418h (1965).
249. Merck and Company, British Patent 834,451 (1960); *Chem. Abstr.*, 55, 17647d (1961).
250. Balyakina and Zhdanovich, *Zh. Obshch. Khim.*, 31, 2983 (1961).
251. Balyakina, Zhukova, Malysheva, and Zhdanovich, U.S.S.R. Patent 213,027, 1968; *Chem. Abstr.*, 69, 43810f (1968).
252. Firestone, *Tetrahedron Lett.*, 2629 (1967).
253. Patzelt, Kugler, Horak, and Dostal, Czechoslovakian Patent 113,723 (1965); *Chem. Abstr.*, 63, 18043a (1965).

254. Perina, Vavrova, Kakac, and Horak, Czechoslovakian Patent 116,694 (1965); *Chem. Abstr.*, **65**, 13666e (1966).
255. Merck and Company, Netherland Patent 6,613,567 (1967); *Chem. Abstr.*, **68**, 68891a (1968).
256. Takeda Chemical, Ind., Japanese Patent 22,886 (1963); *Chem. Abstr.*, **60**, 5466e (1964).
257. Balyakina, Zhdanovich, and Preobrazhenskii, *Zh. Prikl. Khim.*, **35**, 1864 (1962); *Chem. Abstr.*, **58**, 6782d (1963).
258. Balyakina, Zhdanovich, and Preobrazhenskii, *Tr. Vses. Nauchn.-Issled. Vitamin. Inst.*, 7, 8 (1961); *Chem. Abstr.*, **59**, 11417e (1963).
259. Yamanouchi Pharmaceutical Company, Japanese Patent 65 5834 (1965); *Chem. Abstr.*, **62**, 16205a (1965).
260. Yodogawa Pharmaceutical Company, Japanese Patent 65 27813 (1965); *Chem. Abstr.*, **64**, 11182e (1966).
261. Nielsen, Elming, and Clauson-Kaas, *Acta Chem. Scand.*, **14**, 938 (1960).
262. Elming, Carlsten, Lennart, and Ohlsson, British Patent 862,581 (1961); *Chem. Abstr.*, **56**, 11574g (1962).
263. Aktieselskabet, Sadolin, and Holmblad, Danish Patent 80,971 (1956); *Chem. Abstr.*, **50**, 14811c (1956).
264. Balyakina, Zhukova, and Zhdanovich, *Zh. Prikl. Khim.* (Leningrad), **41**, 2324 (1968); *Chem. Abstr.*, **70**, 350 (1969).
265. Sumitomo Chemical Company, Ltd., Japanese Patent 70 00,497 (1970); *Chem. Abstr.*, **72**, 78890w (1970).
266. Takeda Chemical Industries, Ltd., British Patent 1,151,252 (1969); *Chem. Abstr.*, **71**, 49776n (1969).
267. Sumitomo Chemical Company, Ldt., Japanese Patent 70 01,613 (1970); *Chem. Abstr.*, **72**, 100529j (1970).
268. Kondrat'eva, U.S.S.R. Patent 196,854 (1968); *Chem. Abstr.*, **69**, 67240w (1968).
269. Hoffmann-LaRoche and Company, French Patent 1,384,099 (1965); *Chem. Abstr.*, **63**, 4263f (1965).
270. Hoffmann-LaRoche and Company, Netherland Patent 6,403,004 (1964); *Chem. Abstr.*, **62**, 7733h (1965).
271. Hoffmann-LaRoche and Company, Netherland Patent 6,506,703 (1965); *Chem. Abstr.*, **64**, 15851d (1966).
272. Merck and Company, Netherland Patent 6,614,802 (1967); *Chem. Abstr.*, **68**, 87191p (1968).
273. Merck and Company, Netherland Patent 6,614,801 (1967); *Chem. Abstr.*, **68**, 87190n (1968).
274. Osbond, British Patent 1,034,483 (1966); *Chem. Abstr.*, **69**, 106571f (1968).
275. Harris, Rosenburg, and Chamberlin, U.S. Patent 3,381,014 (1968); *Chem. Abstr.*, **69**, 52023d (1968).
276. Pollak, U.S. Patent 3,365,461 (1968); *Chem. Abstr.*, **69**, 35963t (1968).
277. Miki and Matsuo, Japanese Patent 67 25,664 (1964); *Chem. Abstr.*, **69**, 43807k (1968).
278. Murakami and Iwanami, *Bull. Chem. Soc. Jap.*, **41**, 726 (1968).
279. Pfister, Harris, and Firestone, U.S. Patent 3,227,721 (1966); *Chem. Abstr.*, **64**, 9689g (1966).
280. Harris, Firestone, Pfister, Boettcher, Cross, Currie, Monaco, Peterson, and Reuter, *J. Org. Chem.*, **27**, 2705 (1962).
281. Miki and Matsuo, *Yakugaku Zasshi*, **87**, 323 (1967); *Chem. Abstr.*, **67**, 3074 (1967).

282. Takeda Chemical, Ind., Japanese Patent 67 11,745 (1967); *Chem. Abstr.*, **68**, 87177p (1968).
283. Daiichi Seiyaku Company, Japanese Patent 65 22,740 (1965); *Chem. Abstr.*, **64**, 3496g (1966).
284. Takeda Chemical Industries, Ltd., Japanese Patent 69 21,097 (1969); *Chem. Abstr.*, **71**, 124,263y (1969).
285. Doktorova, Ionova, Karpeiskii, Padyukova, Turchin, and Florent'ev, *Tetrahedron*, **25**, 3527 (1969).
286. Colombini and Celon, *Gazz. Chim. Ital.*, **99**, 526 (1969).
287. Argoudelis and Kummerow, *Biochemistry*, **5**, 1 (1966).
288. Nakai, Ohishi, Shimizu, and Fukui, *Bitamin*, **35**, 213 (1967).
289. Sakuragi and Kummerow, *J. Org. Chem.*, **24**, 1032 (1959).
290. Smirnov, Lezina, Bystrov, and Dyumaev, *Izv. Akad. Nauk SSSR, Otd. Khim. Nauk*, 752 (1963).
291. Taborsky, *J. Org. Chem.*, **26**, 596 (1961).
292. McCasland, Gottwald, and Furst, *J. Org. Chem.*, **26**, 3541 (1961).
293. Tomita, Brooks, and Metzler, *J. Heterocycl. Chem.*, **3**, 178 (1966).
294. Paul and Korytnyk, *Chem. Ind.* (London), 230 (1967).
295. Katritzky, Kucharska, Tucker, and Wuest, *J. Med. Chem.*, **9**, 620 (1966).
296. Schmidt, *Ann. Chem.*, **657**, 156 (1962).
297. Seiyaku Co., Japanese Patent 67 9341 (1967); *Chem. Abstr.*, **68**, 68893c (1968).
298. Uno, Funabiki, Irie, and Yoshimura, *Yakugaku Zasshi*, **87**, 1293 (1967); *Chem. Abstr.*, **68**, 6706 (1968).
299. Seiyaku Company, British Patent 1,070,120 (1967); *Chem. Abstr.*, **68**, 78148c (1968).
300. Seiyaku Company, British Patent 1,101,369 (1968); *Chem. Abstr.*, **69**, 10374p (1968).
301. Seiyaku Company, French Patent 1,479,985 (1967); *Chem. Abstr.*, **68**, 114447k (1968).
302. Okumura, Imado, and Oda, *Bitamin*, **35**, 375 (1967).
303. Kuroda, Tanaka, and Moeda, *Bitamin*, **35**, 20 (1967).
304. Société Belge de l'Azote, German Patent 1,228,261 (1966); *Chem. Abstr.*, **66**, 28661c (1967).
305. Sapchim-Fournier-Cimag, French Patent 1,518,970 (1968); *Chem. Abstr.*, **71**, 30365f (1969).
306. Société Belge. de l'Azote, Belgian Patent 640,827 (1964); *Chem. Abstr.*, **63**, 587h (1965).
307. Mizuno, Aoki, and Kamada, *Bitamin*, **38**, 129 (1968); *Chem. Abstr.*, **69**, 58157x (1968).
308. Nitto Surfactant, Ind., Company, Japanese Patent 64 21,846 (1964); *Chem. Abstr.*, **62**, 11825d (1965).
309. Yodogawa Pharmaceutical Company, Ltd., Japanese Patent 68 23,946 (1968); *Chem. Abstr.*, **70**, 57669z (1969).
310. Yodogawa Pharmaceutical Company, Ltd., Japanese Patent 68 23,945 (1968); *Chem. Abstr.*, **70**, 57651n (1969).
311. Mizuno, Aoki, and Kamada, *Bitamin*, **38**, 120 (1968); *Chem. Abstr.*, **69**, 96407m (1969).
312. Mizuno, Aoki, Kamada, and Sumimoto, *Bitamin*, **38**, 125 (1968); *Chem. Abstr.*, **69**, 58156w (1968).
313. Balyokina, Zhdanovich, Zemskova, and Preobrazhenskii, *Zh. Obshch. Khim.*, **32**, 1172 (1962).

314. Korytnyk and Fricke, *J. Med. Chem.*, **11**, 180 (1968).
315. Seiyaku Company, Japanese Patent 62 11,832 (1962); *Chem. Abstr.*, **59**, 10077d (1963).
316. Laboratoires Houde, British Patent 1,109,539 (1968); *Chem. Abstr.*, **69**, 67225v (1968).
317. Reglisse Zan Anciens Etablissements Tersonniere et Kreitman, French Patent M 3341 (1965); *Chem. Abstr.*, **63**, 13222e (1965).
318. Laboratoires d'opochimio-therapie, French Patent 1,459,299 (1966); *Chem. Abstr.*, **67**, 64,247f (1967).
319. Laboratoires d'Opochimio-therapie, French Patent M 5417 (1967); *Chem. Abstr.*, **71**, 61228t (1969).
320. Laboratoires Sohio A.A., French Patent 1,434,375 (1966); *Chem. Abstr.*, **67**, 43686d (1967).
321. Sophymex, S.A., French Patent M 4028 (1966); *Chem. Abstr.*, **67**, 100017n (1967).
322. Laboratoires Cassenne, French Patent M 5464 (1967); *Chem. Abstr.*, **71**, 124266f (1969).
323. Yoshikawa, Kato, and Takenishi, U.S. Patent 3,365,460 (1968); *Chem. Abstr.*, **69**, 27,257a (1968).
324. Research Foundation for Practical Life, Japanese Patent 65 10,155 (1965); *Chem. Abstr.*, **63**, 5609h (1965).
325. Kyowa Fermentation Industry Company, Ltd., German Patent 1,910,035 (1969); *Chem. Abstr.*, **72**, 66838x (1970).
326. Hoffmann-LaRoche and Company, German Patent 1,807,603 (1969); *Chem. Abstr.*, **71**, 91330n (1969).
327. Kyowa Fermentation Industry Company, Ltd., British Patent 1,166,426 (1969); *Chem. Abstr.*, **72**, 43467m (1970).
328. Laboratoire M. Richard S. A., French Patent M 3115 (1965); *Chem. Abstr.*, **63**, 1772h (1965).
329. Société d'Etudes et de Recherches Pharm., French Patent 1,477,452 (1967); *Chem. Abstr.*, **68**, 87594x (1968).
330. Greene and Montgomery, *J. Med. Chem.*, **7**, 17 (1964).
331. Kreisky, *Monatsh. Chem.*, **89**, 685 (1958).
332. Schmidt and Giesselmann, *Ann. Chem.*, **657**, 162 (1962).
333. Kuroda, *Bitamin*, **30**, 431 (1964); *Chem. Abstr.*, **62**, 10402b (1965).
334. Schmidt, U.S. Patent 3,075,987 (1963); *Chem. Abstr.*, **59**, 9999f (1963).
335. Koch, Klemna, and Seiter, *Arzneim.-Forsch.* (Weinheim), **10**, 683 (1960).
336. Merck, A.-G., German Patent 1,186,064 (1965); *Chem. Abstr.*, **62**, 10417g (1965).
337. Greene, Williams, and Montgomery, *J. Med. Chem.*, **7**, 20 (1964).
338. Merck, A.-G., British Patent 1,156,769 (1969); *Chem. Abstr.*, **71**, 71321k (1969).
339. Darge, Liss, and Oeff, *Arzneim.-Forsch.* (Weinheim), **19**, 9 (1969).
340. Darge, Liss, and Oeff, *Arzneim.-Forsch.* (Weinheim), **19**, 5 (1969).
341. Merck, A.-G., German Patent 1,238,473 (1967); *Chem. Abstr.*, **68**, 39749s (1968).
342. Merck, A.-G., German Patent 1,238,474 (1967); *Chem. Abstr.*, **68**, 39480k (1968).
343. Merck, A.-G., U.S. Patent 3,086,023 (0000); *Chem. Abstr.*, **59**, 9995 (1963).
344. Deguchi, Japanese Patent 68 06,790 (1968); *Chem. Abstr.*, **69**, 87010z (1968).
345. Hikawa, Aki, Fushimi, and Yumgi, Japanese Patent 67 25,668 (1967); *Chem. Abstr.*, **69**, 52014b (1968).
346. Merck, A.-G., Netherland Patent 6,412,891 (1965); *Chem. Abstr.*, **64**, 8154c (1966).
347. Yamanouchi Pharmaceutical Company, Ltd., Japanese Patent 69 11,369 (1969); *Chem. Abstr.*, **71**, 70497g (1969).

348. Takeda Chemical Industries, Ltd., Japanese Patent 69 28,095 (1969); *Chem. Abstr.*, **72**, 43475n (1970).
349. Petrova and Bel'tsova, *Zh. Obshch. Khim.*, **32**, 274 (1962).
350. Seiyaku Co., Japanese Patent 67 2706 (1967); *Chem. Abstr.*, **67**, 73531s (1967).
351. Daiichi Seiyaku Company, Ltd., Japanese Patent 69 16,655 (1969); *Chem. Abstr.*, **71**, 124258a (1969).
352. Iwanami, Osawa, and Murakami, *J. Vitaminol. (Kyoto)*, **14**, 326 (1968); *Chem. Abstr.*, **71**, 316 (1969).
353. Takeda Chemical Industries, Ltd., Japanese Patent 69 16,654 (1969); *Chem. Abstr.*, **71**, 91314k (1969).
354. Takeda Chemical Industries, Ltd., Japanese Patent 68 26,176 (1968); *Chem. Abstr.*, **70**, 57600v (1969).
355. Iwanami, Osawo, and Nurekami, *Bitamin*, **36**, 122 (1967); *Chem. Abstr.*, **68**, 12818g (1968).
356. Merck, A.-G., British Patent 927,666 (1963); *Chem. Abstr.*, **60**, 5467h (1964).
357. Iwanami, Osawa and Murakami, *Bitamin*, **36**, 119 (1967); *Chem. Abstr.*, **68**, 1222 (1968).
358. Hirano, Masuda, and Fushimi, Japanese Patent 68 02,712 (1968); *Chem. Abstr.*, **69**, 86827j (1968).
359. Korytnyk, *J. Org. Chem.*, **27**, 3724 (1962).
360. Seiyaku Company, Japanese Patent 67 9348 (1967); *Chem. Abstr.*, **68**, 68892b (1968).
361. Korytnyk and Paul, *Tetrahedron Lett.*, 777 (1966).
362. Korytnyk and Wiedeman, *J. Chem. Soc.*, 2531 (1962).
363. Korytnyk and Paul, *J. Org. Chem.*, **32**, 3791 (1967).
364. Argoudelis and Kummerow, *J. Org. Chem.*, **29**, 2663 (1964).
365. Huettenranch and Matthey, *Z. Chem.*, **6**, 421 (1966).
366. Korytnyk and Singh, *J. Amer. Chem. Soc.*, **85**, 2813 (1963).
367. Korytnyk and Paul, *J. Heterocycl. Chem.*, **2**, 481 (1965).
368. DeJongh, Perricone, and Korytnyk, *J. Amer. Chem. Soc.*, **88**, 1233 (1966).
369. Abbott and Martell, *J. Amer. Chem. Soc.*, **91**, 6931 (1969).
370. Bridges, Creaven, Davies, and Williams, *Biochem. J.*, **88**, 65p (1963).
371. Ikeda, Oku, Ohishi, and Fukui, *Bitamin*, **38**, 109 (1968); *Chem. Abstr.*, **69**, 9015 (1968).
372. Efimovsky, Fourneau, Jacquier, and Stoven, *Chim. Ther.*, **4**, 89 (1969).
373. Ogata, Uchida, Kurihara, Tani, and Tochikura, *J. Vitamino. (Kyoto)*, **15**, 160 (1969); *Chem. Abstr.*, **71**, 512 (1969).
374. Harris, Zabriskie, Chamberlin, Crane, Peterson, and Reuter, *J. Org. Chem.*, **34**, 1993 (1969).
375. Korytnyk and Paul, *J. Med. Chem.*, **13**, 187 (1970).
376. Conference, "Vitamin B_6 in Metabolism of the Nervous System," in *Acad. Sci.*, **166**, 1–364 (1969), New York, art. 1.
377. Melius and Marshall, *J. Med. Chem.*, **10**, 1157 (1967).
378. Dreizen, Goodrich, and Stern, *Int. Z. Vitaminforsch.*, **38**, 210 (1968); *Chem. Abstr.*, **69**, 6993 (1968).
379. Foukas, *Zentralbl. Gynaekol.*, **89**, 831 (1967); *Chem. Abstr.*, **68**, 173 (1968).
380. Korytnyk, Paul, Bloch, and Nichol, *J. Med. Chem.*, **10**, 345 (1967).
381. Société Agrologique Française, Belgian Patent 621,423 (1962); *Chem. Abstr.*, **59**, 12768g (1963).
382. Hansson, *Acta Soc. Med. Upsal.*, **73**, 19 (1968); *Chem. Abstr.*, **69**, 3976 (1968).

383. Schunk, *Strahlentherapie*, **131**, 445 (1966); *Chem. Abstr.*, **66**, 3458 (1967).
384. Johansson, Lindotedt, and Tiselius, *Biochemistry*, **7**, 2327 (1968).
385. Koerner and Nowak, *Arzneim.-Forsch.* (Weinheim), **17**, 572 (1967).
386. Korytnyk and Paul, "Pyridoxal Catalogue: Enzymes Model Systems, Proceedings of the International Symposium," 2nd ed., (Snell, Ed.), Interscience, New York, 1966, p. 615.
387. Efremov, Goryachenkova, and Gvozdova, "Pyridoxal Catalogue: Enzymes Model Systems, Proceedings of the International Symposium, 2nd ed., Snell, Ed., Interscience, New York, 1966, p. 713.
388. Korytnyk and Kris, *Chem. Ind.* (London), 1834 (1961).
389. Korytnyk, Kris, and Singh, *J. Org. Chem.*, **29**, 574 (1964).
390. Houston, Laakso, and Metzler, *J. Heterocycl. Chem.*, **3**, 126 (1966).
391. Chugai Pharm. Co., Japanese Patent 64 26,728 (1964); *Chem. Abstr.*, **62**, 11788b (1965).
392. Heinert and Martell, *J. Amer. Chem. Soc.*, **81**, 3933 (1959).
393. Morozov, Bazhulina, Cherkashina, and Karpeiskii, *Biofizika*, **12**, 773 (1967); *Chem. Abstr.*, **68**, 1795 (1968).
394. Gansow and Holm, *Tetrahedron*, **24**, 4477 (1968).
395. Merck, A.-G., German Patent 1,244,786 (1967); *Chem. Abstr.*, **68**, 49466z (1968).
396. Takanashi, Tamura, Yoshino, and Iidaka, *Chem. Pharm. Bull.* (Tokyo), **16**, 758 (1968).
397. Ahrens and Korytnyk, *J. Heterocycl. Chem.*, **4**, 625 (1967).
398. Argondelis and Kummerow, *Biochim. Biophys. Acta*, **74**, 568 (1963).
399. Yamanouchi Pharmaceutical Company, Japanese Patent 66 6907 (1966); *Chem. Abstr.*, **65**, 7151e (1966).
400. Iwata and Metzler, *J. Heterocycl. Chem.*, **4**, 319 (1967).
401. Florent'ev, Drobinskaya, Ionova, and Karpeiskii, *Tetrahedron Lett.*, 1747 (1967).
402. Okumura, Oda, and Nishihara, *Bitamin*, **35**, 384 (1967); *Chem. Abstr.*, **67**, 4097 (1967).
403. Okumura, Oda, and Nishihara, *Bitamin*, **35**, 380 (1967); *Chem. Abstr.*, **67**, 4096 (1967).
404. Duhault, Gonnard, and Fenard, *Bull. Soc. Chim. Biol.*, **49**, 177 (1967).
405. Hata, Yamakawa, Anraku, and Sano, *Yakuzaigaku*, **26**, 294 (1966); *Chem. Abstr.*, **69**, 9015 (1968).
406. Yodogawa Pharmaceutical Company, Japanese Patent 63 2533 (1963); *Chem. Abstr.*, **59**, 11445f (1963).
407. Seiyaku Company, Japanese Patent 66 5095 (1966); *Chem. Abstr.*, **65**, 2230g (1966).
408. King, *Sci. Progr.* (London), **52**, 645 (1964).
409. Fischer, Seiler, Thobe, and Werner, *Naturwissenschaften*, **55**, 445 (1968).
410. Sennello and Argondelis, *J. Org. Chem.*, **33**, 3983 (1968).
411. Muhlradt, Morino, and Snell, *J. Med. Chem.*, **10**, 341 (1967).
412. Muhlradt and Snell, *J. Med. Chem.*, **10**, 129 (1967).
413. Hullar, *J. Med. Chem.*, **12**, 58 (1969).
414. Florent'ev, Drobinshaya, Ionova, Karpeiskii, and Turchin, *Dokl. Akad. Nauk SSSR*, **177**, 617 (1967).
415. Zenker, *J. Med. Chem.*, **9**, 826 (1966).
416. Iwanami, Numata, and Murakami, *Bull. Chem. Soc. Jap.*, **41**, 161 (1968).
417. Merck, A.-G., British Patent 1,111,876 (1968); *Chem. Abstr.*, **69**, 67396b (1968).
418. Seiyaku Co., French Patent 1,473,663 (1967); *Chem. Abstr.*, **68**, 87195t (1968).
419. Yamanouchi Pharmaceutical Company, Japanese Patent 67 3305 (1967); *Chem. Abstr.*, **67**, 82109n (1967).

420. Seiyaku Company, Japanese Patent 66 21,031 (1966); *Chem. Abstr.*, **66**, 65401p (1967).
421. Seiyaku Company, British Patent 1,076,910 (1967); *Chem. Abstr.*, **68**, 59443d (1968).
422. Wakamoto Pharmaceutical Company, Japanese Patent 63 19739 (1963); *Chem. Abstr.*, **60**, 4116a (1964).
423. Yamada and Saito, *J. Vitaminol.* (Kyoto), **11**, 192 (1965); *Chem. Abstr.*, **64**, 6955h (1966).
424. Dempsey and Christensen, *J. Biol. Chem.*, **237**, 1113 (1962).
425. Ueno, Ishikawa, and Naito, *Tetrahedron Lett.*, 1283 (1969).
426. Hullar, *Ann. N. Y. Acad. Sci.*, **166**, 191 (1969).
427. Hullar, *J. Med. Chem.*, **12**, 58 (1969).
428. Daiichi Seiyaku Company, Ltd., French Patent 1,530,842 (1968); *Chem. Abstr.*, **71**, 81203j (1969).
429. Daiichi Seiyaku Company, Ltd., Japanese Patent 68 26,296 (1968); *Chem. Abstr.*, **70**, 57667x (1969).
430. Nakai, Masugi, Okeshi, and Fukui, *Bitamin*, **38**, 189 (1968); *Chem. Abstr.*, **70**, 348 (1969).
431. Daiichi Seiyaku Company, Ltd., Japanese Patent 68 15,435 (1968); *Chem. Abstr.*, **70**, 57660q (1969).
432. Daiichi Seiyaku Company, Ltd., Japanese Patent 69 30,272 (1969); *Chem. Abstr.*, **72**, 55279q (1970).
433. Kyowa Fermentation Industry Company, Ltd., British Patent 1,166,425 (1969); *Chem. Abstr.*, **72**, 12576s (1970).
434. Kyowa Fermentation Industry Company, Ltd., French Patent 1,533,562 (1968); *Chem. Abstr.*, **71**, 38811u (1969).
435. Nakagawa and Matsui, *Nippon Nogei Kagaku Kaishi*, **42**, 300 (1968); *Chem. Abstr.*, **70**, 344 (1969).
436. Pocker and Fischer, *Biochemistry*, **8**, 5181 (1969).
437. Dainippon Pharmaceutical Company, Ltd., German Patent 1,912,602 (1969); *Chem. Abstr.*, **72**, 21612c (1970).
438. Testa, Bonati, and Pagani, *Chimia*, **15**, 314 (1961).
439. Lepetil S.P.A., British Patent 924,514 (1963); *Chem. Abstr.*, **59**, 11440d (1963).
440. Vitali, Mossini, and Bertaccini, *Farmaco, Ed. Sci.*, **20**, 634 (1965).
441. Okumura, Japanese Patent 68 02,716 (1968); *Chem. Abstr.*, **69**, 59107t (1968).
442. Heinert and Martell, *J. Amer. Chem. Soc.*, **85**, 183 (1963).
443. Heinert and Martell, *J. Amer. Chem. Soc.*, **85**, 1334 (1963).
444. Heinert and Martell, *J. Amer. Chem. Soc.*, **85**, 1488 (1963).
445. Danno, Matsuoka, Nurimoto, and Hayashi, *Bitamin*, **37**, 25 (1968); *Chem. Abstr.*, **68**, 4647 (1968).
446. Yamanouchi Pharmaceutical Company, Japanese Patent 65 26,820 (1965); *Chem. Abstr.*, **64**, 8154b (1966).
447. Loo and Whittaker, *J. Neurochem.*, **14**, 997 (1967).
448. Matsushima and Martell, *J. Amer. Chem. Soc.*, **89**, 1322 (1967).
449. Matsushima and Martell, *J. Amer. Chem. Soc.*, **89**, 1331 (1967).
450. Iwanami and Murakami, *Bitamin*, **36**, 301 (1967); *Chem. Abstr.*, **68**, 4717 (1968).
451. Hill and Mann, *Biochem. J.*, **99**, 454 (1966).
452. Bruice and Topping, *J. Amer. Chem. Soc.*, **85**, 1480 (1963).
453. Bruice and Topping, *J. Amer. Chem. Soc.*, **85**, 1493 (1963).
454. Heinert and Martell, *J. Amer. Chem. Soc.*, **84**, 3257 (1962).

455. Davis, Roddy, and Metzler, *J. Amer. Chem. Soc.*, **83**, 127 (1961).
456. Bruice and Topping, *J. Amer. Chem. Soc.*, **84**, 2448 (1962).
457. Okumura, Kotera, Oda, Kondo, Masukawa, Ueba, and Inoue, *Bitamin*, **37**, 575 (1968); *Chem. Abstr.*, **69**, 7219 (1968).
458. Schirch, "Pyridoxal Catalogue: Enzymes Model Systems, Proceedings of the International Symposium," 2nd ed., Snell, Ed., Interscience, New York, 1966, p. 203.
459. Johnson, Walter, and Aisner, *J. Heterocycl. Chem.*, **6**, 579 (1969).
460. Kobayashi and Makino, *Jikeikai Med. J.*, **15**, 249 (1968); *Chem. Abstr.*, **71**, 250 (1969).
461. Kyowa Fermentation Industry Company, Ltd., German Patent 1,929,272 (1969); *Chem. Abstr.*, **72**, 31624x (1970).
462. Abbot and Martell, *J. Amer. Chem. Soc.*, **92**, 1754 (1970).
463. Skirkh, *Khim. Biol. Piridoksal. Kotal.*, 128 (1968); *Chem. Abstr.*, **71**, 360 (1969).
464. Malsushima, *Chem. Pharm. Bull.* (Tokyo), **16**, 2046 (1968).
465. Kyowa Fermentation Industry Company, Ltd., French Patent 1,525,504 (1968); *Chem. Abstr.*, **71**, 3281u (1969).
466. Malsushima and Hino, *Chem. Pharm. Bull.* (Tokyo), **16**, 2277 (1968).
467. Nakai, Ohishi, Shimizu and Fukui, *Bitamin*, **36**, 521 (1967); *Chem. Abstr.*, **68**, 5737 (1968).
468. Iwata, *Biochem. Prep.*, **12**, 117 (1968).
469. Sattsangi and Argoudelis, *J. Org. Chem.*, **33**, 1337 (1968).
470. Wampler and Churchich, *Biochem. Biophys. Res. Commun.*, **32**, 629 (1968).
471. Wampler and Churchich, *J. Biol. Chem.*, **244**, 1477 (1969).
472. Paul and Korytnyk, *Tetrahedron*, **25**, 1071 (1969).
473. Tanabe Seiyaku Company, Ltd., German Patent 1,811,054 (1969); *Chem. Abstr.*, **72**, 31635b (1970).
474. Tanabe Seiyaku Company, Ltd., Japanese Patent 69 28,307 (1969); *Chem. Abstr.*, **72**, 43471h (1970).
475. Tanabe Seiyaku Company, Ltd., Japanese Patent 69 28,305 (1969); *Chem. Abstr.*, **72**, 78887a (1970).
476. Florent'ev, Drobenskaya, Ionova, and Karpeiskii, *Khim. Geterotsikl. Soedin.*, 1028 (1969); *Chem. Abstr.*, **72**, 381 (1970).
477. Takeda Chem. Ind., Belgian Patent 648,226 (1964); *Chem. Abstr.*, **63**, 18036h (1965).
478. Merck, A.-G., British Patent 915,451 (1963); *Chem. Abstr.*, **59**, 2840f (1963).
479. Takeda Chemical Ind., Japanese Patent 65 17,590 (1965); *Chem. Abstr.*, **63**, 18047c (1965).
480. Sen and Chatterjee, *J. Indian Chem. Soc.*, **42**, 695 (1965).
481. Larrouquere, *Bull. Soc. Chim. Fr.*, 329 (1968).
482. Dannenberg and Iglesias, *Z. Physiol. Chem.*, **349**, 1077 (1968).
483. Dempsey and Snell, *Biochemistry*, **2**, 1414 (1963).
484. Ayling and Snell, *Biochemistry*, **7**, 1616 (1968).
485. Ayling, Dunathan, and Snell, *Biochemistry*, **7**, 4573 (1968).
486. Dunathan, Davis, Kury, and Kaplon, *Biochemistry*, **7**, 4532 (1968).
487. Banks, Diamantis, and Vernon, *J. Chem. Soc.*, 4235 (1961).
488. Argoudelis and Kummerow, *J. Org. Chem.*, **26**, 3420 (1961).
489. Vasak and Kopecky, "Toxicol. Carbon Disulfide, Proceeding of the Symposium," Prague, 35 (1966); *Chem. Abstr.*, **68**, 5624 (1968).
490. Dregval and Rybak, *Khim. Geterolsikl. Svedin., Sb. 1: Azolsoderzhashchie Geterolsikly*, 227, 1967; *Chem. Abstr.*, **70**, 305 (1969).
491. CIBA Corp., U.S. Patent 3,471,505 (1969); *Chem. Abstr.*, **72**, 12585u (1970).

492. Banyu Pharmaceutical Company, Ltd., Japanese Patent 69 22,288 (1969); *Chem. Abstr.*, **72**, 3385b (1970).
493. Shimamoto, Ishikawa, Ishikawa, Inoue and Shimamoto, South African Patent 68 00,677 (1968); *Chem. Abstr.*, **71**, 70504g (1969).
494. Nepera Chemical Company, U.S. Patent 3,438,993 (1969); *Chem. Abstr.*, **71**, 30368j (1969).
495. Yoshitomi Pharmaceutical Industries, Ltd., Japanese Patent 68 14,466 (1968); *Chem. Abstr.*, **70**, 57653q (1969).
496. Lilly and Company, French Patent 1,512,567 (1968); *Chem. Abstr.*, **70**, 106392q (1969).
497. Nielsen and Platt, *J. Heterocycl. Chem.*, **6**, 891 (1969).
498. Mallinckrodt Chemical Works, U.S. Patent 3,478,038 (1969); *Chem. Abstr.*, **72**, 31630w (1970).
499. Wykret and Kuczynski, *Diss. Pharm. Pharmacol.*, **21**, 407 (1969); *Chem. Abstr.*, **72**, 306 (1970).
500. Shimamoto, Ishikawa, Ishikawa, Inoue, and Shimamoto, French Patent 1,546,425 (1968); *Chem. Abstr.*, **72**, 55272g (1970).
501. Bristol-Myers Company, U.S. Patent 3,418,328 (1968); *Chem. Abstr.*, **70**, 57662s (1969).
502. Banyu Pharmaceutical Company, Ltd., Japanese Patent 69 30,271 (1968); *Chem. Abstr.*, **72**, 25278p (1970).
503. Sumitomo Chemical Company, Ltd., Japanese Patent 69 22,494 (1969); *Chem. Abstr.*, **72**, 12578u (1970).
504. Takeda Chemical Industries, Ltd., Japanese Patent 69 26,103 (1969); *Chem. Abstr.*, **72**, 12590s (1970).
505. Hoffmann-LaRoche, Swiss Patent 448,085 (1968); *Chem. Abstr.*, **70**, 28830q (1969).
506. Sumitomo Chemical Company, Ltd., Japanese Patent 69 22,495 (1969); *Chem. Abstr.*, **72**, 12591t (1970).
507. Tanabe Seiyaku Company, Ltd., Japanese Patent 69 22,289 (1969); *Chem. Abstr.*, **72**, 12575r (1970).
508. Tanabe Seiyaku Company, Ltd., Japanese Patent 68 23,944 (1968); *Chem. Abstr.*, **70**, 77808x (1969).

CHAPTER XIV

Pyridine Aldehydes and Ketones

RENAT H. MIZZONI

Ciba Pharmaceutical Company,
Division of the Ciba-Geigy Corporation
Summit, New Jersey

115

I. Pyridine Aldehydes

1. Preparation

A. *Conversion of Picolines to Aldehydes and Derivatives*

Bredereck and co-workers (1) have described the preparation of isonicotinalde-hyde from 4-picoline by way of an enamine intermediate (**XIV-1**). The product was isolated as the 2,4-dinitrophenylhydrazone. The same intermediate was also

XIV-1

prepared by Arnold (2) in another reaction from 4-picoline (**XIV-2**).

Arnold (2) has successfully applied the Vilsmeier-Haack reaction to the preparation of various pyridine side-chain aldehydes and derivatives, which also proceed *via* enamine intermediates (**XIV-3**).

2- And 4-picolines are nitrosated in the presence of strong bases to give the corresponding pyridine aldoximes (**XIV-4**) (3). Similarly, nitrosation of 2-picoline methobromide in aqueous alkali yields picolinaldoxime methobromide (4).

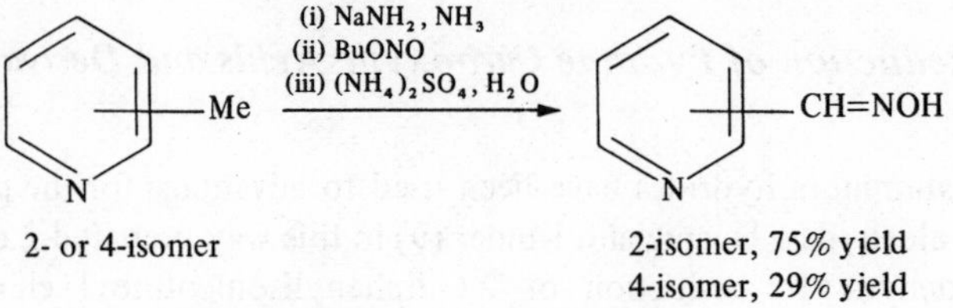

2- or 4-isomer   2-isomer, 75% yield
                                      4-isomer, 29% yield

XIV-4

B. *Preparation of Pyridine Aldehydes and Derivatives by Thermal Reactions*

Picolinaldehyde can be prepared by heating 2-hydroxymethylpyridine-1-oxide *in vacuo* at elevated temperatures (5). The reaction is an extension of the process in which 2-picoline-1-oxide and acetic anhydride gives picolinaldehyde in a stepwise process (**XIV-5**).

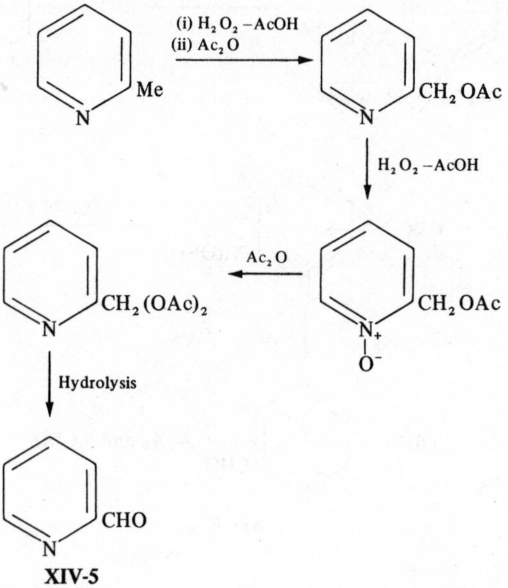

XIV-5

4-Hydroxymethylpyridine-1-oxide reacts with phenylhydrazine and with hydroxylamine to give isonicotinaldehyde phenylhydrazone and isonicotinaldoxime, respectively (6). Analogous results are obtained with 2-hydroxymethylpyridine, which also reacts with arylamines and nitrosobenzene in the presence of base to give anils of picolinaldehyde (7).

Mixtures of cyanopyridines, formic acid, and water give pyridine aldehydes with thorium oxide-aluminum oxide catalysis at elevated temperatures (8).

C. Reduction of Pyridine Carboxylic Acids and Derivatives

Complex aluminum hydrides have been used to advantage for the preparation of pyridine aldehydes. Huenig and Ruider (9) in this way prepared 2,6-diphenylisonicotinaldehyde by reduction of 2,6-diphenylisonicotinoyl chloride with lithium tributoxyaluminum hydride.

Reducing agents of this kind are especially valuable for the preparation of pyridine dialdehydes which are not readily available by other means. Quéguiner and his co-workers (10–13) have carried out extensive work along these lines. The stepwise reduction of pyridinedicarboxylic esters under controlled conditions leads to selectively deuterated pyridine dialdehydes and derivatives **(XIV-6)**. The use of lithium aluminum hydride is less satisfactory than that of diisobutylaluminum hydride in the reaction (13).

XIV-6

D. *Preparation of 2-Pyridylglyoxal*

Oxidation of 2-acetylpyridine by conventional reactions employing selenium dioxide does not give 2-pyridylglyoxal (14). The reaction product is 2-pyridylglyoxylic acid, which undergoes subsequent decarboxylation to give picolinaldehyde (XIV-7).

$$2\text{-PyCOMe} \xrightarrow{\text{SeO}_2} (2\text{-PyCOCHO})$$

$$2\text{-PyCHO} + CO_2 \longleftarrow (2\text{-PyCOCO}_2\text{H})$$
$$\mathbf{XIV\text{-}7}$$

2-Pyridylglyoxal dimethylacetal is produced in the reaction of 2-pyridyllithium with dimethoxyacetylpiperidide (XIV-8) (14).

$$2\text{-PyLi} + (MeO)_2\text{CHCON} \bigcirc$$

$$\downarrow \begin{array}{c} -16° \\ (77\%) \end{array}$$

$$2\text{-PyCOCH(OMe)}_2$$
$$\mathbf{XIV\text{-}8}$$

E. *Pyridoxal and Related Compounds*

Hullar (15) has studied the effect on enzymic properties of replacing the ester linkage of pyridoxal phosphate by a carbon-to-phosphorus bond. The synthesis of the resultant phosphonic acids is shown in XIV-9 (see also Chapter XIII).

XIV-9

Brooks and his collaborators (16) have synthesized isopropylidene isopyridoxal and the corresponding aldehydophenolic alcohol as outlined in **XIV-10**.

4,5-Diformyl-3-hydroxy-2-picoline, which is related to pyridoxal, has been prepared by the method shown in **XIV-11** (17).

XIV-11

2. Reactions

Bergmann and Paul (18) have subjected mixtures of nicotinaldehyde and substituted benzaldehydes to conditions of the benzoin condensation. The reaction gives mixed benzoins in varying yields (**XIV-12**). Isonicotinaldehyde does not participate in the reaction under these conditions.

R = Me$_2$N, 92% yield
R = OH, 48% yield

XIV-12

The reaction of pyridine dialdehydes with diazomethane gives various pyridine aldehydoketones (**XIV-13**) (19).

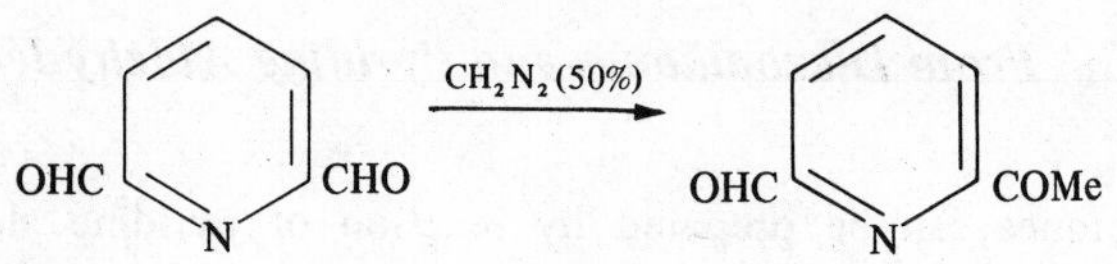

XIV-13

II. Pyridine Ketones

1. Preparation

A. *From Acyclic Components*

The synthesis of 2- and 2,6-disubstituted-3-pyridyl ketones by condensation of appropriate acetylenic compounds and 2-amino-2-pentene-4-one with thermal cyclization is described in a recent German patent **(XIV-14)** (20).

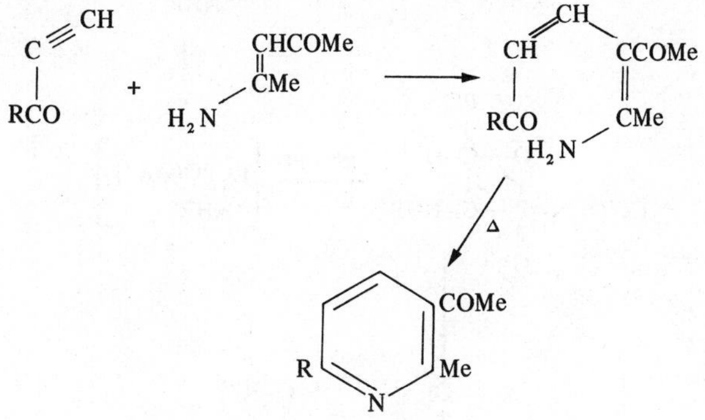

XIV-14

B. *By Acylation of Methylpyridines*

The preparation of mono- and diacylpyridines can be affected by vapor phase acylation of appropriate methylpyridines with vanadium pentoxide-aluminum oxide catalysis (21). In other reactions, various pyridine carboxylic esters react with aliphatic acids at elevated temperatures and similar catalysis to give the respective acylpyridines (22). In an analogous reaction, mono- and dicyanopyridines and aliphatic nitriles react to give the corresponding acylpyridines (23).

C. *From Diazoalkanes and Pyridine Aldehydes*

Pyridine ketones can be prepared by reaction of pyridine aldehydes with diazoalkanes in the presence of aluminum chloride. The concurrent formation of 1,2-epoxides has also been noted for this type of reaction (24). The reaction is also successful in the absence of catalyst (25).

The reaction of pyridine dialdehydes with diazomethane gives pyridine aldehydoketones (previously cited) and pyridine diketones (19).

D. *From Pyridine Aldehydes via Enamine Intermediates*

Zimmer and Bercz (26) have described a general method for the conversion of aromatic and heterocyclic aldehydes into ketone derivatives by way of enamine intermediates. The reaction has been employed to advantage in the synthesis of pyridyl ketones **(XIV-15).**

$$RCH=NR' + HPO(OR'')_2 \longrightarrow \underset{\underset{NHR'}{|}}{RCHPO(OR'')_2}$$

$$R = \text{pyridyl}$$
$$R' = \text{alkyl, phenyl}$$
$$R'' = \text{phenyl}$$
$$R''' = \text{alkyl, aryl, heterocycl.}$$

$$R'''CH_2COR + R'NH_2$$

XIV-15

E. *α,β-Unsaturated Ketones from Pyridine Aldehydes*

Isonicotinaldehyde and 2,6-disubstituted isonicotinaldehydes react with acetophenone in the presence of aqueous alkali to give α,β-unsaturated ketones (9).

The formation of 1,5-diketones as side products was noted earlier by Marvel and his co-workers (XIV-16) (27).

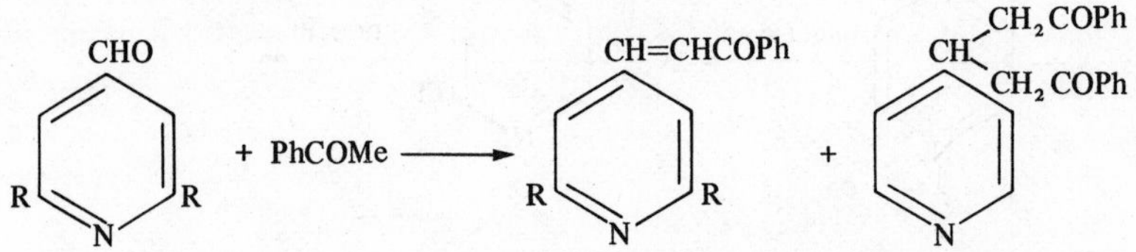

XIV-16

Acetylpyridines react with aromatic and heterocyclic aldehydes, with diethylamine catalysis, to give pyridyl (β-substituted vinyl) ketones (XIV-17) (28).

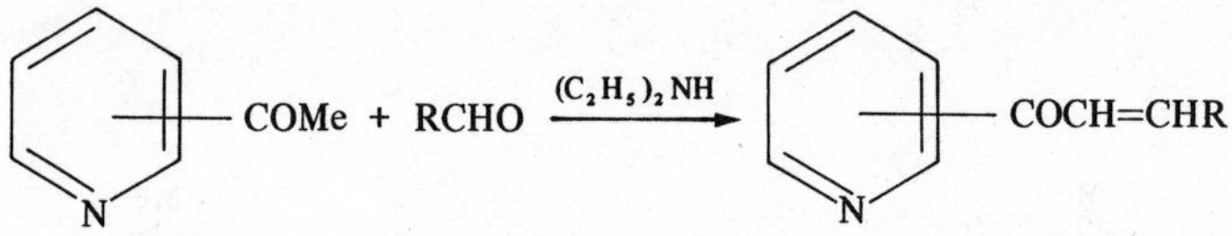

XIV-17

F. *From Pyran Derivatives*

4-Pyrones have long been known to give pyridine derivatives on reaction with ammonia. An extension of this reaction is found in the conversion of 4-benzoylmethylenepyrans to 4-phenacylpyridines by a similar process (XIV-18) (31).

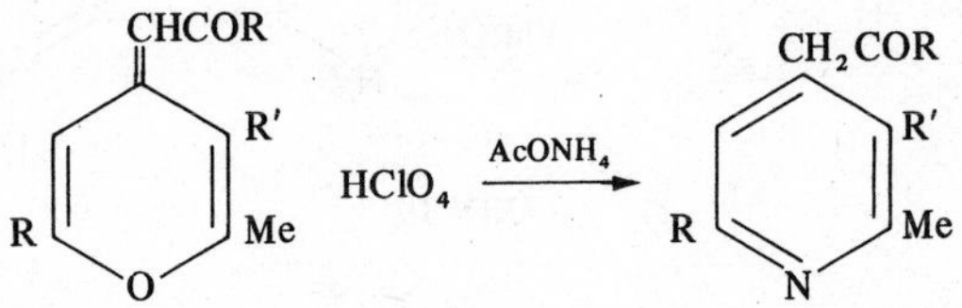

XIV-18

G. *By Acylation of Pyridines and their 1-Oxides*

Abramovitch and Smith (30) have investigated the reaction of aliphatic esters with various 2-pyridyllithium and 1-oxido-2-pyridyllithium derivatives. The products of these reactions are substituted 2-acylpyridines and 1-oxido-2-acyl-pyridines. A similar reaction with dimethylacetamide and benzonitrile gives acylated bimolecular condensation products (XIV-19) (31).

$$2\text{-PyLi} + \text{RCO}_2\text{Et} \longrightarrow 2\text{-PyCOR}$$

XIV-19

2. Reactions

Abramovitch and Tertzakian (32) have studied the Pschorr cyclization of 3-(o-aminobenzoyl)pyridine under a variety of conditions. The products of the reaction were 4-azafluorenone, 2-azafluorenone, 3-benzoylpyridine, and phenol, in varying proportions, depending on the method employed.

Chang (33) investigated the reactions of 2-, 3-, and 4-acetylpyridines with nitric acid. 3-Acetylpyridine gives the nitrate salt and nicotinic acid. 4-Acetylpyridine behaves in an analogous manner. 2-Acetylpyridine, however, gives the nitrate salt, picolinic acid, and the complex diketone **XIV-20**, in 28% yield.

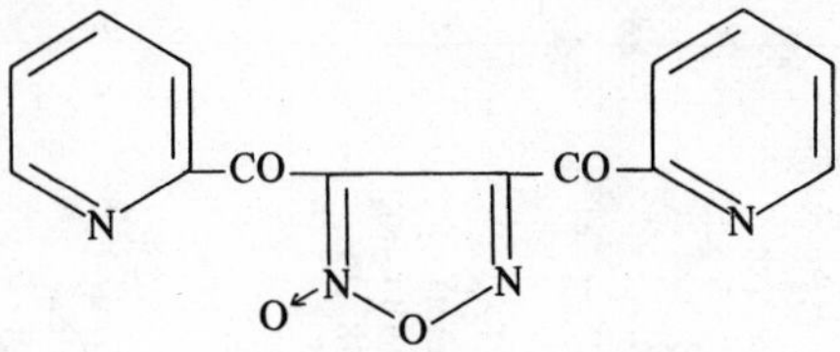

XIV-20

III. Tables

Abbreviations.

PH = Phenylhydrazone
DNPH = 2,4-Dinitrophenylhydrazone
SC = Semicarbazone
TSC = Thiosemicarbazone

TABLE XIV-1. Pyridine Aldehydes

Compound	Method of preparation	Yield (%)	Properties	Ref.
(pyridine-2-CHO)	(2-methylpyridine) + (i) t-BuOCH(NMe$_2$)$_2$ (ii) 5N HCl	80	Isolated as the 2,4-DNPH	1
(Me, CHO pyridine)	(Me, CH$_2$OH pyridine), Selenous acid		B.p. 134–136° (10–12 mm), 89° (0.3 mm), 82° (0.18 mm); osazone, m.p. 157–158°; PH, m.p. 138°	14
2-PyCOCHO	2-PyLi + (MeO)$_2$CHCON(cyclohexyl)	77		49
(Ph, CHO, Ph pyridine)	(Ph, COCl, Ph pyridine), LiAlH(OBu)$_3$	64	M.p. 95–96°	9
(CHO, CHO, CHO pyridine)	CO$_2$Me, CO$_2$Me, (isoBu)$_2$AlH	64	N-Oxide, m.p. 128°	10, 12, 13

Dialdehyde	Diester, Reagent	Yield (%)	Derivatives	References
(pyridine) CHO, CHO	(pyridine) CO_2Me, CO_2Me , $(isoBu)_2AlH$	68	Dioxime, m.p. 240°; Bis-2,4-DNPH, m.p. 331°	10
OHC, CHO (pyridine)	(pyridine) CO_2Me, CO_2Me , $LiAlH_4$	30	Bis-PH, m.p. 212°; N-oxide, m.p. 148°	11, 12
OHC, CHO	MeO_2C, CO_2Me , $(isoBu)_2AlH$	59	Bis-2,4-DNPH, m.p. 339°; N-oxide, m.p. 208°	10–12
OHC, CHO	MeO_2C, CO_2Me , $(isoBu)_2AlH$	69	N-Oxide, m.p. 188°	10, 12, 13
OHC, CHO, CHO	CO_2Me, CO_2Me , $(isoBu)_2AlH$		Bis-DNPH, m.p. 279°; N-oxide, m.p. 117°	10, 13
OHC, CHO	MeO_2C, CO_2Me , $(isoBu)_2AlH$	17	N-Oxide, m.p. 224°; bis-Et acetal, b.p. 135–140°; oxime, m.p. 250°	10, 13

TABLE XIV-1. Pyridine Aldehydes (*Continued*)

Compound	Method of preparation	Yield (%)	Properties	Ref.
(pyridine: Me, HO, N, CHO, CH₂OH)	(pyridine: HO, Me, N, CH₂OH, CH₂OH) "active" MnO₂, H⁺		Oxime, m.p. 212° (decomp.)	86
(pyridine: RO, Me, N, CHO, CH₂OR')	R = H, R' = palmitoyl R = H, R' = benzoyl R = Ac, R' = palmitoyl R = Ac, R' = benzoyl R = R' = palmitoyl	85	M.p. 68–78° M.p. 106–108° M.p. 65–67° M.p. 108–109° M.p. 68–69°	85
(Me₂ pyrano-pyridine, Me, N, CHO)	(Me₂ pyrano-pyridine, Me, N, CH₂OH), "active" MnO₂	86.7	M.p. 62–63°; diethylmercaptal, m.p. 99–100°	16, 17
(pyridine: AcO, Me, N, CH₂OAc, CH(SEt)₂)	(Me₂ dioxino-pyridine, Me, N, CH(SEt)₂) Ac₂O, C₅H₅N	94	M.p. 132–133°	17

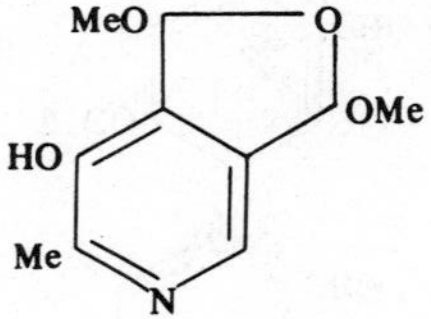

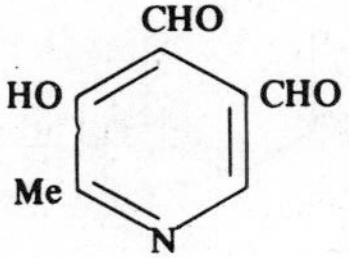

MeO | O OMe bicyclic

HO Me — CHO

·HCl + (i) MeOH, 50–60°
(ii) NH$_3$

44.8

M.p. 164–165°

17

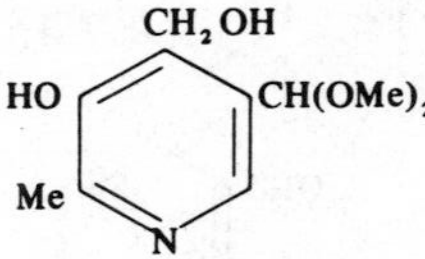

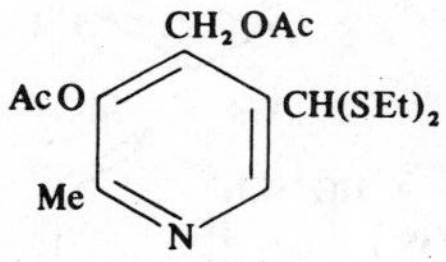

HO Me — CHO / CHO

HO Me — CH$_2$OH

HO Me — CHO / CHO

AcO Me — CH(SEt)$_2$, HgCl$_2$, HgO, MeOH

89.6

M.p. 167–168°

17

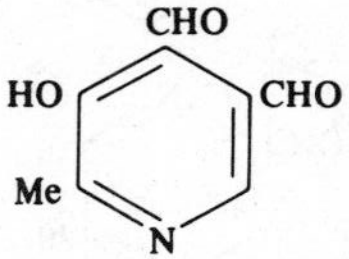

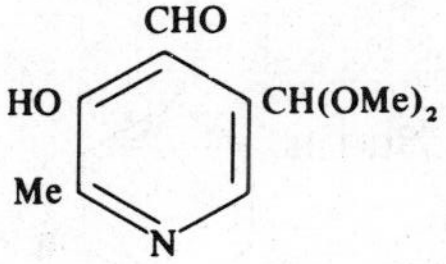

Me HO N — CHO / CHO

Me HO N — CH$_2$OH

Me AcO N — CH$_2$OAc

HO Me — CHO / CH(OMe)$_2$

HO Me — CHO / CH(OMe)$_2$, N-HCl

90

M.p. 158–161° (decomp.);
bis-TSC, m.p. 172–174°

17

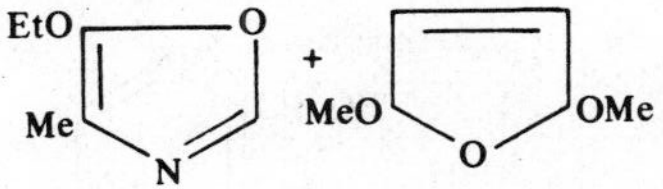

Me HO N — CHO / CH(OMe)$_2$

EtO Me N + O / MeO O OMe

(i) Heat
(ii) KOH-MeOH
(iii) HCl

HCl salt, m.p. 158–160°

34

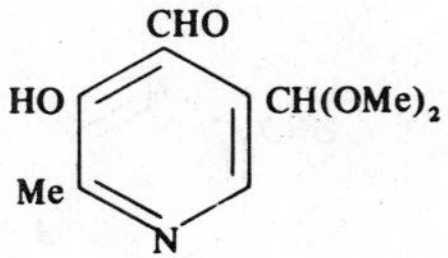

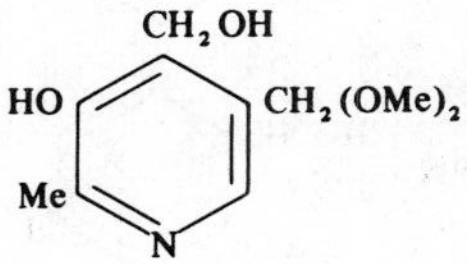

Me HO N — CH$_2$OH / CH(OMe)$_2$, "active" MnO$_2$

68.2

M.p. 58–59°;
oxime, m.p. 195–197°

17

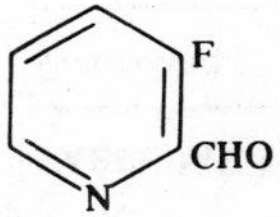

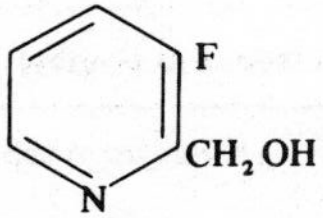

F N — CHO

F N — CH$_2$OH , MnO$_2$

41

B.p. 95–98° (18 mm);
TSC, m.p. 222.5–223.5°

25

TABLE XIV-1. Pyridine Aldehydes (Continued)

Compound	Method of preparation	Yield (%)	Properties	Ref.
F-pyridine-CHO	F-pyridine-CH_2OH , MnO_2	39	M.p. 31–33°; b.p. 72–74° (20.8 mm); TSC, m.p. 236–237°	25
Cl-pyridine-CHO	Cl-pyridine-CH_2OH , MnO_2	67	M.p. 60–62°; TSC, m.p. 235–236°	25
Br-pyridine-CHO	Br-pyridine-CH_2OH , MnO_2	76	M.p. 85–87°; TSC, m.p. 241–242	25
I-pyridine-CHO	I-pyridine-CH_2OH , MnO_2	61	M.p. 102–103°; TSC, m.p. 247–247.5°	25
F_3C-pyridine-CHO	F_3C-pyridine-CH_2OH , Pb(OAc)$_4$	52	B.p. 65–68° (21 mm); TSC, m.p. 208–209°	25
F_3CO-pyridine-CHO	F_3CO-pyridine-CH_2OH , Pb(OAc)$_4$	59	B.p. 73–76° (20 mm); TSC, m.p. 206–207°	25

Me$_2$N—⟨pyridine⟩—CHO → Me$_2$N—⟨pyridine⟩—CH$_2$OH , MnO$_2$ 81 M.p. 86–88°; TSC, m.p. 230–231° 25

MeSO$_2$—⟨pyridine⟩—CHO → MeSO$_2$—⟨pyridine⟩—CH$_2$OH , MnO$_2$ 82 M.p. 167–169°; TSC, m.p. 255–255.5° 25

HO—⟨pyridine⟩—CHO → HO—⟨pyridine⟩—CH$_2$OH , MnO$_2$ M.p. 183–184°; TSC, m.p. 236–237° 25

AcO—⟨pyridine⟩—CHO → AcO—⟨pyridine N-oxide⟩—CH(OAc)$_2$, Hydrolysis 52 TSC, m.p. 200–201° 25

AcO—⟨pyridine⟩—CH(OAc)$_2$ → AcO—⟨pyridine N-oxide⟩—CH$_2$OAc , Ac$_2$O M.p. 86–87.5° 25

⟨pyridine⟩ CO$_2$Et, CH(OAc)$_2$ → ⟨pyridine N-oxide⟩ CO$_2$Et, CH$_2$OAc , Ac$_2$O M.p. 100–103° 25

TABLE XIV-1. Pyridine Aldehydes (Continued)

Compound	Method of preparation	Yield (%)	Properties	Ref.
[pyridine with CDO, CDO]	[pyridine with CO₂Et, CO₂Et] + (i) (isoBu)₂AlH (ii) HC(OEt)₃ (iii) LiAlD₄ (iv) H₃O⊕	2-Acetal, 65 Dialdehyde, 30	M.p. 52° N-oxide, m.p. 128° n_D 1.473	13
[pyridine with CDO, CHO]	[pyridine with CO₂Et, CO₂Et] + (i) (isoBu)₂AlH (ii) HC(OEt)₃ (iii) LiAlD₄ (iv) H₃O⊕	4-Acetal, 68 Dialdehyde, 50	M.p. 70° N-oxide, m.p. 148°	13
[pyridine with ODC, CHO]	[pyridine with EtO₂C, CO₂Et] + (i) (isoBu)₂AlH (ii) HC(OEt)₃ (iii) LiAlD₄ (iv) H₃O⊕	2-Acetal, 61 Dialdehyde, 68	M.p. 65° N-oxide, m.p. 208°	13
[pyridine with Me₂NCH=CCHO]	4-PyMe, POCl₃, DMF	68	M.p. 90–92°	2
	4-PyMe, COCl₂, DMF	51		2
[pyridine with CH—CHO, CH—CHO]	4-PyMe, POCl₃, DMF			2
	CHO 4-PyC=CHNMe₂, (i) KOH (ii) H₃O⊕		Did not melt, 320°	2
	4-PyMe, (i) POCl₃, DMF (ii) NaOH	80		2

Starting material	Reagent	Product	Yield	Property	Ref.
pyridine: OHC, CHO, CHO	(isoBu)₂ AlH — *$(isoBu)_2$ AlH*	pyridine: MeO₂C, CO₂Me, CO₂Me		Oxime, m.p. 230°	13
pyridine: COMe, CHO	CH_2N_2	pyridine: CHO, CO₂Me	28	M.p. 105–106°	19
pyridine: OHC, COMe	CH_2N_2	pyridine: OHC, CHO	52	M.p. 64–65°	19
pyridine: MeCO, CHO	CH_2N_2	pyridine: OHC, CHO	15	M.p. 76°	19
pyridine: CHO, COMe	CH_2N_2	pyridine: CHO, CHO	21	M.p. 57°	19

TABLE XIV-2. Pyridine Aldoximes and Derivatives

Compound	Method of preparation	Yield (%)	Properties	Ref.
2-PyCH=$NOCH_2CH_2OH$	2-PyCH=NOH + (i) NaOEt (ii) $ClCH_2CH_2OH$		M.p. 48–50°; b.p. 104–105° (0.2 mm)	35
2-PyCH=$NOCH_2CH_2Cl$	2-PyCH=$NOCH_2CH_2OH$, $SOCl_2$		B.p. 73–75° (0.15 mm)	35
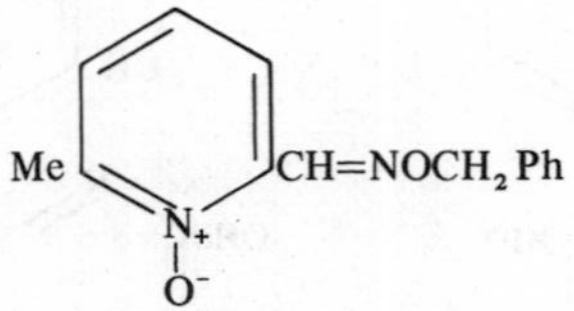			M.p. 61–62°	36
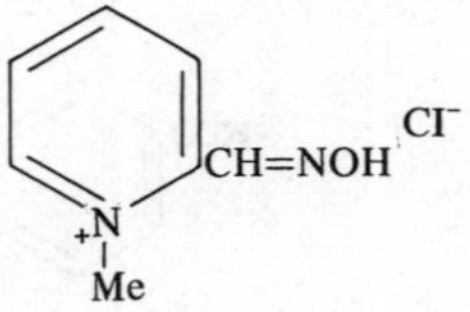	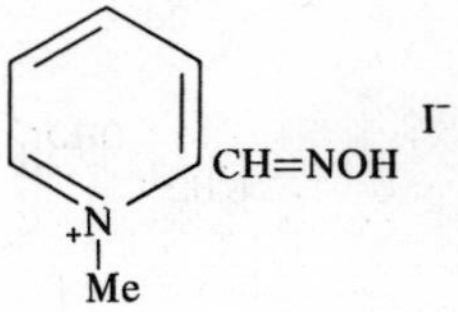 , HCl-MeOH		M.p. 235–237°	37
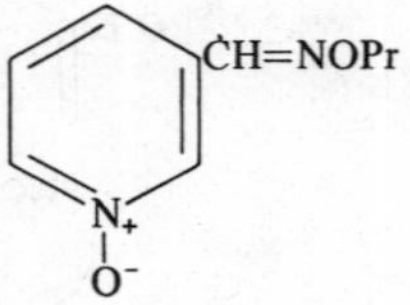			M.p. 65–66.6°	36

CH=NOCHMe₂ ... M.p. 108-109° ... 36

CH=NO(isoC₅H₁₁) ... M.p. 93-94° ... 36

CH=NOMe ... M.p. 118-119° ... 36

CH=NOEt ... M.p. 80-81° ... 36

TABLE XIV-2. Pyridine Aldoximes and Derivatives (*Continued*)

Compound	Method of preparation	Yield (%)	Properties	Ref.
CH=NOPr			M.p. 78–79°	36
CH=NOPr, Me			M.p. 80–81°	36
CH=NOCHMe$_2$			M.p. 93–94°	36
CH=NOCH$_2$CH=CH$_2$			M.p. 62–64°; b.p. 62–63° (0.03 mm); HCl salt, m.p. 120–122°; $n_D^{24} = 1.5458$	36

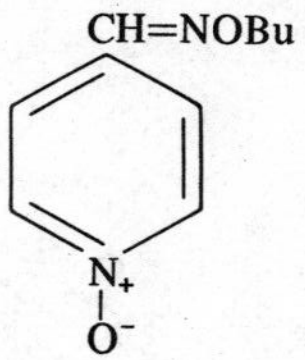

CH=NOBu

M.p. 78–79°

36

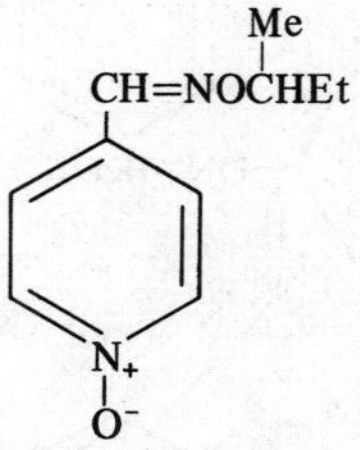

CH=NOCHEt
Me

M.p. 61-62°

36

CH=NOCH$_2$CHMe$_2$

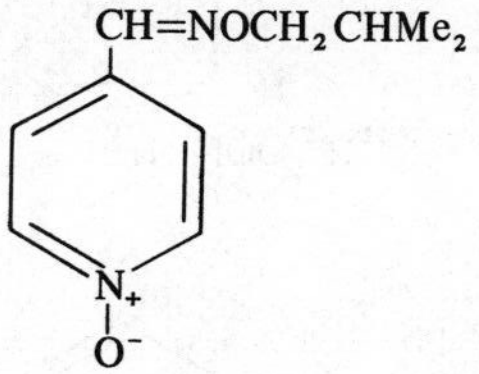

M.p. 128–130°

36

CH=NOC$_5$H$_{11}$

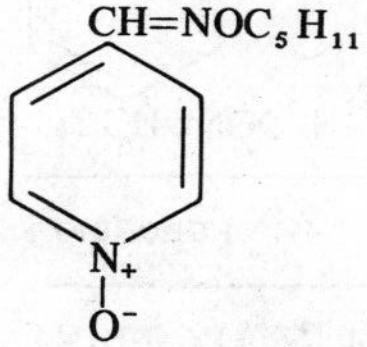

M.p. 81–83°

36

TABLE XIV-2. Pyridine Aldoximes and Derivatives (*Continued*)

Compound	Method of preparation	Yield (%)	Properties	Ref.
CH=NOC$_6$H$_{13}$ (pyridine N-oxide)			M.p. 66–68°	36
CH=NOC$_{12}$H$_{25}$ (pyridine N-oxide)			M.p. 77.5–78°	36
CH=NO–cyclohexyl (pyridine N-oxide)			M.p. 129–131°	36

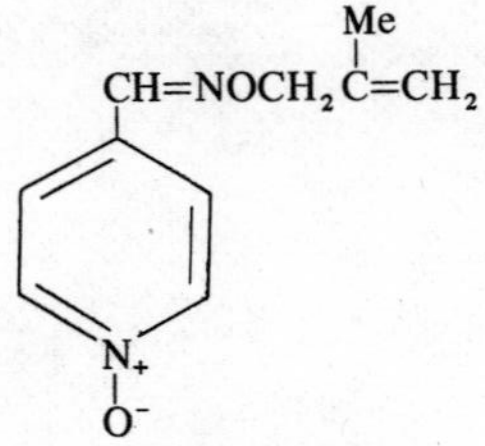

$CH=NOCH_2\overset{Me}{C}=CH_2$

HCl salt, m.p. 104–105°   36

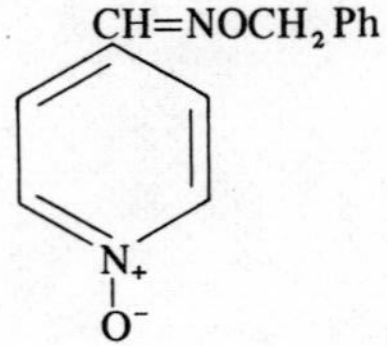

$CH=NOCH_2Ph$

M.p. 161–163°   36

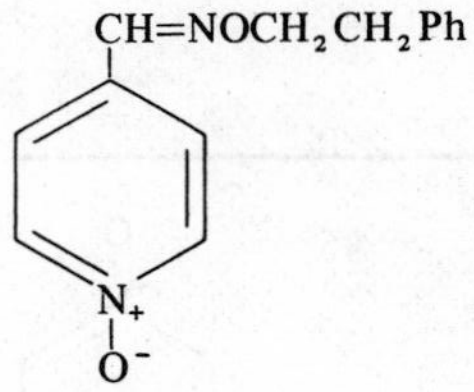

$CH=NOCH_2CH_2Ph$

M.p. 119–121°   36

$CH=NO(CH_2)_3Ph$

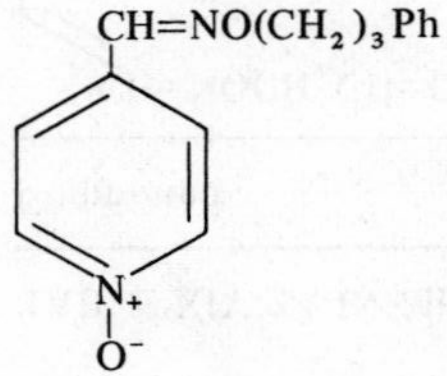

HCl salt, m.p. 145–147°   36

TABLE XIV-2. Pyridine Aldoximes and Derivatives (*Continued*)

Compound	Method of preparation	Yield (%)	Properties	Ref.
CH=NOCH$_2$CH=CHPh on pyridine N-oxide			M.p. 125–127°	36

TABLE XIV-3. Alkyl 2-Pyridyl Ketones

Compound	Method of preparation	Yield (%)	Properties	Ref.
2-PyCOMe			PH, m.p. 102°	38
[structure: Me-pyridine-COMe]	[structure: Me-pyridine-CO$_2$Et] + (i) Na, AcOEt (ii) hydrolysis		B.p. 94–96° (10 mm)	39
C$_4$H$_9$ [structure: pyridine]			TSC, m.p. 118–119° MeNHCSNHNR[a] (HCl salt), m.p. 198–199° PhNHCSNHNR, m.p. 166–167° C$_6$H$_{11}$NHCSNHNR, m.p. 169–170° C$_{12}$H$_{25}$NHCSNHNR (HCl salt), m.p. 187–189°	40
2-PyCOEt	2-PyCHO, MeCHN$_2$ (MeOH)	89	B.p. 94–98° (17 mm) SC, m.p. 163–164°	41
2-PyCOCH$_2$CH$_2$N[morpholine]	2-PyCOMe, H$_2$CO HN[morpholine], HCl	5–8	HCl salt, m.p. 191–192°	31

TABLE XIV-3. Alkyl 2-Pyridyl Ketones (*Continued*)

Compound	Method of preparation	Yield (%)	Properties	Ref.
2-PyCOPr			PH, m.p. 90° *p*-Chlorophenylhydrazone, m.p. 131° *p*-Methoxyphenylhydrazone, m.p. 102° *p*-Benzyloxyphenylhydrazone, m.p. 96–97°	38
			PH, m.p. 112–113°	38
2-PyCOCH₂CH₂OEt	2-PyLi, EtOCH₂CH₂CON⟨O⟩	67	b.p. 79–80°	30
2-PyCO(CH₂)₄OEt	2-PyCN, EtO(CH₂)₄MgBr	54	B.p. 158–162° (10 mm)	42
(−)-2-PyCOCHCH₃ OMe	2-PyLi, (−)-CH₃CHCO₂Et OMe	56	B.p. 66° (0.05 mm)	30
	(i) BuLi (ii) AcOEt	65	M.p. 61–62°	30

(i) BuLi

(ii) AcN 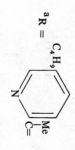   2.8

30

aR = C₄H₉ ... N ... Me C=

TABLE XIV-4. Alkyl 3-Pyridyl Ketones

Compound	Method of preparation	Yield (%)	Properties	Ref.
3-PyCOMe	3-PyCO$_2$Et + (i) AcOEt, NaOEt (ii) HCl		B.p. 107–108° (8 mm), n_D^{20} = 1.5311; oxime, m.p. 116–117°	43, 44
[pyridine ring: N, COMe, Me]	3-PyCO$_2$Et + AcOH, ZrO$_2$–Al$_2$O$_3$, 450°	80	B.p. 112° (16 mm)	22
[pyridine ring: Me, N, Me, COMe]	H$_2$NCH=C–CH=CHCHO, 140° (COMe)		B.p. 122° (13 mm)	20
	MeC=C–CH=CHCOMe, 150° (NH$_2$, COMe)			
[pyridine ring: Ph, N, COMe]	MeCOCH=CMe, PhC≡CCHO, heat (NH$_2$)		B.p. 195–197° (6 mm); m.p. 83°	20
3-PyCOEt	3-PyCHO, MeCHN$_2$ (MeOH)	75	B.p. 115–117° (18 mm); oxime, m.p. 117–118°	41, 43
3-PyCOPr	3-PyCHO, EtCHN$_2$ (MeOH)	81	B.p. 133–136° (23 mm); oxime, m.p. 89–90°	41
3-PyCOPr	3-PyCO$_2$Et + EtCO$_2$H, ZrO$_2$–Al$_2$O$_3$, 450°	30	B.p. 135–137° (26 mm)	22
3-PyCOPr	3-PyCN + PrMgBr		B.p. 117–120° (8 mm); oxime, m.p. 79–80°	43

TABLE XIV-5. Alkyl 4-Pyridyl Ketones

Compound	Method of preparation	Yield (%)	Properties	Ref.
4-PyCOMe	4-PyCO₂Et, AcOH, ZrO₂-Al₂O₃, 450°	22	B.p. 94–98° (10 mm); $n_D^{24} = 1.5233$	22
	4-PyCO₂Et + (i) AcOEt, NaOEt (ii) HCl	44		44
COMe/Me pyridine	CN/Me pyridine , MeMgI	45	B.p. 120–124° (10 mm); picrate, m.p. 174–175°	45
	CN/Me pyridine , MeMgX	30	B.p. 83° (5 mm); picrate, m.p. 169–170°; TSC, m.p. 227–229°; TSC (HCl salt), m.p. 216–218°	46
COMe/Et pyridine (a)	CN/Et pyridine , MgMgX	24		46
COMe/Et pyridine (b)	CO₂Et/Et pyridine + (i) AcOEt, NaOEt (ii) HCl	33	B.p. 95° (5 mm); picrate, m.p. 108°; TSC, m.p. 225°; TSC (HCl salt), m.p. 218–224°	46

4-PyCOEt	4-PyCHO, MeCHN$_2$ (MeOH)	79	B.p. 104–106° (10 mm), 120–123° (22 mm); HBr salt, m.p. 188–190°; picrate, m.p. 100–103°	41, 47
4-PyCOCHBrMe			HBr salt, m.p. 188–190°	48
4-PyCOPr	4-PyCO$_2$Et, EtCO$_2$H, ZrO$_2$-Al$_2$O$_3$, 450°	50	B.p. 137° (30 mm)	39
[pyridine ring, N-Et, COBu]	[pyridine ring, N-Et, CN], BuMgX	58	B.p. 140–144° (9 mm); picrate, m.p. 78°; TSC, m.p. 143°; TSC (HCl salt), m.p. 176–178°	49
[pyridine ring, N-Et, COC$_{10}$H$_{21}$]	[pyridine ring, N-Et, CN], C$_{10}$H$_{21}$MgX	26	B.p. 184–186° (5 mm)	49

TABLE XIV-6. Aryl 2-Pyridyl Ketones

Compound	Method of preparation	Yield (%)	Properties	Ref.
2-PyCOPh	C_5H_5N, PhCO, Et, Al-Hg		B.p. 124-126° (1.5 mm); $n_D^{20} = 1.5940$; $d_4^{20} = 1.1458$; 4-isomer also formed	50
			Phenylhydrazones: more soluble isomer: m.p. 226-227°; less soluble isomer: m.p. 200-201°; mixed isomers: m.p. 200-206°	51
2-PyCO—C₆H₄—Cl			M.p. 52-54°; HCl salt, m.p. 160-164°	52
2-PyCO—C₆H₄—NHCOCH₂Cl	2-PyCO—C₆H₄—NHCOCH₂Cl , NH_4OH		Oxime, m.p. 166.5-168°	52, 53
2-PyCO—C₆H₄—NHCOCH₂NH₂			M.p. 122.5°	52, 53

TABLE XIV-6. Aryl 2-Pyridyl Ketones (*Continued*)

Compound	Method of preparation	Yield (%)	Properties	Ref.
2-PyCO, Cl, NH₂ (phenyl)	2-PyCO, NH₂ (phenyl), Cl₂, AoOH	20	M.p. 99–101°	52–56
2-PyCO, Br, NH₂ (phenyl)	2-PyCO, Br, NHCOPh (phenyl), conc. HCl		M.p. 96–98°, 98–100°; oxime, m.p. 163–166°; HCl salt, m.p. 98–100°	52–56
2-PyCO, Br, NHCOCH₂Cl (phenyl)	2-PyCO, Br, NH₂ (phenyl), ClCH₂COCl		Oxime, m.p. 166.5–168° (decomp.)	55, 56
2-PyCO, Br, NHCOCH₂Br (phenyl)			M.p. 103–106° (decomp.); HBr salt, m.p. 205–206°	52, 53, 56

Starting material	Reaction / product	M.p.	Ref.
2-PyCO / NHCOPh / Br	CrO₃ oxidation → indole (Br, N–H, Ph, Py-2)	M.p. 138–140° (decomp.); m.p. 131.5–133°	54, 55
2-PyCO / NHAc / Br	acetylation, from 2-PyCO / NH₂	M.p. 151–153°	53, 56
2-PyCO / NHAc / NO₂	acetylation, from 2-PyCO / NH₂ / Br	M.p. 131.5–133°	53, 56
2-PyCO / NO₂ / Cl	HNO₃, from 2-PyCO / NH₂ / Cl	M.p. 137–138°; oxime, m.p. 166.5–168°	52, 53, 56
2-PyCO / CF₃ / Cl		M.p. 67–69°	52, 53, 56
2-PyCO / NO₂ / NH₂		M.p. 156–158°	52, 53, 56

TABLE XIV-6. Aryl 2-Pyridyl Ketones (*Continued*)

Compound	Method of preparation	Yield (%)	Properties	Ref.
2-PyCO, NH₂, CF₃ (benzene ring)			M.p. 91.5–93.5°; bromoacetyl derivative, m.p. 198–201°	52, 53, 56
2-PyCO, NH₂, CN (benzene ring)			M.p. 153–155°; bromoacetyl derivative, m.p. 116–121°	52, 53, 56
2-PyCO, CH₃ (naphthalene ring)	2-PyCN, (i) CH₂MgBr (naphthalene); (ii) Hydrolysis	48	B.p. 112–114° (0.12 mm)	57
2-PyCO, NH₂, Br (benzene ring)	2-PyCO, NH₂, Br , Br₂-AcOH (benzene ring)	78	M.p. 133–135°	58

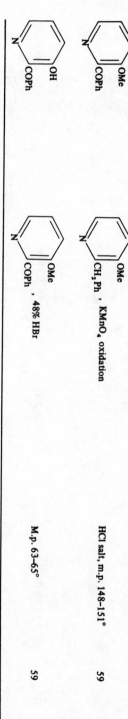

CH₂Ph , KMnO₄ oxidation   HCl salt, m.p. 148–151°   59

COPh , 48% HBr   M.p. 63–65°   59

TABLE XIV-7. Aryl 3-Pyridyl Ketones

Compound	Method of preparation	Yield (%)	Properties	Ref.

3-PyCOPh			Phenylhydrazones:
			more soluble isomer:
			  m.p. 245°;
			less soluble isomer:
			  m.p. 246–248°;
			mixed isomers:
			  m.p. 240–244°
				51

(3-pyridyl Me ketone, COPh, Me)	MeC=CCH=CHCHO, 130° (NH₂, COPh)		B.p. 118° (1.5 mm)	20
3-PyCO—C₆H₄—Cl	3-PyCOCl, PhCl, AlCl₃	60	M.p. 87–88°; b.p. 160–164° (0.4 mm)	
3-PyCO—C₆H₄—NO₂	3-PyLi, (i) (CHO, NO₂) (ii) CrO₃	32	M.p. 60–60.5°	
3-PyCO—C₆H₄—NH₂	3-PyCO—C₆H₄—NO₂, N₂H₄, Pd-C	32	M.p. 247–249°	

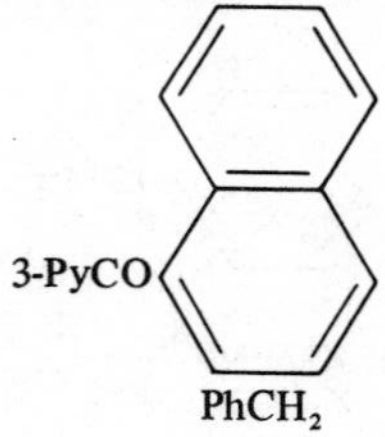

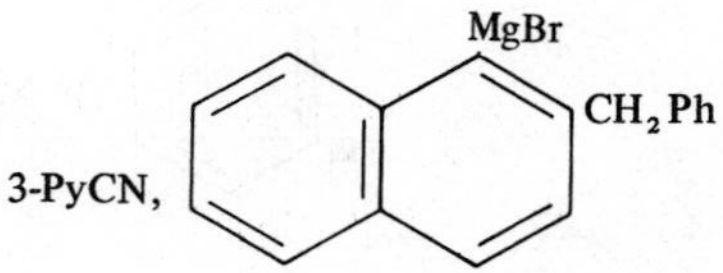

3-PyCO

PhCH₂

3-PyCN,

CH₂Ph

MgBr

64   B.p. 212–219° (0.6 mm);   57
      picrate, m.p. 145–147°

3-PyCO

CH₂

3-PyCN,

CH₂

MgBr

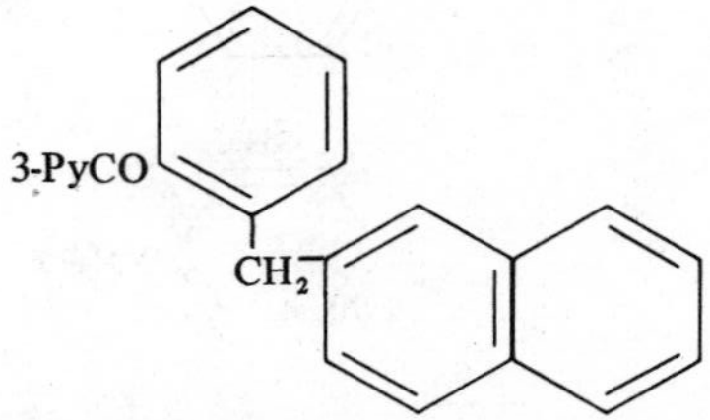

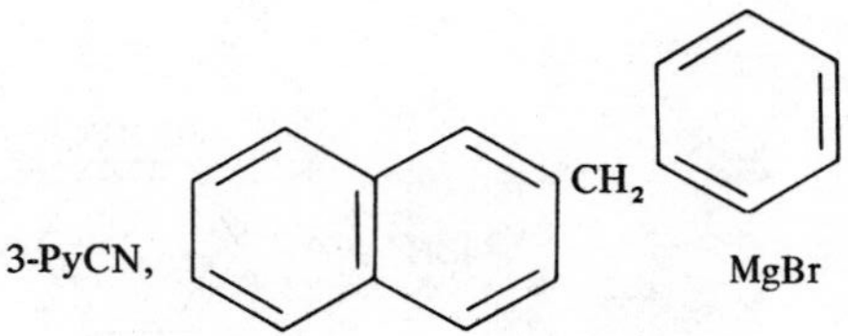

B.p. 207–208° (0.1 mm)   57

TABLE XIV-8. Aryl 4-Pyridyl Ketones

Compound	Method of preparation	Yield (%)	Properties	Ref.
4-PyCOPh	C_5H_5N, $PhCO_2Et$, Al-Hg (2-isomer also formed)		M.P. 72.5–73°; b.p. 126–146° (19 mm); picrate, m.p. 160–161°	50
			Phenylhydrazones: more soluble isomer: m.p. 228°; less soluble isomer: m.p. 204–205° mixed isomers: m.p. 228–229°	51
4-PyCO— (NH₂)	4-PyCO— (NHCOPh) , conc. HCl		M.p. 159–160°; O-benzyl carbamate, m.p. 130–130.5°; N-acetyl derivative, m.p. 161–162°, m.p. 151–153°	52–54, 56
4-PyCO— (NHCOCH₂NH₂)			M.p. 122–125°	52, 53, 56
4-PyCO— (Me)	4-PyCN, (i) Me Li (ii) 20% HCl	65	M.p. 41–42°; b.p. 170–172° (14 mm)	61

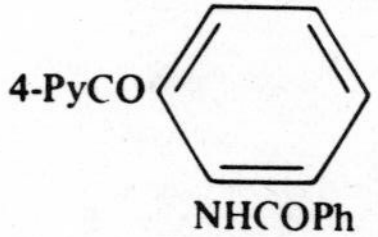

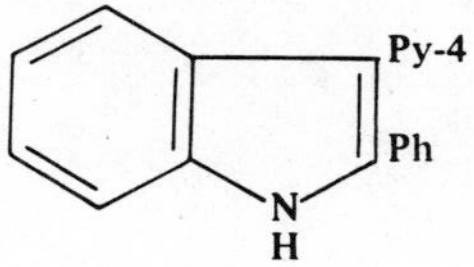

4-PyCO

NHCOPh , CrO₃ oxidation

Py-4
N Ph
H

M.p. 129–130°   54

4-PyCO

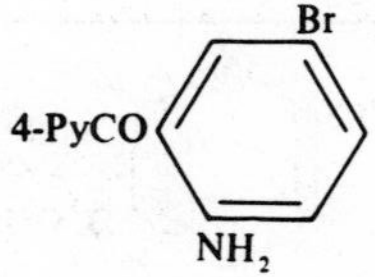

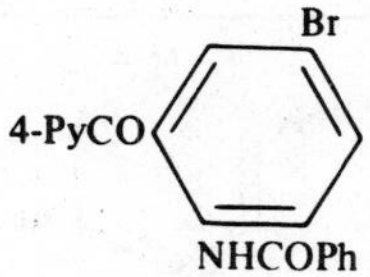

NH₂   Br , conc. HCl

4-PyCO

NHCOPh   Br

M.p. 213–214°   53, 55

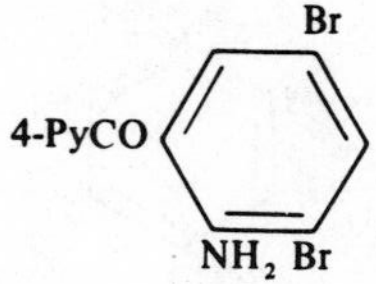

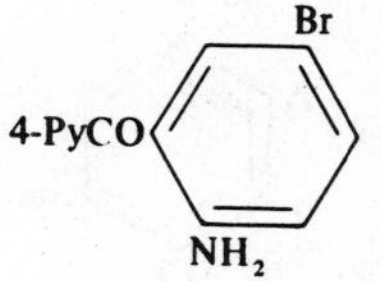

4-PyCO

NH₂   Br , Br₂-AcOH

4-PyCO

NH₂   Br

M.p. 141–142°   54

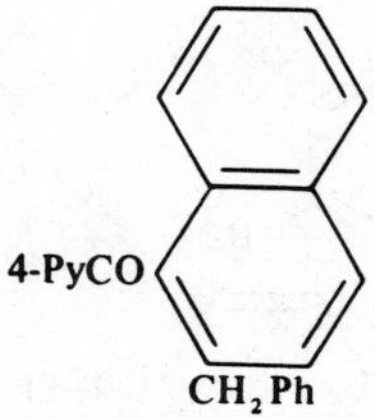

4-PyCO

CH₂Ph

4-PyCN, MgBr

CH₂Ph

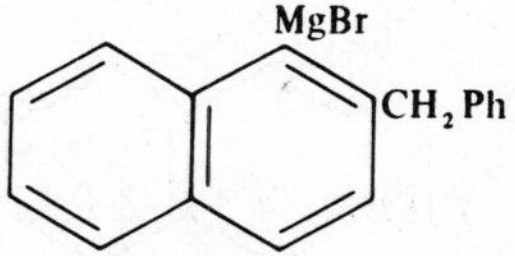

65

M.p. 106–107°;
b.p. 234–250° (0.8 mm)   57

TABLE XIV-8. Aryl 4-Pyridyl Ketones (Continued)

Compound	Method of preparation	Yield (%)	Properties	Ref.
4-PyCO—⟨CH₂—naphthyl⟩	4-PyCN, ⟨CH₂—naphthyl⟩MgBr	52	B.p. 207–209° (20 mm)	57
Me—⟨pyridyl⟩COPh, Me	⟨CH₂Ph pyridyl⟩, Me, KMnO₄, Mg(NO₃)₂, 60°		M.p. 47°; picrate, m.p. 143–144°; methiodide, m.p. 163–165°	62
HO₂C—⟨pyridyl⟩COPh / ⟨pyridyl⟩CO₂H	Me—⟨CH₂Ph pyridyl⟩—Me, KMnO₄ oxidation, 100°		M.p. 218–221°; Di-Et ester, m.p. 92–94°	62

TABLE XIV-9. Pyridyl Aralkyl and Aralkenyl Ketones

Compound	Method of preparation	Yield (%)	Properties	Ref.
2-PyCOCH₂Ph	2-PyCO₂Et, + PhCH₂CN, (i) NaOEt (ii) hydrolysis	74.5	M.p. 44°; SC, m.p. 155°; TSC, m.p. 159°;	63
2-PyCOCH₂[C₆H₄NO₂]	2-PyCH=NR + (ii) O₂N[C₆H₄]CHO, (i) HPO(OR)₂ (iii) H⊕		M.p. 158-159°	26
2-PyCOCH₂[C₆H₄NH₂]	Reduction of nitro compound		M.p. 117.5-118°	26
2-PyCOCHPh (NO)	2-PyCOCH₂Ph, HNO₂	78	M.p. 130-133°	64
3-PyCOCH₂Ph	3-PyCO₂Et + PhCH₂CN, (i) NaOEt (ii) Hydrolysis	87.7	M.p. 63°; PH, m.p. 119-119.5°; SC, m.p. 175-176°; TSC, m.p. 169-170°	63
3-PyCOCH₂Ph	3-PyCOCHPh (CN), conc. HBr		M.p. 53-56°	65
3-PyCOCHPh (CN)	3-PyCO₂Et, PhCH₂CN, NaOEt		M.p. 137-141°	65

TABLE XIV-9. Pyridyl Aralkyl and Aralkenyl Ketones (*Continued*)

Compound	Method of preparation	Yield (%)	Properties	Ref.
3-PyCOCH₂—[C₆H₄—NO₂]	(i) HPO(OR)₂ 3-PyCH=NR + (ii) O₂N—[C₆H₄]—CHO (iii) H⊕		M.p. 158–159°	26
[structure: PhCO, Cl, Me, N⁺O⁻, Cl, Me]	(i) BuLi, (ii) PhCN ; [structure: Cl, Me, N⁺O⁻]	11.5	M.p. 235°	30
3-PyCOCHPh (NO)	3-PyCOCH₂Ph, HNO₂	64	M.p. 164–167°	64
4-PyCOCH₂Ph	4-PyCO₂Et + PhCH₂CN, (i) NaOEt, (ii) hydrolysis		M.p. 96°; PH, m.p. 56°; SC, m.p. 182°; TSC, m.p. 198°	63
	4-PyCHO, PhCHN₂ (i) HPO(OR)₂	50		24
4-PyCOCH₂—[C₆H₄—NO₂]	4-PyCH=NR + (ii) O₂N—[C₆H₄]—CHO (iii) H⊕		M.p. 112–113°	26

NO
4-PyCOCHPh

4-PyCOCH₂Ph, HNO₂ 67 M.p. 171–174° 64

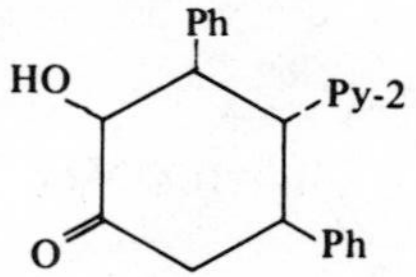

HO
Ph
Ph
Py-2

2-PyCH₂COPh, MeCOCH=CHPh, NaOEt 66

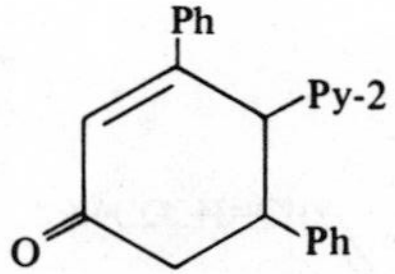

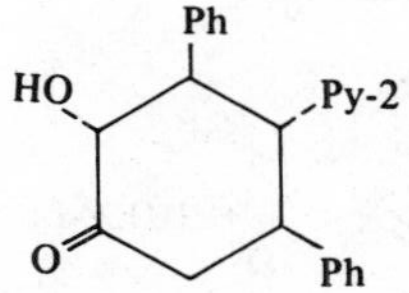

O
Ph
Ph
Py-2

, H₃PO₄, 100° (epimerizes)

Cis-isomer,
m.p. 170.5–171.5° 67

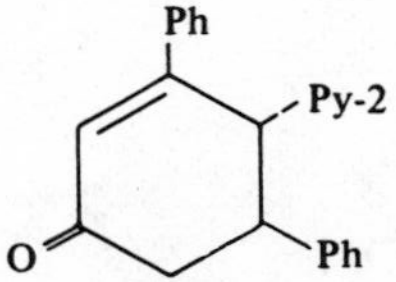

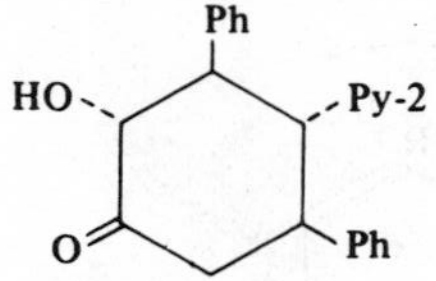

HO
Ph
Ph
Py-2

Cis, cis-isomer,
m.p. 157–158° 67

O
Ph
Ph
Py-2

Trans-isomer,
m.p. 140–141° 67

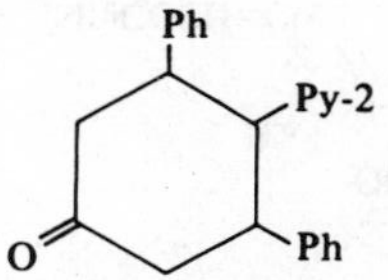

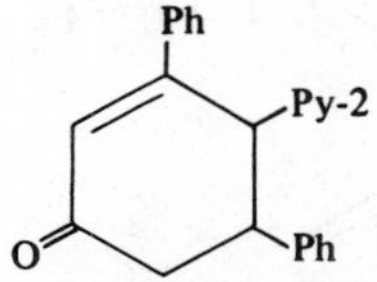

O
Ph
Ph
Py-2

HO
Ph
Ph
Py-2

, H₂ (Pd-C)

2-PyCOCH₂Ph

O
Ph
Py-2

, H₂ (Pd-C)

2-PyCOCH=CHPh

O
Ph
Py-2

2-PyCOMe, Et₂NH M.p. 71° 28

2-PyCOMe + PhCHO, Et₂NH M.p. 84° 28

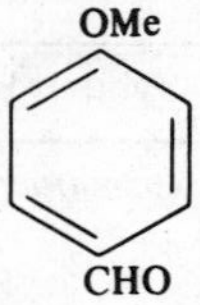

2-PyCOCH=CHPh
OMe

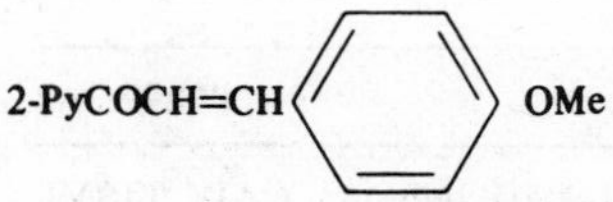

2-PyCOMe +
OMe
CHO
, Et₂NH 28

TABLE XIV-9. Pyridyl Aralkyl and Aralkenyl Ketones (*Continued*)

Compound	Method of preparation	Yield (%)	Properties	Ref.
2-PyCOCH=CH— (C6H4, OH)	2-PyCOMe + (C6H4, OH, CHO), Et2NH		M.p. 147–148°	28
2-PyCOCH=CH— (C6H4, NO2)	2-PyCOMe + (C6H4, NO2, CHO), Et2NH		M.p. 154°	28
2-PyCOCH=CH— (benzodioxole, O–CH2–O)	2-PyCOMe + (benzodioxole, O–CH2–O, CHO), Et2NH		M.p. 153°	28
2-PyCOCH=CH— (C6H3, OMe, OMe)	2-PyCOMe + (C6H3, OMe, OMe, CHO), Et2NH		M.p. 116–117°	28
4-PyCOCH=CHPh	3-PyCOMe + PhCHO, Et2NH		M.p. 84–85°	28

Product	Reactants	M.p.	Ref.
3-PyCOCH=CH-C$_6$H$_4$-NMe$_2$	3-PyCOMe + C$_6$H$_4$(NMe$_2$)CHO , Et$_2$NH	M.p. 74–75°	28
3-PyCOCH=CH-C$_6$H$_4$-OMe	3-PyCOMe, C$_6$H$_4$(OMe)CHO , Et$_2$NH	M.p. 93°	28
3-PyCOCH=CH-C$_6$H$_4$-OH	3-PyCOMe, C$_6$H$_4$(OH)CHO , Et$_2$NH	M.p. 174°	28
3-PyCOCH=CH-C$_6$H$_4$-NO$_2$	3-PyCOMe, C$_6$H$_4$(NO$_2$)CHO , Et$_2$NH	M.p. 186–187°	28
3-PyCOCH=CH-(benzo-1,3-dioxole)	3-PyCOMe, (benzo-1,3-dioxole)CHO , Et$_2$NH	M.p. 154°	28

TABLE XIV-9. Pyridyl Aralkyl and Aralkenyl Ketones (Continued)

Compound	Method of preparation	Yield (%)	Properties	Ref.
3-PyCOCH=CH—C$_6$H$_3$(OMe)(OMe)	3-PyCOMe, C$_6$H$_3$(OMe)(OMe)CHO, Et$_2$NH		M.p. 98°	28
4-PyCOCH=CHPh	4-PyCOMe, PhCHO, Et$_2$NH		M.p. 87–88°	28
4-PyCOCH=CH—C$_6$H$_4$(OMe)	4-PyCOMe, C$_6$H$_4$(OMe)CHO, Et$_2$NH		M.p. 115–116°	28
4-PyCOCH=CH—C$_6$H$_4$(HO)	4-PyCOMe, C$_6$H$_4$(OH)CHO, Et$_2$NH		M.p. 181–182°	28
4-PyCOCH=CH—C$_6$H$_4$(NMe$_2$)	4-PyCOMe, C$_6$H$_4$(NMe$_2$)CHO, Et$_2$NH		M.p. 127–128°	28

4-PyCOCH=CH— (benzodioxole, O–CH₂–O) → 4-PyCOMe, CHO (benzodioxole, O–CH₂–O) , Et₂NH   M.p. 140°   28

4-PyCOCH=CH— (OMe, OMe) → 4-PyCOMe, CHO (OMe, OMe) , Et₂NH   M.p. 222°   28

TABLE XIV-10. Pyridyl Heterocyclic Ketones

Compound	Method of preparation	Yield (%) Properties	Ref.
2-PyCO (thiazole)	2-PyCO$_2$Et, (thiazolyl-Li) (−50°)	M.p. 77.5–79°; oxime, m.p. 169.5–170.5°	68
2-PyCO (indole, N–H)	2-PyMgBr, (indole-2-CHO, N–H), air oxidation	M.p. 173–175°	69
2-PyCOCH=CH (furan)	2-PyCOMe (furan-CHO), Et$_2$NH	M.p. 53–54°	28
3-PyCO (pyrrole, N–CH$_2$CH$_2$NMe$_2$)		B.p. 145° (0.06 mm); citrate, m.p. 128°	70
3-PyCO (pyrrole, N–CH$_2$CH$_2$NEt$_2$)		B.p. 165° (0.1 mm); citrate, m.p. 156°	70

3-PyCOCH=CH (furan)   3-PyCOMe,   95   M.p. 80.9-82.1°; methiodide, m.p. 200-201.5° (decomp.)   44

(pyridine)–C(COMe)(Me)–(pyridine)   3-PyCOMe, (furan)CHO, 10% NaOH   M.p. 84°   28

4-PyCO (thiazole, N–S)   3-Py–C(Me)(OH)–C(Me)(OH)–Py-3, H₂SO₄   M.p. 45–49°   71

4-PyCO (N-Me indole)   4-PyCO₂Et, (thiazole)–Li   M.p. 117–118°; HCl salt, m.p. 203–204°; oxime, decomp. 218–219°; methiodide, m.p. 204–206°   72

4-PyCO₂Et, (N-Me indole)–Li   M.p. 134.5–135°; oxime m.p. 252.5–253°   73

4-PyCO (pyrrole, N-CH₂CH₂NMe₂)   B.p. 127° (0.05 mm)   70

TABLE XIV-10. Pyridyl Heterocyclic Ketones (*Continued*)

Compound	Method of preparation	Yield (%)	Properties	Ref.
4-PyCO, $CH_2CH_2CH_2NMe_2$ (pyrrole)			B.p. 145° (0.07 mm)	70
4-PyCOCH=CH (furan)	4-PyCOMe, CHO, 10% NaOH	77	M.p. 74–75.1°; ethiodide, m.p. 141–143°; ethio-dide (+1H_2O), m.p. 107.5–108°	44
	4-PyCOMe, CHO, Et_2NH		M.p. 77°	28
4-PyCOC—Py-4, Me, Me	4-PyC—C—Py-4, H_2SO_4, OH OH, Me Me		M.p. 75–76°	71

TABLE XIV-11. Pyridyl Diazoketones

Compound	Method of preparation	Yield (%) Properties	Ref.
CCOCH$_3$, N$_2$	COCH$_3$, HClO$_4$	HClO$_4$ salt, dec. 132° (expl.)	74
CCOPr, N$_2$	COPr, HClO$_4$	HClO$_4$ salt, dec. 117–118°	74
CCO(furyl), N$_2$	CO(furyl), HClO$_4$	HClO$_4$ salt, dec. 137°	74
CCOC$_6$H$_5$, N$_2$	COC$_6$H$_5$, HClO$_4$	HClO$_4$ salt, dec. 139° (expl.)	74

2-Py—C—CO(t-Bu)
    ‖
    N₂

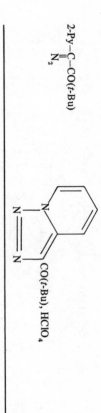

CO(t-Bu), HClO₄

Perchlorate salt, m.p.
117–118° (decomp.)

66

TABLE XIV-12. Pyridine Side-chain Ketones

Compound	Method of preparation	Yield (%)	Properties	Ref.
2-PyCH₂COMe	(2-PyCH₂)₂CHOH + (i) NaNH₂ (ii) H₂O		B.p. 49–50° (0.05 mm), n_D^{20} = 1.5310; picrate salt, m.p. 132–133°	48
2-PyCH₂COMe	2-PyCH₂Li, MeCN		B.p. 101° (10 mm)	75
2-PyCH₂COEt	2-PyCH₂Li, EtCN	80	B.p. 113° (10 mm)	75
2-PyCH₂COPr	2-PyCH₂Li, PrCN	84	B.p. 123° (10 mm)	75
2-PyCH₂COBu	2-PyCH₂Li, BuCN	71	B.p. 136° (10 mm)	75
2-PyCH₂COC₅H₁₁	2-PyCH₂Li, C₅H₁₁CN	77	B.p. 151° (10 mm)	75

	CH₂Li, MeCN		B.p. 86–92° (0.1–0.12 mm) n_D^{25} = 1.5313	76
	+ (i) HClO₄ (ii) AcONH₄		M.p. 164°	29
	+ (i) HClO₄ (ii) AcONH₄		M.p. 161°	29

TABLE XIV-12. Pyridine Side-chain Ketones (*Continued*)

Compound	Method of preparation	Yield (%)	Properties	Ref.
Pyridine: Ph, CH$_2$COPh, CH$_2$Ph, Me	Pyranone: Ph, CHCOPh, CH$_2$Ph, Me, O + (i) HClO$_4$ (ii) AcONH$_4$		M.p. 162°	29
(Br-phenyl)-pyridine: CH$_2$CO(Br-C$_6$H$_4$), CH$_2$Ph, Me	(Br-phenyl)-pyranone: CHCO(Br-C$_6$H$_4$), CH$_2$Ph, Me, O (i) HClO$_4$ (ii) AcONH$_4$		M.p. 152°	29
3-PyCHOHCO-C$_6$H$_4$-NMe$_2$	3-PyCHO, C$_6$H$_4$(NMe$_2$)CHO, KCN	92	M.p. 156–157°	18
3-PyCHOHCO-C$_6$H$_4$-OH	3-PyCHO, C$_6$H$_4$(OH)CHO, KCN	48	M.p. 205–206°	18

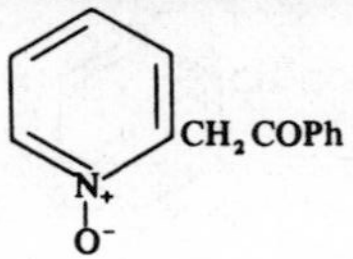

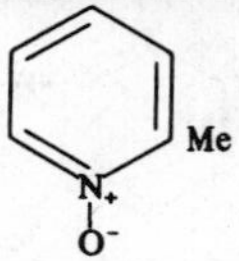

CH₂COPh (pyridine N-oxide)

(i) NaNH₂-NH₃
(ii) PhCO₂Me
(iii) hydrolysis

Me (pyridine N-oxide)

85.3   M.p. 158°   77

CH₂CO- (naphthalene) (pyridine N-oxide)

(i) NaNH₂-NH₃
(ii)
Me (pyridine N-oxide) + naphthalene-CO₂Me

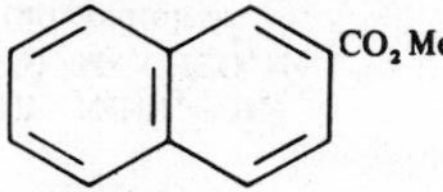

83.9   M.p. 163-164°   77

CH₂CO- (thiophene, S) (pyridine N-oxide)

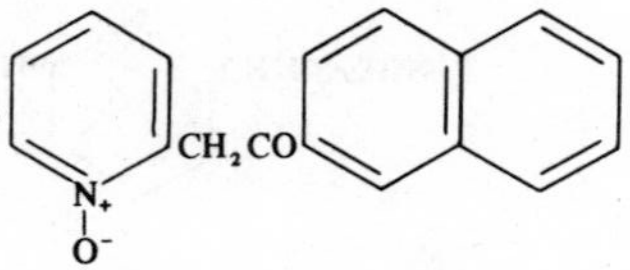

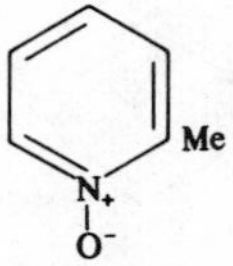

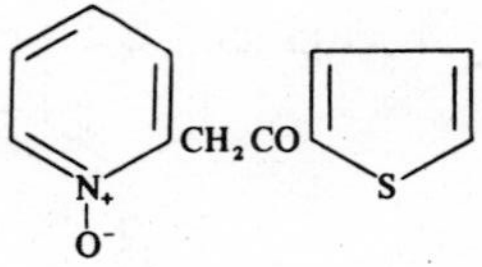

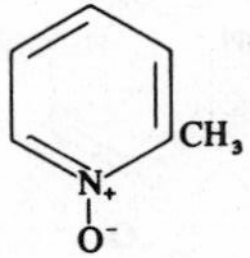

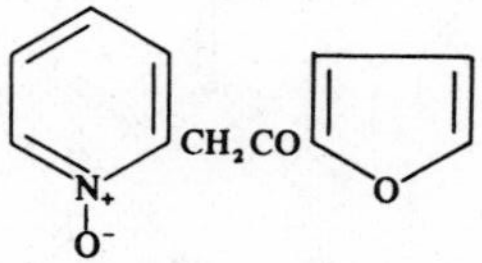

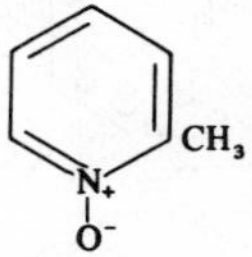

(i) NaNH₂-NH₃
(ii)
(iii) hydrolysis
CH₃ (pyridine N-oxide) + thiophene-CO₂Me

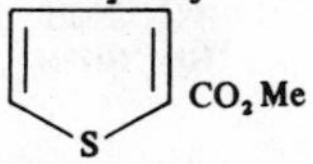

83   M.p. 153-154°   77

CH₂CO- (pyridine) (pyridine N-oxide)

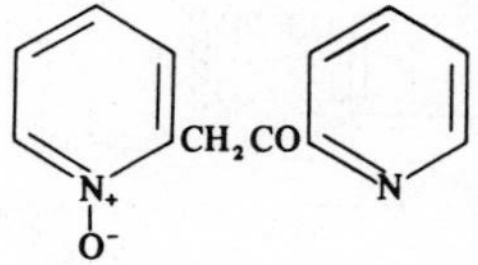

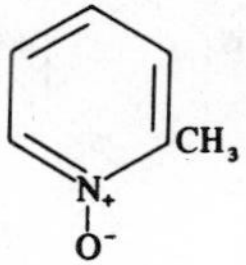

(i) NaNH₂-NH₃
(ii) 2-PyCO₂Me
(iii) hydrolysis
CH₃ (pyridine N-oxide)

74.7   M.p. 141-142°   77

CH₂CO- (furan, O) (pyridine N-oxide)

(i) NaNH₂-NH₃
(ii) + furan-CO₂Me
(iii) hydrolysis
CH₃ (pyridine N-oxide)

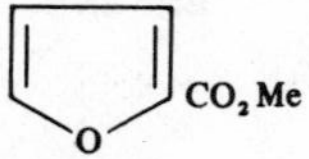

65   M.p. 152-153°   77

TABLE XIV-12. Pyridine Side-chain Ketones

Compound	Method of preparation	Yield (%)	Properties	Ref.
Pyridine N-oxide, 2-$CH_2CO(isoPr)$	(i) $NaNH_2$-NH_3 (ii) Me_2CHCO_2Me (iii) hydrolysis; from pyridine N-oxide, 2-CH_3	55.5	B.p. 153–155° (1 mm); m.p. 53–54°	77
Pyridine N-oxide, CH_2COPh	(i) $NaNH_2$-NH_3 (ii) $PhCO_2Me$ (iii) hydrolysis	79.5	M.p. 141–142°	77
Pyridine N-oxide, CH_2CO-(2-thienyl)	(i) $NaNH_2$-NH_3 (iii) hydrolysis; + thiophene-CO_2Me	73	M.p. 117–118°	77
Pyridine N-oxide, CH_2CO-(pyridyl)	(i) $NaNH_2$-NH_3 (ii) 3-$PyCO_2Me$ (iii) hydrolysis	50	M.p. 133–134°	77
Pyridine N-oxide, Me, $CH_2COCHMe_2$	(i) $NaNH_2$-NH_3 (ii) Me_2CHCO_2Me (iii) hydrolysis	58	B.p. 140–146° (0.5 mm); m.p. 91–92°	77
Pyridine N-oxide, Me, CH_2COCMe_3	(i) $NaNH_2$-NH_3 (ii) Me_3CCO_2Me (iii) hydrolysis	77	B.p. 135–140° (0.5–1 mm); m.p. 96–97°	77

TABLE XIV-13. Pyridyl Ketoximes

Compound	Method of preparation	Yield (%)	Properties	Ref.
Me | 2-PyC=NOH	2-PyEt, $C_5H_{11}ONO$, KNH_2 (NH_3)	26	M.p. 123–124°	78
	Et, $C_5H_{11}ONO$, KNH_2 (NH_3)	77	Isomer A, m.p. 215–217°; isomer B, m.p. 215–217°	78
NOH | 2-PyCCH₂Ph	2-PyCH₂CH₂Ph, $C_5H_{11}ONO$, KNH_2 (NH_3)	2		
		29	M.p. 155–156°	78
	CH_2CH_2Ph, $C_5H_{11}ONO$, KNH_2 (NH_3)	54	M.p. 209–211°	78
NOH | 2-PyCPh	2-PyCH₂Ph, $C_5H_{11}ONO$, KNH_2 (NH_3)	73	M.p. 165–166°; syn-phenyl, m.p. 151–152°; anti-phenyl, m.p. 163–164.5°	78 79

Structure		Conditions	Properties	Ref.
2-Py N-oxide, =NOH, =CPh	CH₂Ph, C₅H₁₁ONO, KNH₂ (NH₃)	86	*Anti*-phenyl isomer, m.p. 219–222°	78, 79
Ph-2-PyC=NOCH₂CH₂NMe₂			B.p. 174–176° (1.5 mm)	80
Ph-2-PyC=NOCH₂CH₂NEt₂			B.p. 190–195° (3 mm)	80
Ph-2-PyC=NOCH₂CH₂CH₂NMe₂			B.p. 190–195° (3 mm)	80
3-Py, =NOH, =CPh			*Syn*-phenyl isomer, m.p. 141–143°; *anti*-phenyl isomer, m.p. 162–163°	79
4-Py N-oxide, =NOH, =CPh			*Syn*-phenyl isomer, m.p. 222–224°; *anti*-phenyl isomer, m.p. 178–180°	79

TABLE XIV-13. Pyridyl Ketoximes (*Continued*)

Compound	Method of preparation	Yield (%)	Properties	Ref.
4-PyC=NOH (Me structure)	Sn-HCl reduction		M.p. 183–185°	81
4-PyCH₂Ph, C₅H₁₁ONO, KNH₂ (NH₃)		93	M.p. 181–183°	78
			Syn-phenyl isomer, m.p. 152–155°; *anti*-phenyl isomer, m.p. 186–188°	79
4-PyEt + C₄H₉ONO, (NaNH₂)			M.p. 147–161°	3
4-PyEt + C₅H₁₁ONO, KNH₂ (NH₃)		16	M.p. 156–158°	78
, C₅H₁₁ONO, KNH₂ (NH₃)		57	M.p. 204–205°	78

PhC=NOH (N-oxide)   *Syn*-phenyl isomer, m.p. 229–230°; *anti*-phenyl isomer, m.p. 222–223°   79

NOH
4-PyCCH₂Ph   4,PyCH₂CH₂Ph, C₅H₁₁ONO, KNH₂ (NH₃)   45   M.p. 196–197°   78

C(=NOH)CH₂Ph (pyridine N-oxide)   (pyridine N-oxide) CH₂CH₂Ph, C₅H₁₁ONO, KNH₂ (NH₃)   32   M.p. 339–341°   78

NOH
4-PyCCH₂CH₂Ph

Me
4-PyC=NOMe   4-PyC=NOH, MeI, NaOMe   M.p. 122–124°   82

Me
4-PyC=NOPr   4-PyC=NOH, PrX, NaOMe   M.p. 58–60° (0.025 mm), n_D^{24} = 1.5178; N-oxide, m.p. 108–112°   82

TABLE XIV-14. Pyridine Polyketones and Derivatives

Compound	Method of preparation	Yield (%)	Properties	Ref.
N₂CHCO–[pyridine]–COCHN₂	ClOC–[pyridine]–COCl , CH₂N₂		M.p. 181° decomp.	83
BrCH₂CO–[pyridine]–COCH₂Br	N₂CHCO–[pyridine]–COCHN₂ , HBr		M.p. 110–111°	83
ClCH₂CO–[pyridine]–COCH₂Cl	N₂CHCO–[pyridine]–COCHN₂ , HCl		M.p. 151–153°	83
AcCH₂CO–[pyridine]–COCH₂Ac	XCH₂CO–[pyridine]–COCH₂X (X = Br, Cl) , KOAc, AcOH		M.p. 99°	83
2-PyCOCH₂COCO₂Et	2-PyCOMe, (CO₂Et)₂, NaOEt		M.p. 172–173°	84
3-PyCOCH₂COCO₂Et	3-PyCOMe, (CO₂Et)₂, NaOEt	40	M.p. 68–69°	84

OH
3-PyC=CHCOCO₂Et

3-PyCOCH₂COCO₂Et + (i) Base
(ii) AcOH

Structure		M.p. / notes	Yield	Ref
3-PyC=CHCOCO₂Et (OH)		M.p. 187–189°, reverts slowly to diketone	84	
Py–COCO–C₆H₄–NMe₂				
Py–COCO–C₆H₄–OH				
Py(COMe)(COMe)				
Py(MeCO)(COMe)				
C₆H₄(COMe)(COMe)				

Structure		Yield	M.p.	Ref
Py–CHOHCO–C₆H₄–NMe₂, NH₄NO₃, Cu(OAc)₂		40	M.p. 132–133°	18
Py–CHOHCO–C₆H₄–OH, NH₄NO₃, Cu(OAc)₂		66	M.p. 178°	18
Py(CHO)(CHO), CH₂N₂		75	M.p. 71°	19
Py(OHC)(CHO), CH₂N₂		57	M.p. 81°	19
C₆H₄(CHO)(CHO), CH₂N₂		12	M.p. 45°	19

TABLE XIV-14. Pyridine Polyketones and Derivatives (*Continued*)

Compound	Method of preparation	Yield (%)	Properties	Ref.
PrCO / OEt pyridine N-oxide, COPr	[OEt pyridine N-oxide], BuLi, PrCO$_2$Et	16.5	M.p. 64–65°	30
MeCO / N / Me, Me bipyridyl N-oxide	[Me, Me pyridine N-oxide], (i) BuLi (ii) MeCONMe$_2$	12.9	M.p. 217°	30

IV. Appendix

The following supplementary table contains references to spectral data on the subject matter covered in this chapter. The studies cited are of a more comprehensive nature.

Compound	Type of study	Ref.
Picolinaldehyde	IR	97, 111, 114, 121
	UV	93
	NMR	96, 99, 109, 119
	Mass spec. of derivatives	122
Nicotinaldehyde	NMR	96, 99, 105, 106, 118
	IR	97, 98, 111, 114
	UV	93
Isonicotinaldehyde	NMR	96, 99, 125
	IR	97, 98, 111, 114, 121
	UV	93
Isonicotinaldehyde-1-oxide	Mass spec.	88
6-Methylpicolinaldehyde	NMR	109
Pyridoxal and derivatives	IR	87, 111
Isopyridoxal and derivatives	NMR	104
5-Deoxypyridoxal	IR	111
Pyridine aldehydes and derivatives	NMR	90
	IR	112
2-PyCOMe	IR	97, 98, 121
	NMR	95, 96, 107, 113, 116 117, 119, 120
	Mass spec.	115
3-PyCOMe	ESR	92
	NMR	105, 106, 108, 116, 119
	IR	97, 121
4-PyCOMe	IR	97, 125
	NMR	100, 116, 125
4-Acetylpyridine-1-oxide	NMR	100
2,6-Diacetylpyridine	IR, NMR	94
Pyridine methyl ketones and derivatives	NMR	90
Picolyl ketones and thioketones	NMR	102, 103
2-PyCOPh	NMR	91, 115, 116
	IR (oximes)	110
3-PyCOPh	NMR	116
4-PyCOPh	NMR	116
2-Phenacylpyridines	IR, UV	89
2-Picolyl isopropyl ketone	NMR	101
2-Picolyl phenyl ketone	NMR	101
2-PyCH(CO-subst.Ph)$_2$	IR, UV	89

Compound	Type of study	Ref.
6-Bromopyridyl methyl ketone	NMR	124
6-Chloropyridyl methyl ketone	NMR	124
2-(6-Picolyl) methyl ketone	NMR	124

2-Py ⬠ CO−(subst.)Ph

IR, NMR   123

3-Py ⬠ CO−(subst.)Ph

IR, NMR   123

4-Py ⬠ CO−(subst.)Ph

IR, NMR   123

References

1. H. Bredereck, G. Semchen, and R. Wahl, *Chem. Ber.*, 101, 4048 (1968).
2. Z. Arnold, *Coll. Czech. Chem. Commun.*, 28, 863 (1963).
3. S. E. Forman, U.S. Patent 3,150,135; *Chem. Abstr.*, 62, 2765g (1965).
4. R. B. Margerison and J. A. Nelson, U.S. Patent 3,385,860; *Chem. Abstr.*, 69, 77122y (1968).
5. W. Mathes and W. Sauermilch, U.S. Patent 3,088,963; *Chem. Abstr.*, 57, 5895g (1962).
6. W. S. Chilton and A. K. Butler, *J. Org. Chem.*, 32, 1270 (1967).
7. S. Miyano, N. Abe, and A. Abe, *Chem. Pharm. Bull.* (Tokyo), 18, 511 (1970).
8. F. E. Cislak, U.S. Patent 3,160,633; *Chem. Abstr.*, 62, 5258g (1965).
9. S. Huenig and G. Ruider, *Tetrahedron Lett.*, 773 (1968).
10. G. Quéguiner and P. Pastour, *Bull. Soc. Chim. Fr.*, 4117 (1968).
11. G. Quéguiner and P. P. Pastour, *C. R. Acad. Sci., Paris, Ser. C*, 258, 5903 (1964).
12. G. Quéguiner, M. Alas, and P. Pastour, *C. R. Acad. Sci., Paris, Ser. C*, 265, 824 (1967).
13. G. Quéguiner and P. Pastour, *Bull. Soc. Chim. Fr.*, 3655, 3659, 3683 (1969).
14. K. Schank, *Chem. Ber.*, 102, 383 (1969).
15. T. L. Hullar, *J. Med. Chem.*, 12, 58 (1969).
16. H. G. Brooks, J. W. Luskso, and D. E. Meltzler, *J. Heterocycl. Chem.*, 3, 126 (1966).
17. P. D. Sattangi and C. J. Argoudelis, *J. Org. Chem.*, 33, 1337 (1968).
18. P. Bergmann and H. Paul, *Z. Chem.*, 6, 339 (1966); *Chem. Abstr.*, 66, 55342f (1967).

19. G. Quéguiner and P. Pastour, *Bull Soc. Chim. Fr.*, 4082 (1969).
20. German Patent 1,207,930; *Chem. Abstr.*, 64, 9694h (1966).
21. F. E. Cislak, U.S. Patent 3,118,899; *Chem. Abstr.*, 60, 9250f (1964).
22. V. I. Yakerson, L. I. Lafer, and A. M. Rubershtein, *Izv. Akad. Nauk SSSR, Ser. Khim.*, 314; *Chem. Abstr.*, 64, 17535g (1966).
23. R. H. Feldhake, U.S. Patent 3,155,676; *Chem. Abstr.*, 62, 2765b (1965).
24. E. Mueller and R. Herschkeil, *Tetrahedron Lett.*, 2809 (1964).
25. E. J. Blantz, F. A. French, J. R. DoAmaral, and D. A. French, *J. Med. Chem.*, 13, 1124 (1970).
26. H. Zimmer and J. P. Bercz, *Ann. Chem.*, 686, 107 (1965); *Chem. Abstr.*, 63, 14731h (1965).
27. C. S. Marvel, L. E. Coleman, and G. P. Scott, *J. Org. Chem.*, 20, 1785 (1955).
28. L. Krasnek, J. Durinda, and L. Szucs, *Chem. Zvesti*, 15, 558 (1961); *Chem. Abstr.*, 56, 12847g (1962).
29. M. Simalty, H. Strzelecka, and M. Dupre, *C. R. Acad. Sci., Paris, Sec. C*, 266, 1306 (1968).
30. R. A. Abramovitch, D. Auld, and E. M. Smith, University of Alabama, personal communication, 1970.
31. R. A. Abramovitch, E. M. Smith, and R. T. Coutts, *J. Org. Chem*, 37, 3584 (1972).
32. R. A. Abramovitch and G. Tertzakian, *Can. J. Chem.*, 43, 940 (1965).
33. M. S. Chang, *J. Org. Chem.*, 28, 3542 (1963).
34. T. Naito, K. Ueno, M. Sano, Y. Omura, I. Itoh, and F. Ishikawa, *Tetrahedron Lett.*, 5767 (1968).
35. British Patent 880,856; *Chem. Abstr.*, 59, 3899b (1963).
36. French Patent M2092; *Chem. Abstr.*, 60, 9252d (1964).
37. R. I. Ellin, U.S. Patent 3,285,927; *Chem. Abstr.*, 66, 28660b (1967).
38. French Patent 71,930; *Chem. Abstr.*, 57, 3419d (1962).
39. J. Wrobel and Z. Dabrowski, *Rocz. Chem.*, 39, 1239 (1965); *Chem. Abstr.*, 64, 15936g (1966).
40. Japanese Patent 68-00,517; *Chem. Abstr.*, 69, 59112r (1968).
41. C. R. Warner, E. J. Walsh, and R. F. Smith, *J. Chem. Soc.*, 1232 (1962).
42. A. Fozard and G. Jones, *J. Org. Chem.*, 30, 1523 (1965).
43. H. Erdtman, F. Haglid, I. Wellings, and U. S. vonEuler, *Acta Chem. Scand.*, 17, 1717 (1963); *Chem. Abstr.*, 60, 496g (1964).
44. A. P. Terent'ev, R. A. Gracheva, N. N. Preobrazhenskaya, and L. M. Volkova, *Zh. Obshch. Khim.*, 33, 4006 (1963); *Chem. Abstr.*, 60, 9242a (1964).
45. T. R. Govindachari, P. S. Santhanan, and V. Sudarsanan, *Ind. J. Chem.*, 4, 398 (1966).
46. H. D. Eilhauer and K. H. Meinicke, *Arch. Pharm. (Weinheim)*, 298, 131 (1965); *Chem. Abstr.*, 63, 1765d (1965).
47. M. Bieganowska and L. Kuczynski, *Acta Polon. Pharm.*, 20, 15 (1963); *Chem. Abstr.*, 61, 8302a (1964).
48. R. Bodalski and J. Michalski, *Rocz. Chem.*, 41, 939 (1967); *Chem. Abstr.*, 67, 73493f (1967).
49. Yu. I. Chumakov, Z. E. Stolyarov, Yu. P. Shapovalova, and V. F. Novikova, USSR Patent 184,454; *Chem. Abstr.*, 66, 10850q (1967).
50. G. B. Bachman and R. M. Schisla, U.S. Patent 3,120,527; *Chem. Abstr.*, 60, 9252g (1964).
51. I. L. Tschetter, *Proc. S. Dakota Acad. Sci.*, 43, 165 (1964); *Chem. Abstr.*, 63, 8165b (1965).

52. R. I. Fryer, R. A. Schmidt, and L. H. Sternbach, U.S. Patent 3,182,067; *Chem. Abstr.*, **63**, 5662h (1965).
53. R. I. Fryer, R. A. Schmidt, and L. H. Sternbach, U.S. Patent 3,182,065; *Chem. Abstr.*, **63**, 7024e (1965).
54. R. I. Fryer, R. H. Schmidt, and L. H. Sternbach, *J. Pharm. Sci.*, **53**, 264 (1964).
55. R. I. Fryer, R. A. Schmidt, and L. H. Sternbach, U.S. Patent 3,100,770; *Chem. Abstr.*, **60**, 1780e (1964).
56. R. E. Fryer, R. A. Schmidt, and L. H. Sternbach, U.S. Patent 3,182,066; *Chem. Abstr.*, **63**, 2989e (1965).
57. F. A. Vingiello and T. J. Delia, *J. Org. Chem.*, **29**, 2180 (1964).
58. L. H. Sternbach, H. Lehr, E. Reeder, T. Hayes, and N. Steiger, *J. Org. Chem.*, **30**, 2812 (1965).
59. L. A. Walter and N. Sperber, U.S. Patent 2,997,478; *Chem. Abstr.*, **56**, 1434i (1962).
60. British Patent 901,700; *Chem. Abstr.*, **58**, 9031g (1963).
61. F. J. McCarty, C. H. Tilford, and M. G. VanCampen, Jr., *J. Org. Chem.*, **26**, 4084 (1961).
62. N. S. Prostakov, L. A. Shakhparonova, and L. M. Kirillova, *Zh. Obshch. Khim.*, **34**, 3231 (1964); *Chem. Abstr.*, **62**, 4002h (1965).
63. L. Kulcznski, Z. Machon, and L. Wykret, *Diss. Pharm. Pharmacol.*, **13**, 299 (1961); *Chem. Abstr.*, **57**, 8540i (1962).
64. L. Kulcznski and S. Respond, *Diss. Pharm. Pharmacol.*, **20**, 163 (1968); *Chem. Abstr.*, **69**, 43750m (1968).
65. F. J. Villani, U.S. Patent 3,357,986; *Chem. Abstr.*, **69**, 27262y (1968).
66. M. Regitz and A. Liedhegener, *Chem. Ber.*, **99**, 2918 (1966).
67. M. M. Robison, W. G. Pierson, L. Dorfman, and B. F. Lambert, *J. Org. Chem.*, **31**, 3213 (1966).
68. V. G. Ermolaeva, I. S. Musatova, and M. N. Shchukina, *Zh. Obshch. Khim.*, **33**, 825 (1963); *Chem. Abstr.*, **59**, 8722a (1963).
69. A. F. Wagner, P. E. Wittreich, A. Lusi, and K. Folkers, *J. Org. Chem.*, **27**, 3236 (1962).
70. M. Pesson, M. Aurousseau, M. Joannic, and F. Roguet, *Chim. Therap.*, 127 (1966); *Chem. Abstr.*, **66**, 2431z (1967).
71. French Patent 1,317,574; *Chem. Abstr.*, **59**, 10,000e (1963).
72. V. G. Ermolaeva and M. N. Shchukina, *Zh. Obshch. Khim.*, **32**, 2664 (1962); *Chem. Abstr.*, **58**, 9057h (1963).
73. V. M. Aryuzina and M. N. Shchukina, *Geterotsikl. Soedin.*, 605 (1966); *Chem. Abstr.*, **66**, 94952z (1967).
74. P. F. Muhlradt, Y. Morino, and E. E. Snell, *J. Med. Chem.*, **10**, 341 (1963).
75. J. Buechi, F. Kracher, and G. Schmidt, *Helv. Chim. Acta*, **45**, 729 (1962).
76. Belgian Patent 667,078; *Chem. Abstr.*, **65**, 7153c (1966).
77. D. R. Osborne and R. Levine, *J. Heterocycl. Chem.*, **1**, 138 (1964).
78. T. Kato and Y. Goto, *Yakugaku Zasshi*, **85**, 451 (1965); *Chem. Abstr.*, **63**, 5596f (1965).
79. T. Kato, Y. Goto, and T. Chiba, *Yakugaku Zasshi*, **86**, 1022 (1966); *Chem. Abstr.*, **66**, 85680g (1967).
80. F. J. Villani, U.S. Patent 3,290,320; *Chem. Abstr.*, **66**, 46340b (1967).
81. H. Rubinstein, G. Hazen, and R. Zerfing, *J. Chem. Eng. Data*, **12**, 149 (1967).
82. E. L. Schumann, U.S. Patent 3,205,234; *Chem. Abstr.*, **63**, 14826a (1965).
83. Japanese Patent 67-22,616; *Chem. Abstr.*, **69**, 19033j (1968).

84. J. Janculev and B. Podolesov, *Glasnik Hem. Drustva Beograd*, **27**, 415 (1962); *Chem. Abstr.*, **61**, 638e (1964).

85. L. T. Senello and C. J. Argoudelis, *J. Org. Chem.*, **33**, 3983 (1968).

86. H. Ahrens and W. Korytnyk, *J. Heterocycl. Chem.*, **4**, 625 (1967).

87. F. J. Anderson and A. E. Martell, *J. Amer. Chem. Soc.*, **86**, 715 (1964).

88. N. Bild and M. Hesse, *Helv. Chim. Acta*, **50**, 1885 (1967); *Chem. Abstr.*, **67**, 120957x (1967).

89. R. F. Branch, A. H. Beckett, and D. B. Covell, *Tetrahedron*, **19**, 401 (1963).

90. W. Bruegel, *Z. Elektrochem.*, **66**, 159 (1962); *Chem. Abstr.*, **57**, 307b (1962).

91. S. Castellano and A. A. Bothner-By, *J. Chem. Phys.*, **41**, 3863 (1964); *Chem. Abstr.*, **62**, 2384a (1965).

92. P. T. Cottrell and P. H. Reger, *Mol. Phys.*, **12**, 149 (1967); *Chem. Abstr.*, **67**, 69314f (1967).

93. E. P. Crowell, W. A. Powell, and C. J. Varsel, *Anal. Chem.*, **35**, 184 (1963).

94. J. D. Curry, M. A. Robinson, and D. A. Busch, *Inorg. Chem.*, **6**, 1570 (1967).

95. G. G. Dvoryantseva, V. P. Lezina, V. F. Bystrov, R. N. Ul'yanova, G. P. Syrova, and Yu. N. Scheinker, *Izv. Akad. Nauk SSSR., Ser. Khim.*, 994 (1968); *Chem. Abstr.*, **69**, 76367v (1968).

96. D. Herbison-Evans and R. E. Richards, *Mol. Phys.*, **8**, 19 (1964); *Chem. Abstr.*, **61**, 12813 (1964).

97. R. Isaac, F. F. Bentley, H. Sternglanz, W. C. Coburn, Jr., C. V. Stephenson, and W. S. Wilcox, *Appl. Spectroscopy*, **17**, 90 (1963); *Chem. Abstr.*, **59**, 7077f (1963).

98. R. Joeckle and R. Mecke, *Ber. Bunsenges Phys. Chem.*, **71**, 165 (1967); *Chem. Abstr.*, **66**, 70445k (1967).

99. G. J. Karabatsos and F. M. Vane, *J. Amer. Chem. Soc.*, **85**, 3886 (1963).

100. A. R. Katritzky and J. M. Lagowski, *J. Chem. Soc.*, 43 (1961).

101. G. Klose and K. Arnold, *Mol. Phys.*, **11**, 1 (1966); *Chem. Abstr.*, **65**, 18467e (1966).

102. G. Klose and E. Uhlemann, *Nucl. Mag. Res. Chem., Proc. Symp.*, Cagliari, Italy, 237 (1964); *Chem. Abstr.*, **66**, 6931u (1967).

103. G. Klose and E. Uhlemann, *Tetrahedron*, **22**, 1373 (1966).

104. W. Korytuyk, E. J. Kris, and R. P. Singh, *J. Org. Chem.*, **29**, 574 (1963).

105. V. J. Kowalewski and D. G. deKowalewski, *Bol. Acad. Nacl. Cienc.*, **42**, 153 (1961); *Chem. Abstr.*, **58**, 2044c (1963).

106. V. J. Kowalewski and D. G. deKowalewski, *J. Chem. Phys.*, **36**, 266 (1962); *Chem. Abstr.*, **57**, 4207e (1962).

107. V. J. Kowalewski and D. G. deKowalewski, *J. Chem. Phys.*, **37**, 2603 (1962); *Chem. Abstr.*, **58**, 1072f (1963).

108. V. J. Kowalewski, D. G. deKowalewski, and E. C. Ferra, *J. Mol. Spectrosc.*, **20**, 203 (1966); *Chem. Abstr.*, **65**, 8216e (1966).

109. V. J. Kowalewski and D. G. deKowalewski, *Nucl. Magn. Resonance Chem., Proc. Symp.*, Cagliari, Italy, 159 (1964); *Chem. Abstr.*, **65**, 18467b (1966).

110. G. C. Kulasingam, W. R. McWhinnie, and R. R. Thomas, *Spectrochim. Acta*, **22**, 1365 (1966); *Chem. Abstr.*, **65**, 4849e (1966).

111. A. E. Martell, *I. U. B. Symp. Ser.*, **30**, 13 (1962); *Chem. Abstr.*, **62**, 6011f (1965).

112. Y. Matsui and T. Kubota, *Nippon Kagaku Zasshi*, **83**, 985 (1962); *Chem. Abstr.*, **58**, 10872g (1963).

113. N. K. Mahrotra and M. C. Saxena, *Bull. Chem. Soc. Jap.*, **40**, 19 (1967); *Chem. Abstr.*, **67**, 37269d (1967).

114. R. A. Miller, W. G. Fateley, and R. E. Witkowski, *Spectrochim. Acta*, **23A**, 891 (1967); *Chem. Abstr.*, **66**, 109835q (1967).

115. N. Neuner-Jehle, *Tetrahedron Lett.*, 2047 (1968).
116. G. A. Olah and M. Calvin, *J. Amer. Chem. Soc.*, **89**, 4736 (1967).
117. H.-H. Perkampus and U. Krueger, *Chem. Ber.*, **100**, 1165 (1967).
118. H. L. Retcofsky and R. A. Friedel, *J. Phys. Chem.*, **72**, 290 (1968); *Chem. Abstr.*, **68**, 44592c (1968).
119. H. L. Retcofsky and R. A. Friedel, *J. Phys. Chem.*, **72**, 2619 (1968); *Chem. Abstr.*, **69**, 43252a (1968).
120. H. L. Retcofsky and F. R. McDonald, *Tetrahedron Lett.*, 2575 (1968).
121. E. K. Schmid and R. Joeckle, *Spectrochim. Acta*, **22**, 1645 (1966); *Chem. Abstr.*, **65**, 16247a (1966).
122. E. Schumacher and R. Taubenest, *Helv. Chim. Acta*, **49**, 1455 (1966); *Chem. Abstr.*, **65**, 10560c (1966).
123. L. Szucs, J. Durinda, A. Nagy, J. Heger, and L. Drasnek, *Chem. Zvesti*, **22**, 347 (1968); *Chem. Abstr.*, **69**, 96556j (1968).
124. T. K. Wu, *J. Chem. Phys.*, **71**, 3089 (1967); *Chem. Abstr.*, **67**, 86338j (1967).
125. T. K. Wu and B. P. Dailey, *J. Chem. Phys.*, **41**, 3307 (1964); *Chem. Abstr.*, **62**, 1228d (1965).

CHAPTER XV

Sulfur and Selenium Compounds of Pyridine

HARRY L. YALE

The Squibb Institute for Medical Research
New Brunswick, New Jersey

I. Pyridinethiones and Pyridinethiols

1. Preparation

A. *By Synthesis of the Pyridine Ring*

Several novel syntheses involving formation of the pyridine heterocycle have led to a number of pyridinethiones, some of which are not readily accessible by conventional preparative procedures. One of these methods involves heating an aqueous solution of cyanothioacetamide, sodium formylacetone, and piperidine acetate to give 3-cyano-6-methyl-2-pyridinethione **(XV-1)**; the latter by the sequence of reactions shown, affords the bicyclic compound, 3-hydroxy-6-methylthieno[2,3-*b*]pyridine **(XV-2)** (366).

Another generally applicable reaction is the one between acylisothiocyanates and 1,3-dienes which gives 40 to 50% yields of 1-acyl-1,2,3,6-tetrahydro-2-pyridinethiones **(XV-3)** (5); these yields are obtained only after a reaction period of 6 to 7 months at room temperature.

The reaction of 2-cyano-3-ethyl-2-pentenonitrile and carbon disulfide in *N,N*-dimethylformamide gives the intermediate thiopyranthione **XV-4**, which is then rearranged in aqueous base to yield the 2-pyridinethione **(XV-5)** (130). A series of related 4-pyridyl sulfides (e.g., **XV-5a**), have also been synthesized *via* pyridine-ring formation (461).

4*H*-Pyran-4-thiones and alkyl- or aralkylamines readily form 1-alkyl or 1-aralkyl-4-pyridinethiones **(XV-6)** (97); with hydrazine, 4,5,6-triphenyl-2-thiopyrone yields the 1-amino-2-pyridinethione derivative **(XV-7)** (98, 397). 2-Cyano-3-mercapto-(3-methylthio)acrylamide, malononitrile, and aqueous methylamine gave **XV-7a** (481).

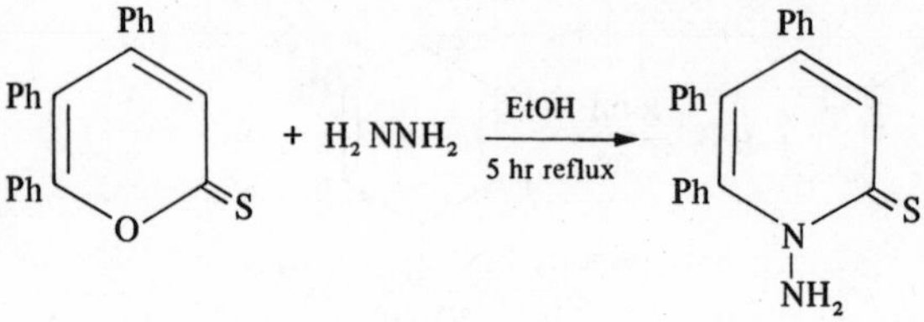

XV-7 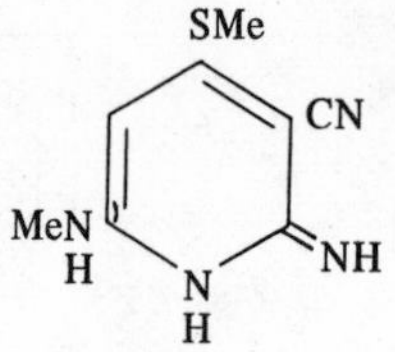   XV-7a

B. From Halopyridines

The conversion of halopyridines and halopyridine-1-oxides to the corresponding pyridinethiones has continued to be the most useful procedure for preparing this class of compounds. The most general methods involve reaction with an alkali hydrosulfide (6, 37, 44, 79, 227, 275, 284, 288, 329, 358, 373, 434, 455) or ammonium dithiocarbamate (221). A halogen at the 4-position is most easily displaced by the nucleophilic mercaptide ion. The reaction of ethanolic sodium hydrosulfide with 2,3,4,5-tetrachloropyridine for example, gives only **XV-8** (378). Several other related reactions provide confirmation for this preferential substitution: 2,3,4,5,6-pentachloropyridine and ethanolic hydrogen sulfide (150) or phosphorus pentasulfide in pyridine (434) give only 2,3,5,6-tetrachloro-4-pyridinethione **(XV-9)**; 3-chloro-2,4,5,6-tetrafluoropyridine yields **XV-10**. In none of these reactions, presumably, does the next most reactive halogen, the one at C_2 participate. The halogen at position 3 is least reactive, almost resembling the halobenzenes. Minor amounts of the symmetrical dipyridyl

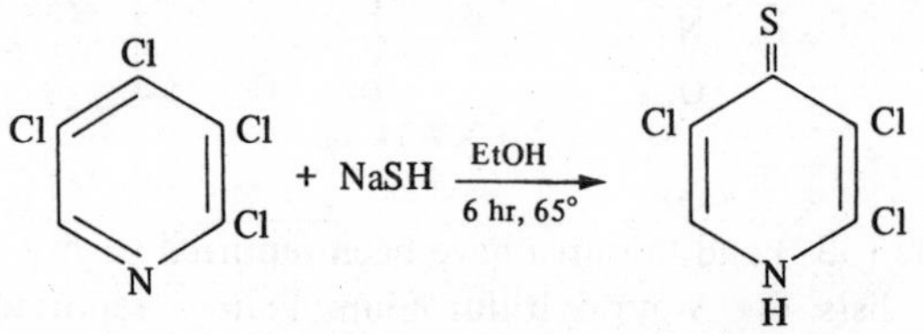

XV-8

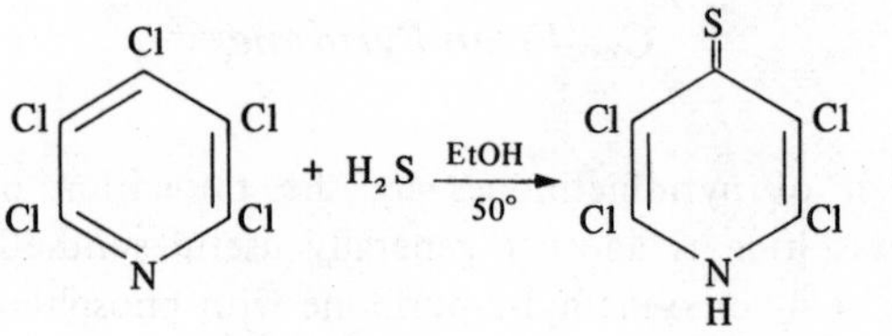

XV-9

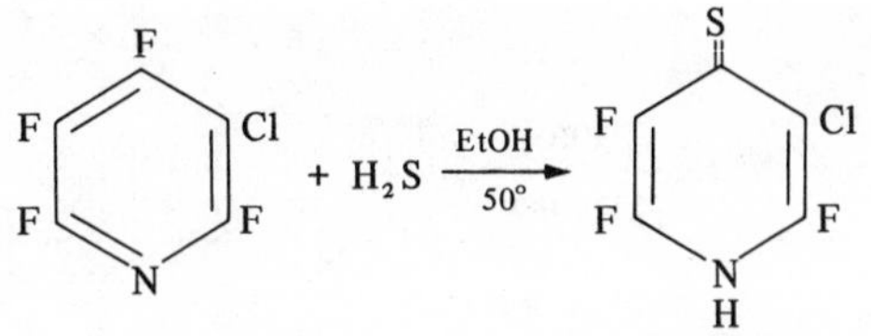

XV-10

sulfide are usually found as by-products in these reactions (79, 267, 275). The decomposition of *S*-pyridylthiuronium halides with base is a well-known preparative method for pyridinethiones and several additional examples have been reported (37, 138, 376). Here, too, the reaction has oftentimes given the sulfide as a minor by-product. In several instances, prolonged heating of the thiuronium halide in methanol or acetone in the absence of base gave the sulfide: diethyl 4-chloro-2,6-pyridinedicarboxylate (221) and 3-fluoro-4-nitro-

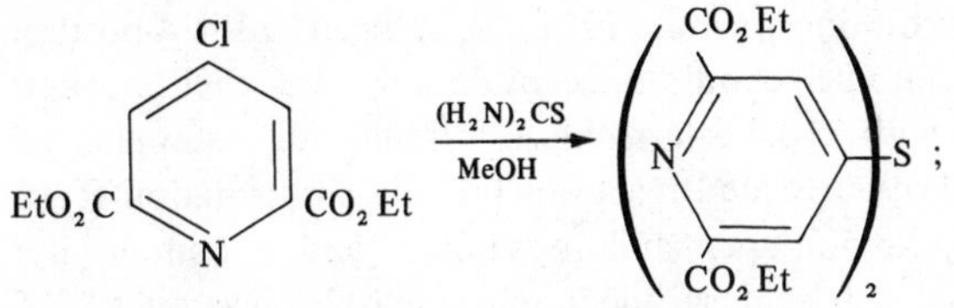

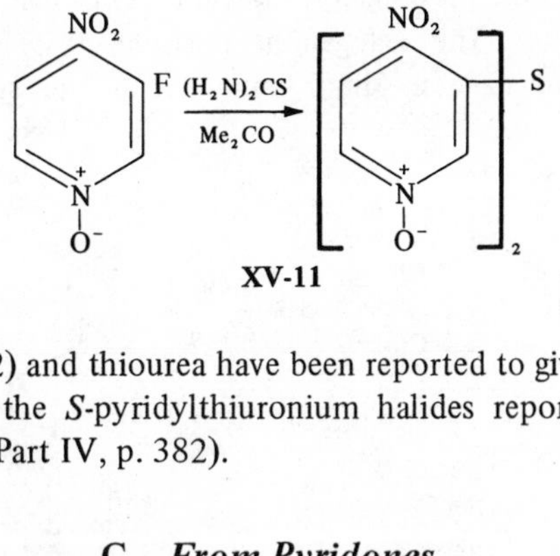

XV-11

pyridine-1-oxide (332) and thiourea have been reported to give only the sulfides.

Table XV-47 lists the *S*-pyridylthiuronium halides reported since the first compilation (1st Ed. Part IV, p. 382).

C. *From Pyridones*

The preparation of pyridinethiones by the thionation of pyridones with phosphorus pentasulfide is another generally useful synthetic procedure. For example, fusion of 4- or 6-methyl-2-pyridone with phosphorus pentasulfide at 140 to 160° gives essentially quantitative yields of the thiones (**XV-12**) (79).

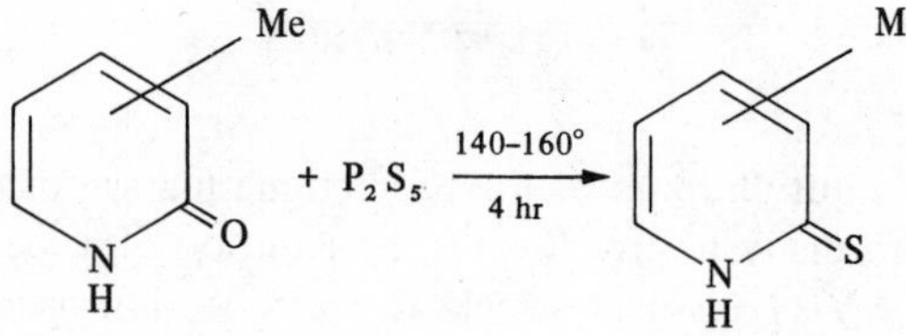

XV-12

Salts of Guareschi imides **(XV-13)** and the same reactant, in xylene, have led to a series of 4-alkyl-3,5-dicyano-6-mercapto-2-pyridinethiones; these dithiono derivatives are stable only as their anions **(XV-14)** (44, 377). Similarly, 1-alkyl-4-

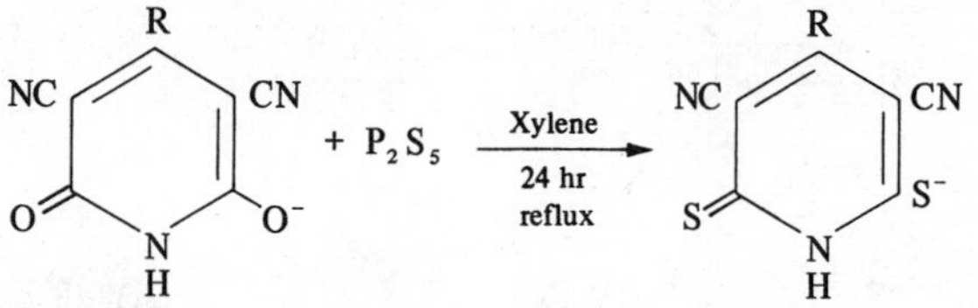

XV-13   **XV-14**

pyridones give 1-alkyl-4-thiopyridinones **(XV-15)** (97, 397). In contrast, an anomalous behavior is seen with 2,6-di-*t*-butyl-4-pyridone, where only the sulfide **XV-16** is obtained. It is significant, however, that 2,6-di-*t*-butyl-4-chloropyridine and potassium hydrosulfide also react in this manner to give **XV-16** rather than the thione (267).

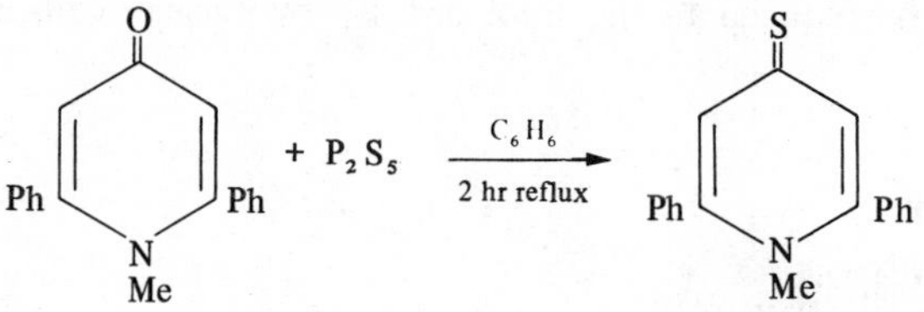

XV-15

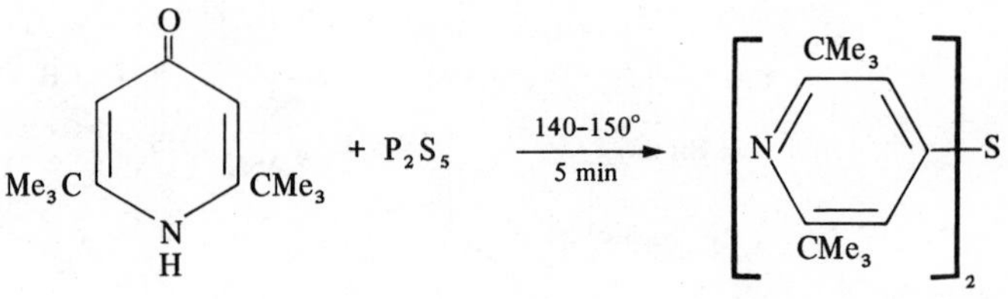

XV-16

D. *From Sulfides*

The addition of *t*-butyllithium to the azomethine linkage of methyl 4-pyridyl sulfide proceeds normally to give **XV-17**; the addition of a second molecule of *t*-butyllithium to **XV-17** involves attack at the S–Me linkage, and attack by a third molecule leads to the formation of the thione **(XV-18)**. In contrast, under the same conditions, *t*-butyllithium and 4-ethoxypyridine give 2,6-di-(*t*-butyl)-4-ethoxypyridine (267).

XV-17   XV-18

E. *From Aminopyridines*

3-Aminopyridines form diazonium cations in the normal fashion and these react with nucleophiles such as potassium ethyl xanthate or sodium sulfide to give 3-pyridinethiols along with lesser amounts of the corresponding disulfide (15, 80, 81, 118, 525); with the former reactant, the xanthyl intermediate **(XV-19)** may be reduced to the thiol directly by means of lithium aluminum hydride (133).

XV-19

F. From Pyridinesulfonyl Chlorides

The reduction of 3-pyridinesulfonyl chloride to 3-pyridinethiol has been carried out with stannous chloride in concentrated hydrochloric acid (15) and with red phosphorus and iodine in glacial acetic acid (118). In the latter procedure, the yield of 3-pyridinethiol is about 25%; the by-products include the disulfide and the sulfonic acid (118).

G. From Pyridylpyridinium Chlorides

1-(4-Pyridyl)pyridinium chloride and hydrogen sulfide in pyridine at 140° give 4-pyridinethione (133, 159).

H. From Organometallic Compounds of Pyridine-1-oxides and Elemental Sulfur

Pyridine-1-oxides, treated first with lithium, sodium, or potassium hydrides, or n-butyllithium, and then with an excess of elemental sulfur gave 1-hydroxy-2-pyridinethiones (433, 471, 537).

I. From Thiocyanates

Diethyl 4-thiocyanopyridine-2,6-dicarboxylate and ammonium sulfide in aqueous ethanol under reflux yield the 4-thione (221).

J. From Carboxylic Acids

Picolinic acid-1-oxide and sulfur, heated in diglyme or pyridine at 110 to 120°, gave 1-hydroxy-2-pyridinethione in 44% yield (489, 495).

2. Properties

The 2- and 4-pyridinethiols and 2- and 4-pyridinethiones are tautomers. Ultraviolet spectral data and potentiometric titrations have furnished evidence that at equilibrium in neutral solution, the species favored is the tautomer with the mobile proton on the nitrogen atom. Below are shown the approximate ratios of tautomers having the proton on nitrogen to those having the proton on sulfur. While far smaller than those for the 2- and 4-isomers, the ratio for the 3-isomer is surprisingly large, particularly when it is realized that a thione structure for the latter would necessarily involve a "dihydro" structure for which no evidence exists. Alternatively, a zwitterionic structure may be the species in equilibrium with the 3-thiol. The ratios for the oxygen analogs are shown for comparison (398).

	Sulfur derivatives	Oxygen derivatives
2-	49,000: 1	340: 1
3-	150: 1	1: 1
4-	35,000: 1	2200: 1

The interpretation of the infrared spectra of these pyridinethiones and pyridinethiols has led to the conclusion that there exists in these compounds strong intermolecular hydrogen bonding (399). It should be noted that the x-ray spectra of these compounds (1st Ed., Part IV, p. 353) have been interpreted to indicate weak hydrogen bonding.

The pK_a for the 2-, 3-, and 4-pyridinethiones are 9.97, 7.01, and 8.83, respectively (398).

The mass spectra of 2-pyridinethione and seventeen homologs have recently been reported (408, 409). Two modes of fragmentation of the molecular ions that are scarcely detectable with the ions from the corresponding 2-pyridones, that is, loss of HO, loss of H, and loss of HCN, are observed with the sulfur derivatives, where H, HS, HCN, and CS are significant products. These observed differences might be interpreted as a reflection of the relative abundance of the tautomeric forms in each class of compounds. This is not true, however, since the molecular ion from 1-($^2H_3$)methyl-2-pyridinethione also loses the fragment ^2HS. In addition to the fragments above, the ions from the methylated pyridinethiones also yield CH_2S, and 3-methyl-2-pyridinethione is unique in also fragmenting to yield H_2S. Separate experiments demonstrated that when kept at 150° for several days, the several 1-alkyl-2-pyridinethiones did not undergo any detectable change, thus ruling out any possibility of rearrangement under the conditions of the mass spectra determination.

3. Reactions

A. *Oxidation and Reduction*

The procedures whereby pyridinethiones and pyridinethiols are oxidized to disulfides, sulfonyl chlorides, and sulfonic acids shall be discussed in later sections of this chapter (pp. 244, 236, 229). Two interesting anomalies should be noted, however: (1) that the betaine **XV-21** could not be prepared from the reaction between hydrogen peroxide and 3,5-dibromo-1-methyl-4-pyridine-thione; only the hydrolysis product, **XV-20**, was obtained, due probably to the steric effects of the two bulky bromine atoms, and (2) that with 0.5M nitric acid, the 4-thione group is eliminated (586), while with 100% nitric acid, it is replaced by the nitroso group (568) **(XV-21a)** or the 4-nitro group (545).

XV-20   XV-21

XV-21a

B. *Acylation, Alkylation, and General Reactions*

Pyridinethiones and acyl halides yield only *S*-acyl derivatives (61, 387); 1-hydroxy-2(1*H*)-pyridinethione **(XV-22)** behaves similarly in reactions with acid halides, chloroformates, and chlorothioformates (43, 72). 4-Pyridinethione and acetic anhydride give 1-acetyl-4-pyridinethione (67).

Alkali metal salts of 4-pyridinethiones react with *p*-toluenesulfonyl chloride and with acyl halides to give the corresponding *S*-sulfonate and thiolacetate, respectively (151). The sodium salt of **XV-22** and acetic anhydride yield the 2-thiolacetate (246), while with trichloromethyl mercaptan **XV-22** gives the disulfide **XV-23** (248, 287); similar unsymmetrical disulfides are also prepared from 2- and 4-pyridinethiones and alkane- and arenesulfenyl chlorides (438, 439).

2-, 3-, And 4-pyridinethiones and their 1-oxides, in neutral solvent or with base, react with alkyl, aralkyl, and activated aryl halides to give predominantly

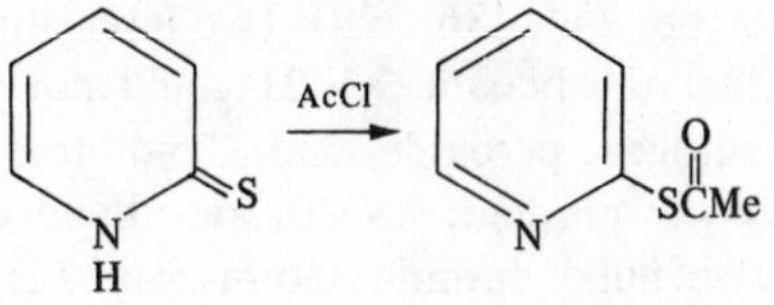

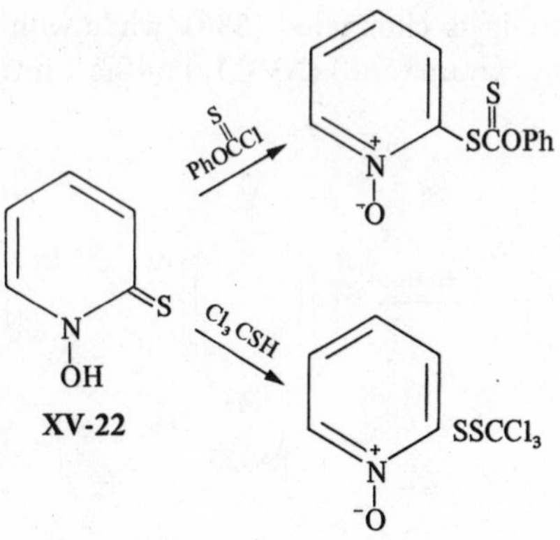

XV-22

XV-23

S-substituted derivatives (37, 73, 152, 248, 273, 278, 279, 287, 378). The chloro-
acetate ion behaves similarly, but 3-chloropropionitrile and 2-pyridinethione
gave significant amounts of both *S*- (**XV-24**) and *N*- alkylation products (**XV-25**)

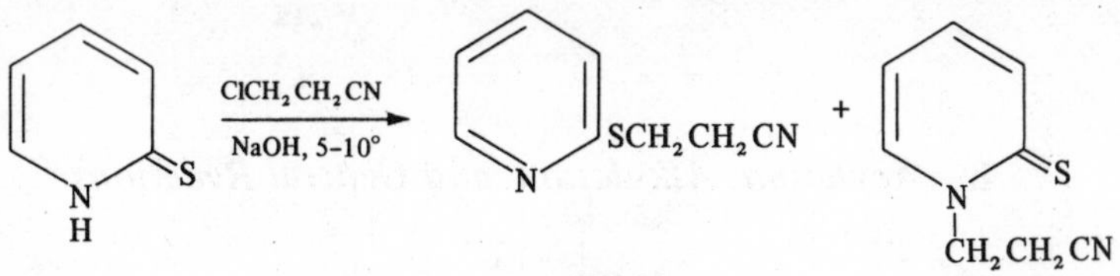

XV-24   **XV-25**

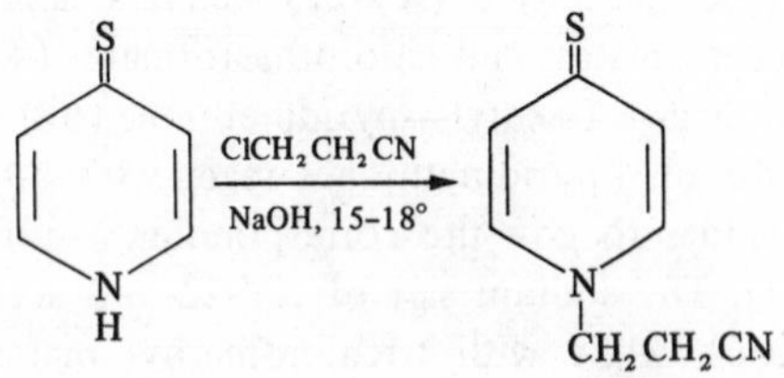

XV-26

in the presence of base (107, 108), while 4-pyridinethione yielded the
N-alkylation product (**XV-25**) principally. The base-catalyzed addition of
4-pyridinethione to acrylonitrile or methyl acrylate has given both *S*- and
N-substituted derivatives (106).

2-Bromoethylchloride and 2-pyridinethione in base give the quaternary compound (**XV-27**), which gave the *N*-alkyl-2-pyridinethione (**XV-28**) with

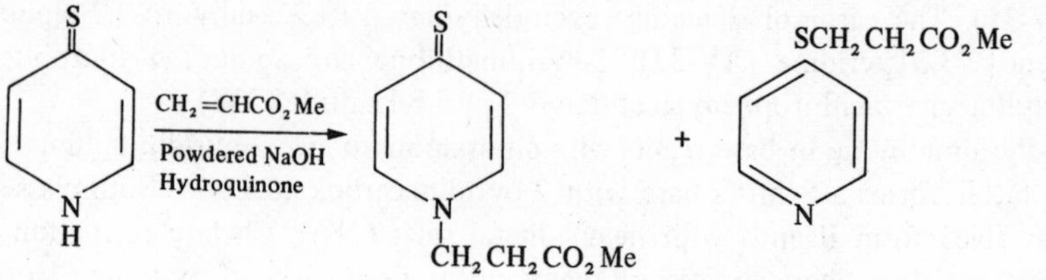

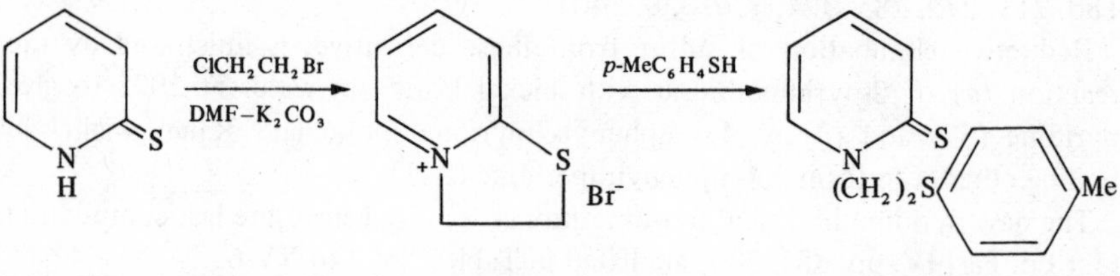

XV-27   **XV-28**

p-thiocresol. The reaction of 2-pyridinethiones with α-haloketones or β-halo-acetals has yielded the novel α-(2-pyridylthio)ketones and β-(2-pyridylthio)-acetals (37). 2-Pyridinethiones are converted to 2-methylthiopyridines by means of diazomethane or dimethyl sulfate (130).

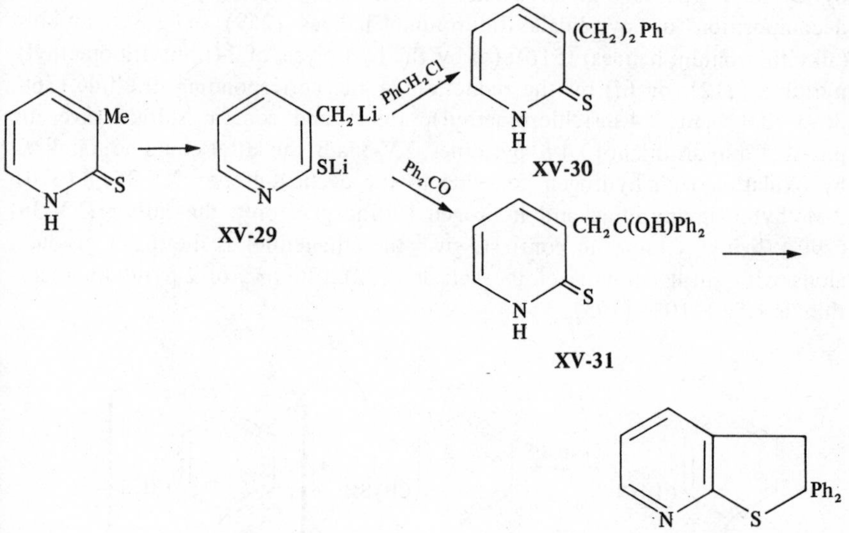

XV-29

XV-30

XV-31

XV-32

3-Methyl-2-pyridinethione and *n*-butyllithium give the dilithio species (XV-29). These organolithium compounds react with aralkyl halides to give derivatives such as XV-30, and with benzophenone give the tertiary carbinol (XV-31). The carbinol is readily cyclodehydrated to 2,3-dihydro-2,2-diphenyl-thieno[2,3-*b*]pyridine (XV-32). 2-Pyridinethione and sodium *p*-chlorophenyl-acetylide give *p*-chlorophenylacetylenyl 2-pyridyl sulfide (338).

2-Pyridinethione in base reacts with chloramine to give 2-pyridinesulfenamide; the latter forms a Schiff's base with 2-pyridinecarboxaldehyde. Both classes of derivatives form ligands with heavy metal salts (286). Chelate formation is a rather common characteristic of many sulfur derivatives of pyridine (46, 169, 180, 211, 272, 285, 324, 326, 339, 340).

Reductive elimination of sulfur from these derivatives is illustrated by the reaction (*a*) of 2-pyridinethione with nickel boride in water at 200° to give pyridine (57) and (*b*) of 4,6-diphenyl-2-pyridinethione and Raney nickel in boiling ethanol to form 2,4-diphenylpiperidine (201).

The new pyridinethiols and pyridinethiones reported since the last compilation (1st Ed. Part IV, pp. 383–386) are listed in Tables XV-2 to XV-6.

II.   Pyridylalkanethiols

Representatives of this class of compounds have been prepared (*a*) from pyridylalkyl halides and an alkali hydrosulfide (172); (*b*) by the alkaline decomposition of pyridylalkylthiuronium halides (259) or 2,6-pyridyl-bis-(alkylthiuronium halides) (516); (*c*) by the hydrolysis of 2-(*t*-butylthiomethyl)-pyridine (212); or (*d*) by the reduction of the corresponding disulfide (368, 369). 2,3- and 3,4-Bis(chloromethyl)pyridine and sodium sulfide gave the unstable dihydrothieno[3,4-*b*]pyridines, XV-35a,b; the latter could be stabilized by oxidation with hydrogen peroxide to the cyclic sulfones, XV-35c,d (576). 2-Methyl-5-vinylpyridine and hydrogen sulfide give only the sulfide (XV-36) (299); 2-vinylpyridine, in contrast, gives the ethanethiol as the major product, along with small amounts of the sulfide (272). The pK_a of 2-pyridylmethane-thiol is 1.52×10^{-9} (195).

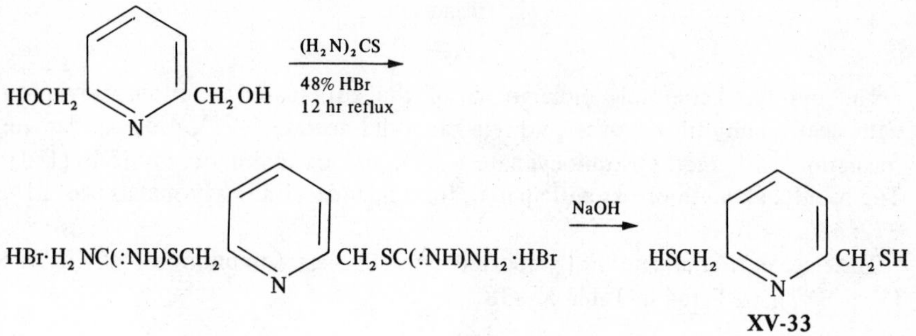

$(H_2N)_2CS$
48% HBr
12 hr reflux

$HOCH_2$ —[pyridine]— CH_2OH

$HBr \cdot H_2NC(:NH)SCH_2$ —[pyridine]— $CH_2SC(:NH)NH_2 \cdot HBr$

NaOH

$HSCH_2$ —[pyridine]— CH_2SH

XV-33

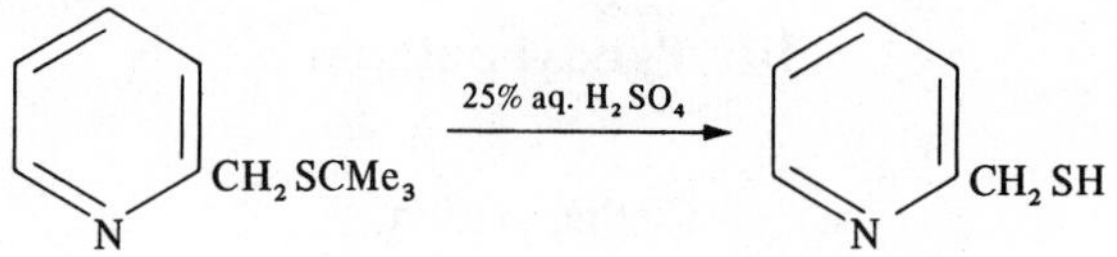

[pyridine]—CH_2SCMe_3

25% aq. H_2SO_4

[pyridine]—CH_2SH

XV-34

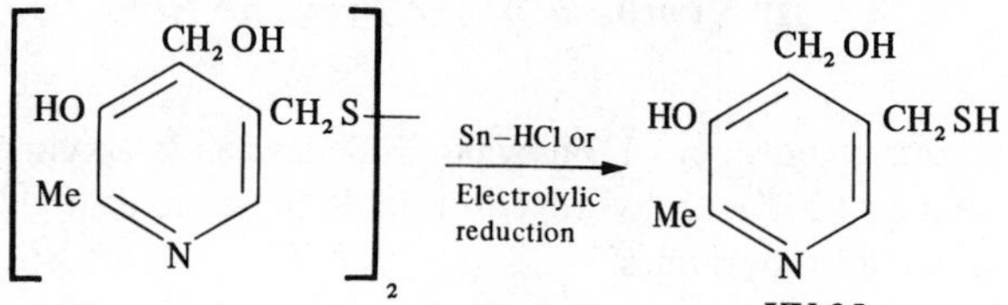

$$\left[\begin{array}{c} CH_2OH \\ HO\text{—[pyridine]—}CH_2S \\ Me \end{array} \right]_2$$

Sn–HCl or
Electrolylic
reduction

CH_2OH
HO—[pyridine]—CH_2SH
Me

XV-35

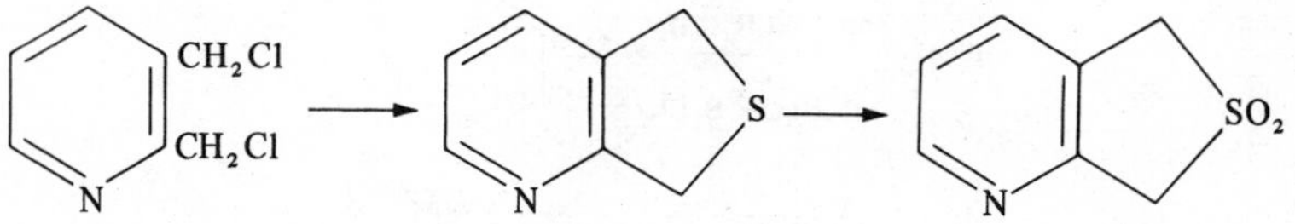

[pyridine]
CH_2Cl
CH_2Cl

→ **XV-35a** → **XV-35c**

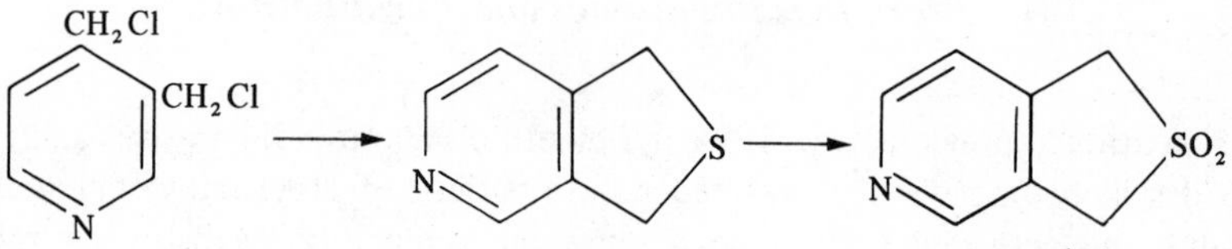

[pyridine]
CH_2Cl
CH_2Cl

→ **XV-35b** → **XV-35d**

[pyridine]—CH:CH_2
Me

H_2S
135°

$$\left[\begin{array}{c} Me\text{—[pyridine]—}CHMe\text{—S} \end{array} \right]_2$$ [No thiol]

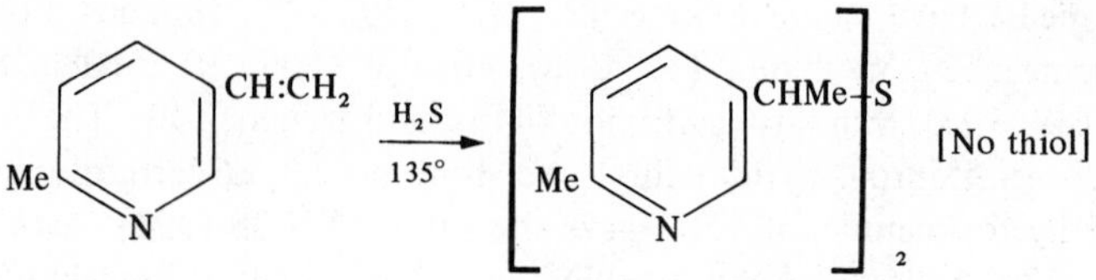

XV-36

1. Reactions

The pyridylalkanethiols undergo all of the anticipated reactions: acylation with acetic anhydride to the pyridylalkanethiol acetate (172), dithiocarbamate formation with methylisothiocyanate (259), and oxidation to disulfide (172). The pyridylalkanethiols, as anticipated, form ligands with heavy metals (46, 211, 272, 326, 340).

The new pyridylalkanethiols reported since the last compilation (1st Ed. Part IV, p. 387), are listed in Table XV-38.

III. Pyridyl Sulfides

1. Preparation

A. *By Synthesis of the Pyridine Ring*

1,3-Dicyano-2-propanols or 1,3-dicyanopropenes and alkylmercaptans or arylthiols heated at 120° with hydrogen bromide in acetic acid give 2-amino-6-(alkylthio) (or arylthio) pyridines.

$$\text{RSH} \xrightarrow[\text{NCCH: CHCH}_2\text{CN}]{\overset{\text{HOCH(CH}_2\text{CN)}_2}{\text{or}}}$$

XV-37

B. *From Pyridinethiones and Pyridinethiols*

As mentioned previously (p. 199), pyridinethiols and pyridinethiones and their 1-oxides in base, have been reacted with a variety of alkyl, alkylene, aralkyl, pyridyl, and activated aryl halides to give predominantly S-substituted, rather than N-substituted, derivatives, although several exceptions were noted. The reaction media have included water (73, 87, 105a, 373), aqueous dioxane (58, 89), dioxane (53), methanol (37, 288), ethanol (105, 107, 108, 289, 347), pyridine (49, 394), dimethyl sulfoxide (446), and toluene (49). The copper salt of 2,3,5,6-tetrafluoro-4-pyridinethiol and 4-bromo-2,3,5,6-tetrafluoroaniline in N,N-dimethylformamide, at 150°, gave the sulfide XV-38 (394). A tin chloride complex of 3-pyridinethiol and octyl iodide yielded 3-octylthiopyridine (14).

XV-38   XV-38a

The product obtained from the reaction of 6-chloro-2(1*H*)-pyridinethione with phosphorus pentasulfide at 130° was shown by X-ray crystallographic analysis to be **XV-38a** (549).

C. *From Halopyridines*

The halopyridines and their 1-oxides have been converted to pyridyl sulfides by nucleophilic displacement of the halogen by mercaptide ion; occasionally, either copper bronze or cuprous oxide has been used as catalyst (1, 265, 331, 334). The solvents have included water (26, 329, 373), methanol (163, 236, 288, 289, 335), chloroform (62), ethanol (1, 105, 138, 140, 219, 265, 267), pyridine (364), lutidine (280), triethylamine (41), *N,N*-dimethylacetamide (47), and *N*-methylpyrrolidone (48).

A halogen at position 4 is far more reactive than one at position 2 toward mercaptide ion; a 2-(trichloromethyl) group is inert. Thus 2,4-dichloro-3-nitropyridine or 3,4,5-trichloropicolinotrichloride and sodium methylmercaptide in methanol gave only the respective 4-(methylthio) derivatives (163, 236, 335).

In an unusual reaction, diethyl 4-methylthiopyridine-2,6-dicarboxylate is reported to be formed by the reaction of diethyl 4-chloropyridine-2,6-dicarboxylate and potassium methyl xanthate (221).

As noted previously (p. 194), the decomposition with base of *S*-pyridylthiuronium halides frequently gives the sulfide as a by-product; in two instances, heating the thiuronium chloride in methanol or acetone has only given the sulfide (221, 332).

1-Methyl-2-pyridinethione and polymethylene dibromides yield bis(sulfonium bromide) derivatives (281).

The pyridyne, generated from either 3-bromopyridine or 4-chloropyridine and sodamide, reacts with methyl mercaptan to give similar mixtures of methyl 3-pyridyl sulfide and methyl 4-pyridyl sulfide (446).

D. *From Aminopyridines*

Diazotized 3-aminoisonicotinic acid reacts with 5-chloro-2-thiocresol to give the sulfide **(XV-39)** (29, 166, 449).

XV-39

E. *From 1-Pyridinium and 1-Picolinium Cations*

By complex and not too well understood reactions, the 1-pyridinium and 1-picolinium cations **(XV-40)** and alkylmercaptide anions give, along with other products, alkylthiopyridines **(XV-41)** (14–17, 498). The products formed, and in particular, the ratios of isomeric sulfides produced, depend largely upon whether

a pyridinium or picolinium cation is involved and the nature of E, X, and R. Table XV-1 summarizes some of the data. The sulfur-free products of these reactions include pyridine (or picoline), unreacted *N*-oxide, and oxidation products, for example, acetaldehyde from the ethosulfates (see also Chapter IV).

Whether substitution of mercaptide ion occurs at the methyl group of a 1-picolinium cation is dependent largely on R. Thus thiophenoxide ion reacts to give (arylthiomethyl)pyridines **(XV-43)** while alkylmercaptide ion gives products involving substitution at the methyl group as well as the pyridine nucleus **(XV-44)**. In contrast, in reactions with alkyl mercaptans in neutral ethanol, the picolinium cation undergoes almost randomized substitution in the pyridine nucleus **(XV-45)**.

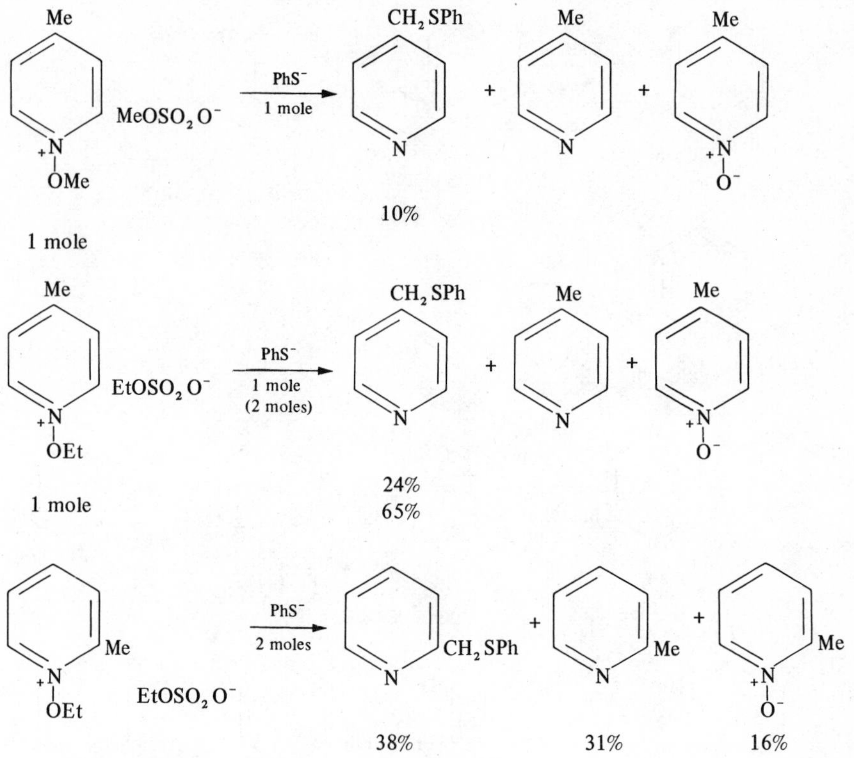

Me

MeOSO₂O⁻

1 mole

PhS⁻
1 mole

CH₂SPh + Me + Me

10%

Me

EtOSO₂O⁻

1 mole

PhS⁻
1 mole
(2 moles)

CH₂SPh + Me + Me

24%
65%

Me

EtOSO₂O⁻

PhS⁻
2 moles

CH₂SPh + Me + Me

38%   31%   16%

XV-43

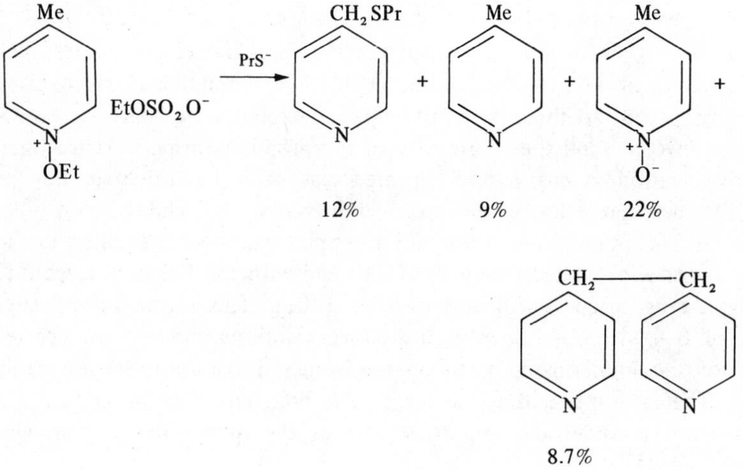

Me

EtOSO₂O⁻

PrS⁻

CH₂SPr + Me + Me +

12%   9%   22%

CH₂——CH₂

XV-44

8.7%

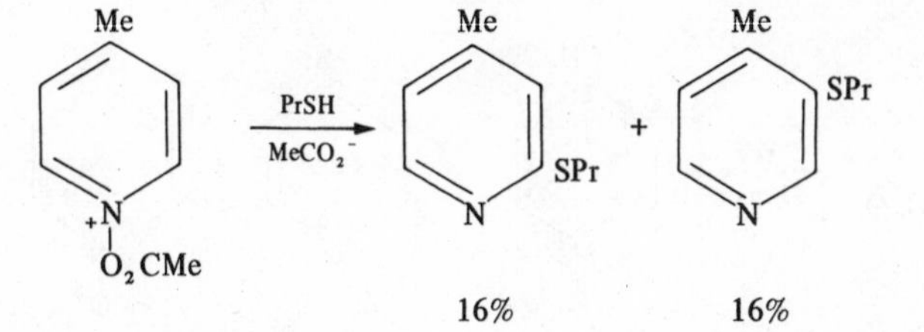

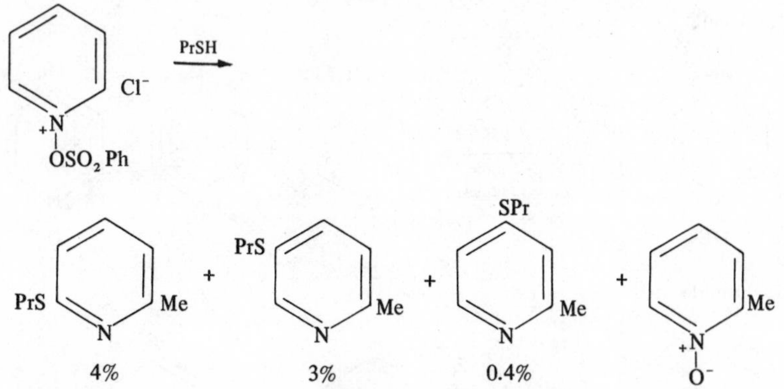

16%   16%

4%   3%   0.4%

XV-45

3%   1%   0.7%

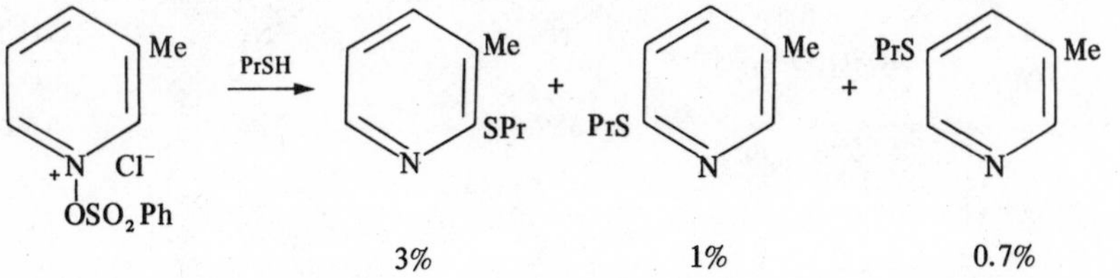

While most of these reactions are not practical preparative procedures, they provide a wealth of data for mechanism elucidation. A number of mechanisms have been proposed but these have not helped to explain adequately the variety of reactions involved and the complexity of the products formed. At the onset of this investigation, it might have appeared reasonable to anticipate that the nucleophilic mercaptide ion would attack only positions 2 and 4, as do other nucleophiles. The pyridinium cation, for example, reacts with cyanide ion to give only 2- and 4-pyridinecarbonitriles (400) and with the Grignard reagent to give only 2-substituted pyridine derivatives (401). The formation of large amounts of 3-alkylthiopyridines in the above reactions and the unexpected influence of the quaternizing agent on the isomer distribution are not easily explained. The assumption that the reaction is both electrophilic and nucleophilic does not consider the important role of the quaternizing agent. One possible explanation, namely, that the 3-sulfide is formed by the rearrangement of the 4-isomer *via* the sequence **XV-46** was not put to test; no apparent attempt was made to determine whether the 4-(alkylthio)pyridine would rearrange under the experimental conditions to the 3-isomer (14). (See Ch. IV for further discussion.)

The mechanism suggested for the formation of (alkylthiomethyl)pyridines involves the formation of an intermediate anhydro base which undergoes nucleophilic attack by mercaptide ion **(XV-47)** (16); this concept is based, in effect, on a similar mechanism proposed for the reaction of 4-picoline-1-oxide with acetic anhydride to give 4-(acetoxymethyl)pyridine (402).

It is important to note that only the quaternary cations are attacked by nucleophilic agents; pyridine and picoline-1-oxides do not react (1st Ed., Part II, p. 125).

F. By Reaction with Sulfur Chloride

2-Pyridone undergoes electrophilic substitution at the 5-position when heated under reflux for 1 week with sulfur dichloride and gives the sulfide (XV-48).

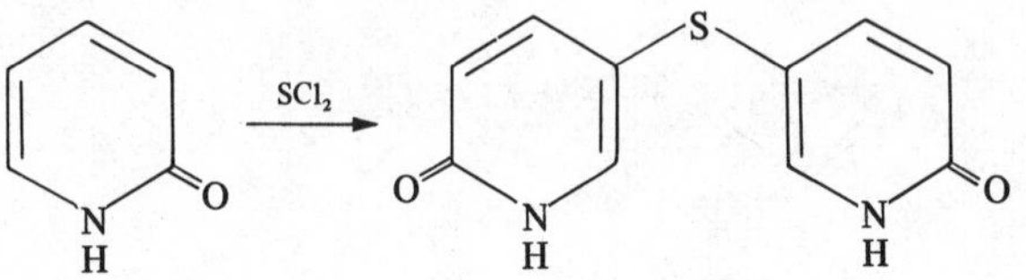

XV-48

G. From 4-Phenoxypyridine

The preparation of 4-(alkylthio)pyridines by the nucleophilic displacement of the phenoxy group in 4-phenoxypyridine by alkyl mercaptide ion is a general preparative method (XV-49) (233, 345). An alternative procedure involves heating 4-phenoxypyridine hydrobromide and benzenethiol, when a quantitative yield of 4-(phenylthio)pyridine is reported (345).

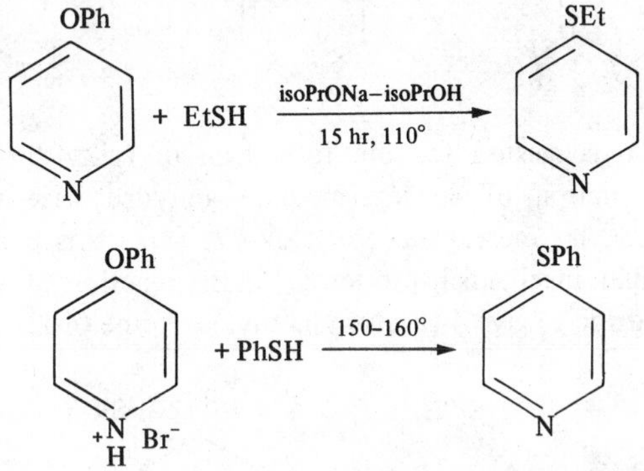

XV-49

H. By Cleavage of a Disulfide with an Organolithium Compound

2-Pyridyllithium cleaves the disulfide linkage in 2,2'-dithiodiacetaldehyde bis(dimethyl acetal) to give 2-(2-pyridylthio)acetaldehyde dimethyl acetal (XV-50) (257).

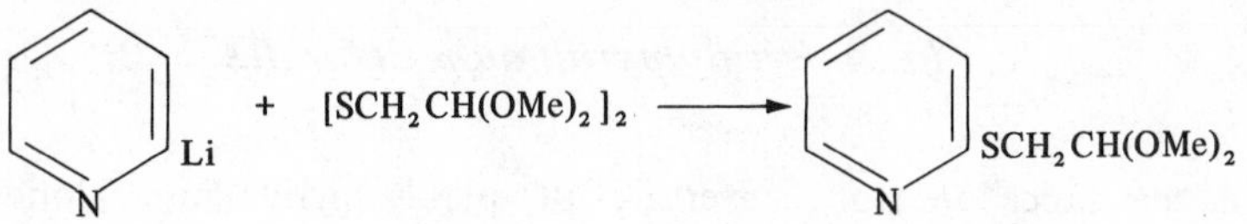

$$\text{(pyridyl-Li)} \quad + \quad [SCH_2CH(OMe)_2]_2 \longrightarrow \text{XV-50} \quad SCH_2CH(OMe)_2$$

XV-50

I. By the Alkylation of Aryl 4-Methyl-2-pyridyl Sulfides

The lateral metalation of 4-methyl-2-pyridyl phenyl sulfide by potassium amide in liquid ammonia gives **XV-44**, which is alkylated with *n*-hexyl bromide to give the 4-*n*-heptyl derivative (**XV-51**) (2). In striking contrast, the sulfone **XV-52** does not react under the same conditions.

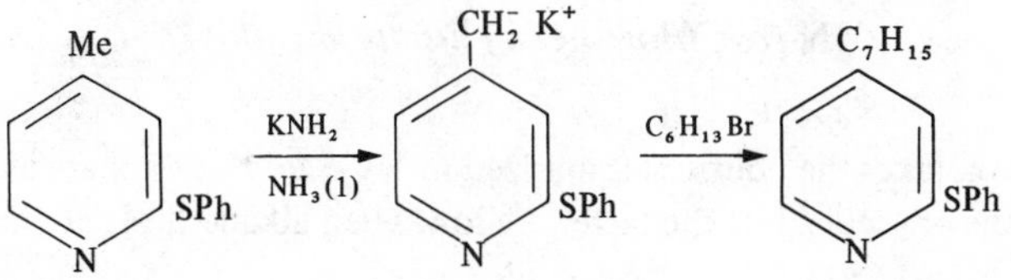

XV-44   **XV-51**

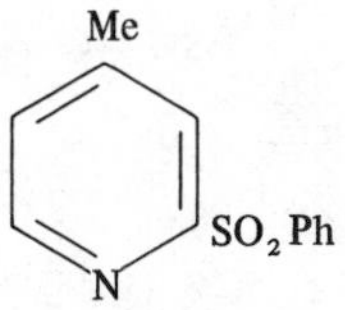

XV-52

J. By Mercaptide Ion Displacement of a Nitro Group in 4-Nitropyridine-1-oxide

One of the reactions frequently encountered with 4-nitropyridine-1-oxides is displacement of the 4-nitro group by various nucleophiles; the reaction is usually not accompanied by reduction of the 1-oxide (1st Ed., Part II, p. 513 ff). Thus 4-nitropyridine-1-oxide gives (4-pyridylthio)acetic acid-1-oxide when heated in ethanolic sodium hydroxide with thioglycollic acid (247); similarly, 4-nitro-3-pyridinecarboxamide-1-oxide and benzenethiol yield 4-(phenythio)-3-pyridine-carboxamide-1-oxide (120).

K. From Pyridylpyridinium Chlorides

When the procedure for converting 1-(4-pyridyl)pyridinium chlorides to 4-pyridinethiones (p. 197) is employed with a mixture of pyridine, hydrogen sulfide, and allyl chloride, the product is 4-(allylthio)pyridine (159). 4-(1-Pyridyl)pyridinium chloride and mercaptoacetic acid give (4-pyridylthio)-acetic acid (485).

L. From Alkyl 4-Pyridinethione-1-carboxylates

4-(Methylthio)pyridine is obtained by heating methyl 4-pyridinethione-1-car-boxylate at 100° (67).

M. From Pyridine-1-oxides, a Thiol, and N-(α-Chlorobenzylidene)aniline

This reaction takes the course summarized in **XV-52a**. No data are available yet for conditions or yields, but the latter are low when alkanethiols are used instead (449).

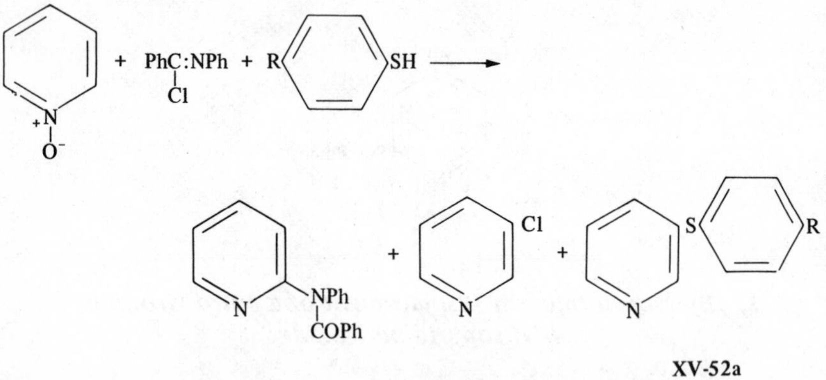

R = H, OMe

XV-52a

2. Reactions

The oxidation of pyridylsulfides to the corresponding sulfoxides and sulfones are discussed in later sections. The sulfide linkage is stable toward cleavage during catalytic or chemical reduction of nitro groups to amines (1, 53, 55, 105, 288, 289) or pyridine-1-oxides to pyridines (140, 273). Reduction of diethyl 4-methylthio-2,6-pyridinedicarboxylate with sodium borohydride is without effect on the sulfide linkage and yields 4-methylthio-2,6-pyridinedimethanol (153, 156). A recent report claims that 2-alkyl and 2-arylsulfides are cleaved by sodium borohydride, with displacement of the sulfur atom from the pyridine heterocycle (472). However, prolonged heating with Raney nickel in ethanol under reflux (130, 481) or nickel boride in water at 200° (57), results in the elimination of alkylthio groups from (alkylthio)pyridines. The carbon sulfur linkage is cleaved by *n*-butyllithium (434). The carbon–sulfur linkage in 4-(benzylthio)pyridine-1-oxide undergoes chlorinolysis in aqueous acetic acid; the pyridine product has not been identified, but is presumed to be

$$\text{XV-53}$$

$$(?) + PhCH_2Cl + PhCH_2O_2CMe + SO_4^=$$

pyridine-1-oxide (207). An attempt to quaternize **XV-53a** with methyl iodide in acetonitrile or *N,N*-dimethylformamide gave only **XV-53b** (459). When the

XV-53a   **VX-53b**

sulfide linkage is *ortho* to an amino group, it is cleaved by either acid or base to give a thiol or thione derivative **(XV-54)**. This reaction, the Smiles rearrangement, has been studied extensively (403). When the rearrangement is effected in ethanolic hydrogen chloride, oxidation is rapid, and the product isolated is the disulfide (214–217, 219); however, under basic conditions, addition of methyl iodide usually traps the thiol or thione as the methylthio derivative. An unusual phenomenon is observed when a 6-alkoxy-2-pyridyl sulfide is involved; with these, the rearranged amine appears to be thermally sensitive to cleavage by base.

XV-54

When the rearrangement occurs under alkaline conditions and the intermediate thiol or thione possesses an *ortho* nitro group, intramolecular displacement by the mercaptide ion of the nitro group may result in cyclization to an aza- or diazaphenothiazine (58, 59, 214-217, 288, 289, 297, 361, 362, 466). The role played by the acetamide group in the sulfide or in the rearranged amine has not

XV-56

been too well understood. With only one exception (XV-55), the products of these reactions are the deacetylated azaphenothiazines; XV-55 is surprisingly stable to base, and deacetylation occurs only at the boiling point in aqueous ethanolic hydrochloric acid. It should be noted that there are a significant number of examples where rearrangement and cyclization to azaphenothiazine occurs without prior acylation of the amino group in the sulfide (see 1st Ed., Part IV, p. 362). That the rate of rearrangement of the acetamido derivative under acidic conditions is slower than that of the amine, and that the rearranged derivative is deacetylated (288) is not relevant since no cyclization has apparently ever occurred under acidic conditions. Insofar as the azaphenothiazines are concerned, the isolation of XV-55 in 80% yield is good evidence that the acetyl group does not interfere in either the rearrangement or the cyclization. Furthermore, in the alkaline rearrangement of diaryl sulfides, there is ample evidence that acylation of the amino group is a necessary prerequisite for obtaining phenothiazines in good yield (404). In addition, there is some indication from recent syntheses of 1,6-diazaphenothiazines, that the acyl group must be retained by the rearranged amine for cyclization to occur. In the deacetylated rearranged amine, hydrogen bonding between the NH proton and the leaving nitro group would protect the latter from attack by mercaptide ion; thus dimethyl sulfoxide, which competes strongly with the nitro group and preferentially hydrogen bonds the NH proton of the rearranged amine, makes possible the synthesis of the hitherto unavailable 1,6-diazaphenothiazines (289).

2-Azaphenothiazine-1-oxide has been synthesized by heating 3-amino-4-(2-bromophenylthio)pyridine-1-oxide, potassium carbonate, copper powder, and pyridine at 130 to 150° (XV-57) (120).

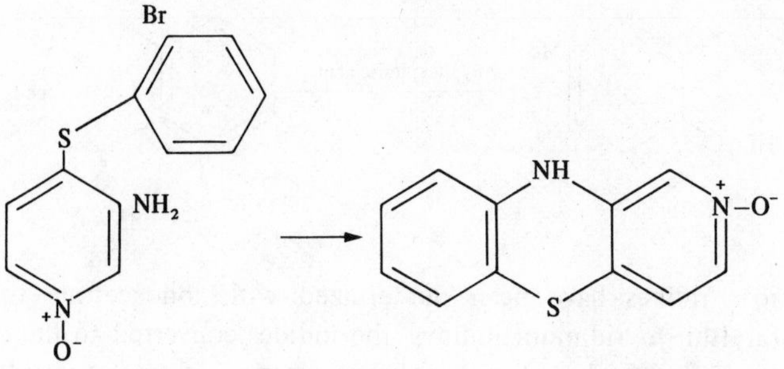

XV-57

6-Chloro-9-methyl-5*H*-[1*H*] benzothiopyrano[2,3-*b*] pyridin-5-one **(XV-58)**, 6-chloro-9-methyl-5*H*-[1*H*] benzothiopyrano[2,3-*c*] pyridin-5-one **(XV-59)**, and 2-methyl-5*H*-[1*H*] benzothiopyrano[2,3-*c*] pyridin-5-one **(XV-60)** have been prepared by the sequences shown below (29, 294).

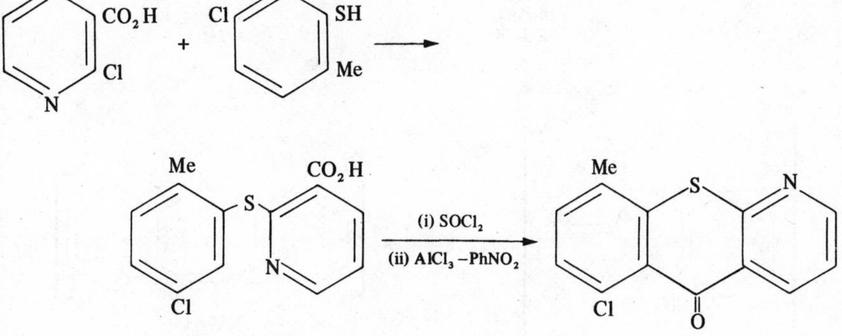

XV-58

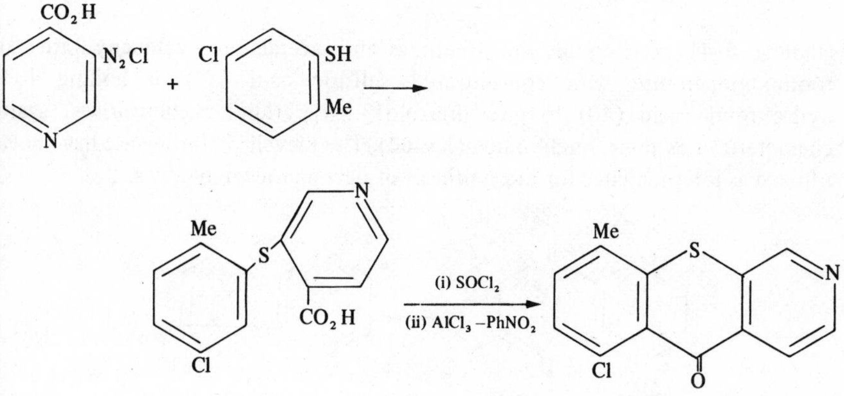

XV-59

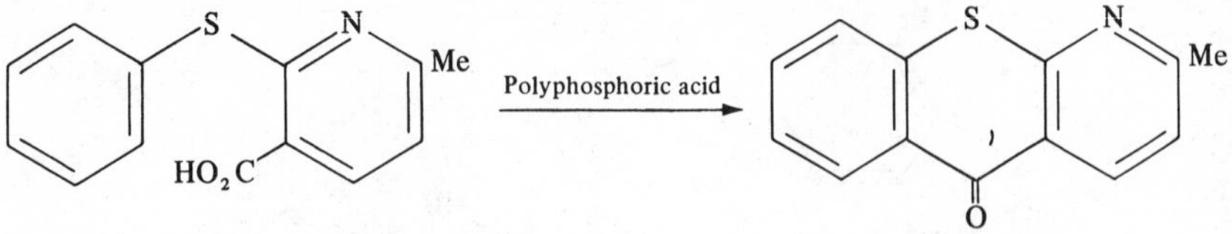

XV-60

2-(Arylthio)pyridines have been quaternized with iodoacetone to give 1-acetonyl-2-(arylthio)pyridinium iodides, the iodides converted to the chlorides, and the latter cyclodehydrated with polyphosphoric acid to give pyrido[2,1-*b*]-benzo[*f*]-1,3-thiazepinium salts **(XV-61)**; of these, the perchlorates are the most

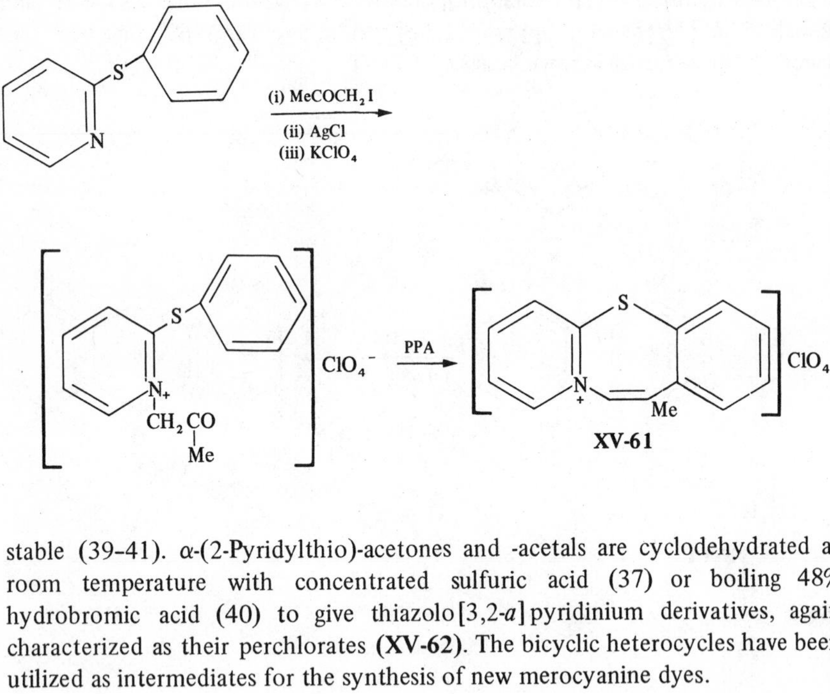

XV-61

stable (39–41). α-(2-Pyridylthio)-acetones and -acetals are cyclodehydrated at room temperature with concentrated sulfuric acid (37) or boiling 48% hydrobromic acid (40) to give thiazolo[3,2-*a*]pyridinium derivatives, again characterized as their perchlorates **(XV-62)**. The bicyclic heterocycles have been utilized as intermediates for the synthesis of new merocyanine dyes.

XV-62

2,3-Diamino-4-(methylthio)pyridine and diethoxymethyl acetate react at room temperature to give 7-methylthio-3H-imidazo[4,5-b]pyridine. 2-(Phenacylthio)-3-acetamidopyridines have been cyclized to give pyrido[2,3-b][1,4]thiazines (482).

The mass spectra of eleven 2-(alkylthio)pyridines have recently been reported. Loss of SH from the molecular ions is a major pathway; thus deuterium labeling in the (ethylthio) derivative indicates that ~75% of the SH loss occurs with a proton from the methyl group, and the remainder from the methylene portions of the ethyl group. The (M^+–SH) ion subsequently loses ethylene, to generate a pyridyne ion. Separate experiments have been employed to demonstrate that, when kept several days at 150°, the 2-(alkylthio)pyridines do not undergo any detectable change (409).

The new pyridyl sulfides reported since the last compilation (1st Ed., Part IV, pp. 388–401) are listed in Tables XV-7 to XV-12, XV-49, and XV-50.

IV. Pyridylalkyl- and Pyridylalkenyl Sulfides

Pyridylmethyl chlorides have been converted to sulfides by reaction with sulfide (155), arylthiolate (16, 122), and alkylmercaptide ions (134, 169, 212). 2-Vinylpyridine has been utilized to prepare 2-pyridylethyl sulfides; addition occurs so as to place the sulfur on the β-carbon atom. Thus the addition of (a) hydrogen sulfide in acetic acid or pyridine gives small amounts of the sulfide but the major product is the ethanethiol (272), (b) benzenethiol in Skellysolve B gives phenyl 2-(2-pyridyl)ethyl sulfide (155), and (c) 2-pyridylmethanethiol in acetic acid yields 2-(2-pyridylethyl)pyridylmethyl sulfide (45). Aminoalkane-thiols also add to vinylpicolines to give aminoalkyl picolinealkyl sulfides (528). In the absence of solvent at 135°, hydrogen sulfide adds in the reverse manner to 6-methyl-3-vinylpyridine and forms only di[1-(3-pyridyl)ethyl]sulfide (299). It is not surprising that α-substituted 2-vinylpyridines, in solution, also add sulfhydryl compounds in the normal fashion to give 2-(2-pyridyl)ethyl deriva-tives; α-phenyl-2-vinylpyridine adds n-butylmercaptan in benzene to give 2-phenyl-2-(2-pyridyl)ethyl n-butyl sulfide (188) while 2-(2-pyridyl)-2-(methyl-ene)ethanol and 2-pyridinethione yield 2[(2-pyridylthio)methyl]-2-(2-pyridyl)-ethanol (XV-63). Distillation of XV-63 in vacuo results in loss of water and the formation of the olefin (XV-64). A remarkable reaction occurs when ethyl

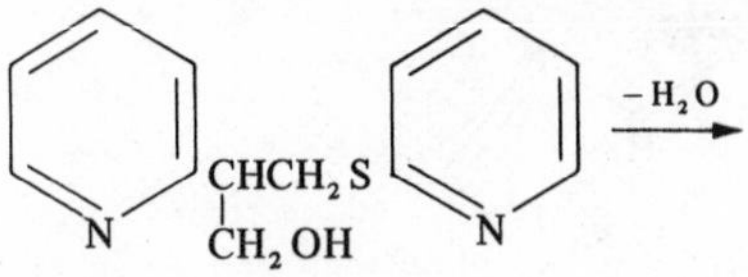

XV-63

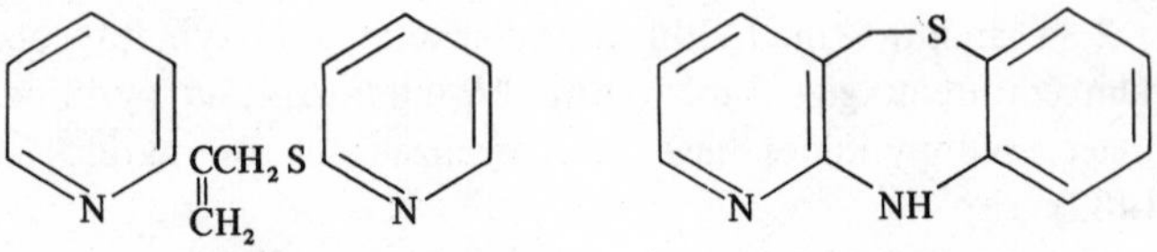

XV-64   **XV-66**

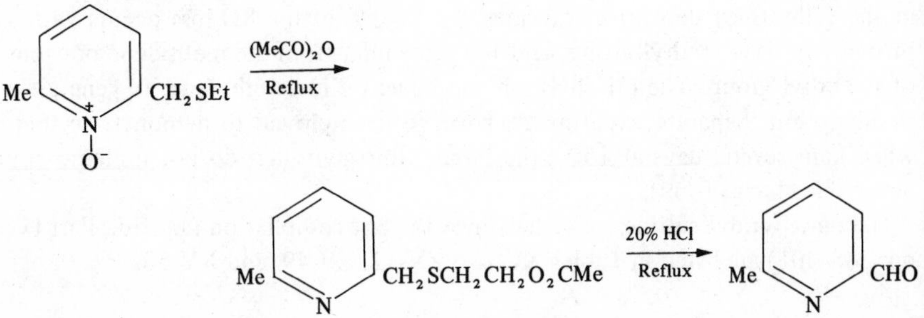

XV-65

6-methyl-1-oxido-2-pyridylmethyl sulfide is heated under reflux with acetic anhydride; the product **XV-65** is reported to have formed by an unusual rearrangement. It is of interest, also, that on heating with 20% hydrochloric acid, **XV-65** is converted to 6-methyl-2-pyridinecarboxaldehyde (172). The carbon-sulfur linkage in di-[2-(2-pyridyl)ethyl] sulfide is unstable to reducing agents, for example, with sodium in ethanol some di-[2-(2-piperidyl)ethyl] sulfide is

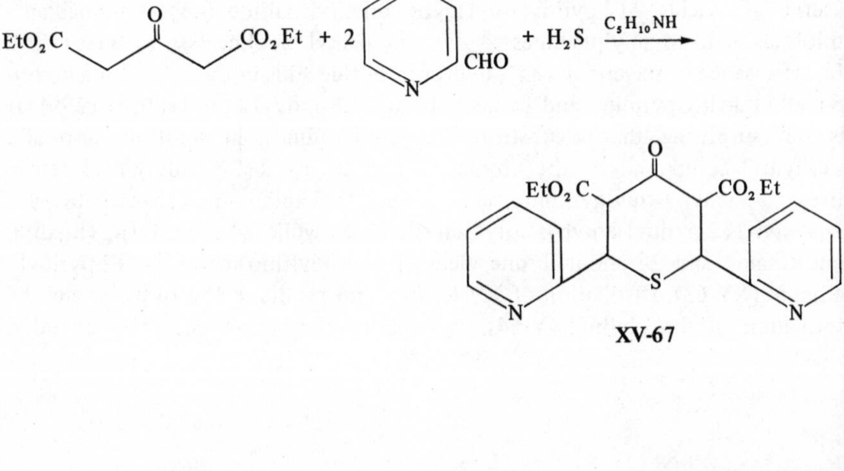

XV-67

formed along with considerable amounts of 2-ethylpiperidine (272). Pyrolysis of these pyridylethyl sulfides regenerates the vinylpyridine and the thiol (272). When 2-chloro-3-pyridylmethyl 2'-aminophenyl sulfide is heated in xylene with dimethylaniline, cyclization occurs, and 5,11-dihydrobenzo[b] pyrido [2,3-e] - 1,4-thiazepine **(XV-66)** is formed (122).

The piperidine catalyzed condensation of 2-pyridinecarboxaldehyde, diethyl acetonedicarboxylate, and hydrogen sulfide gives **XV-67**, where the sulfide portion of the dipyridylalkyl sulfide is part of a cyclic system; since **XV-67** still

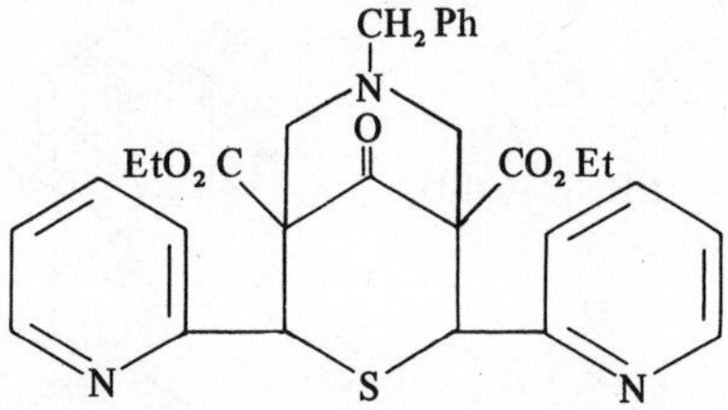

XV-68

retains two functional protons of the acetonedicarboxylate, a Mannich reaction with formaldehyde and benzylamine gives the interesting bicyclic system **XV-68**. In contrast, when 4-pyridinecarboxaldehyde is substituted for the 2-isomer, the only product isolated is bis(4-pyridylmethyl)disulfide. Hydrolysis of **XV-67** with 15% hydrochloric acid at 95° also involves decarboxylation to give 1-thia-2,6-di-2-pyridyl-4-oxocyclohexane **(XV-69)**, while oxidation of **XV-69** with ceric sulfate in cold 2N sulfuric acid gives the dienone **(XV-70)** (137). A related

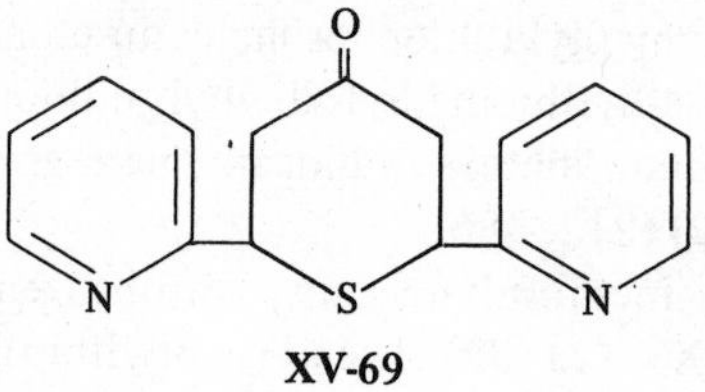

XV-69

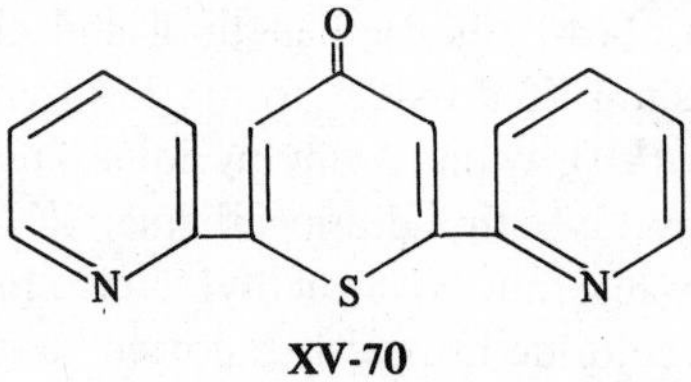

XV-70

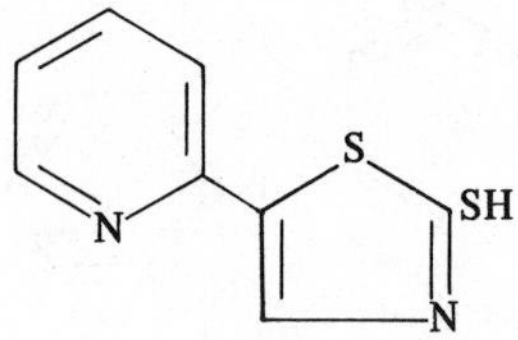

XV-71

product **XV-71** is formed from the base-catalyzed reaction between 2-pyridine-carboxaldehyde, carbon disulfide, and hydrazine (389).

2-Pyridylmethyl chloride reacts with alkali metal or ammonium salts of diethyldithiophosphate to give 2-pyridylmethyl dithiophosphates (99, 104, 375); higher homologs are prepared by the reaction with 2-(β-nitrovinyl)pyridine

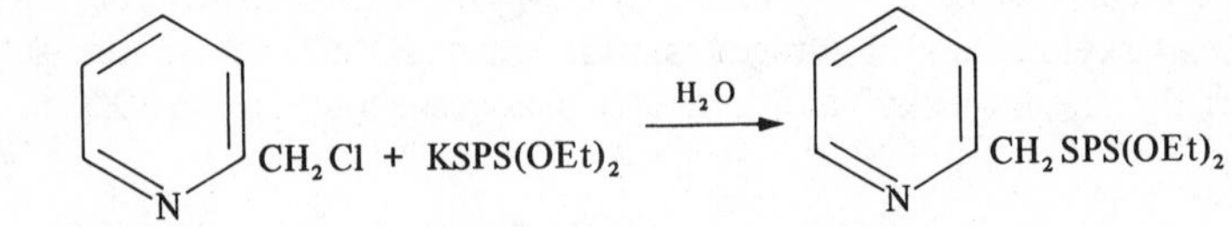

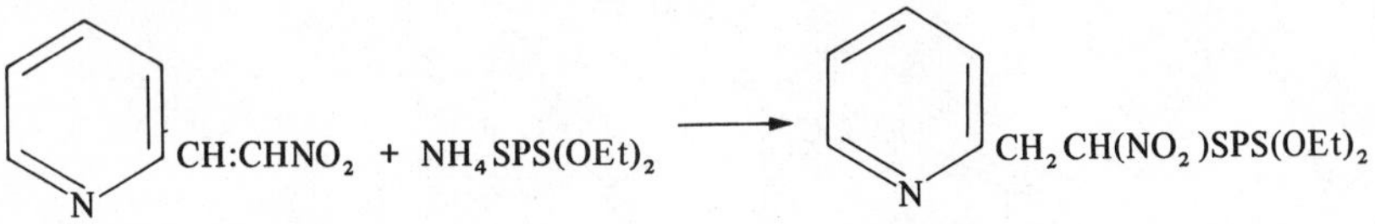

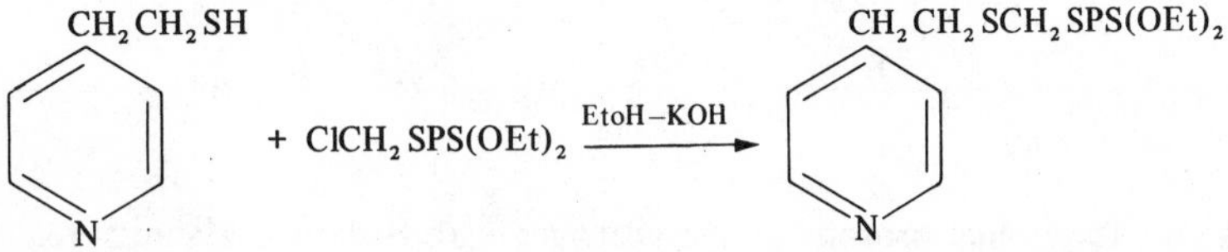

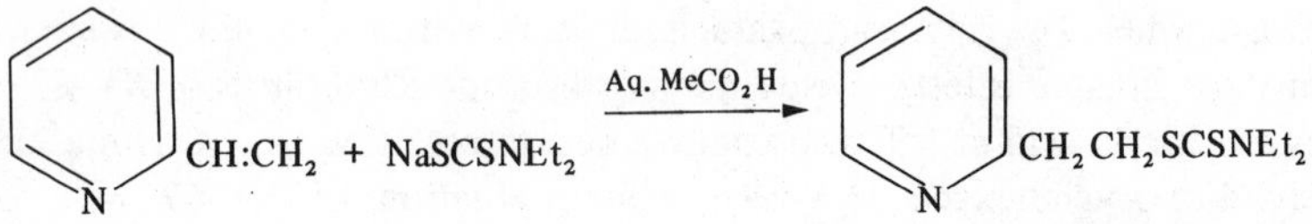

XV-72

(357). 2-(4-Pyridyl)ethanethiol and chloromethyldithiophosphates in ethanolic potassium hydroxide form 2-(4-pyridyl)ethylthiomethyl diethyl dithiophosphate (99), while 2-vinylpyridine and sodium diethyl dithiocarbamate give the 2-(2-pyridyl)ethyl dithiocarbamate **(XV-72)** (212, 386).

The reaction of 1-methyl 2-picolinium methosulfate and o-nitrobenzenesulfenyl chloride in base is reported to give **XV-72a** (82). The di-(methylthio)ketal

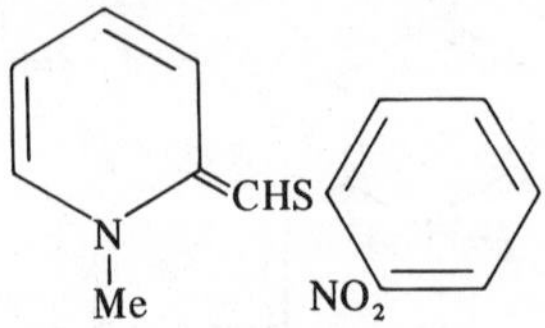

XV-72a

XV-73 and dimethyl sulfate react to give the unusual sulfonium cation **XV-74**, which with hot aqueous sodium chloride yields the first example of a pyridylalkenyl sulfide **(XV-75)** (135).

Pyridylalkyl sulfides are oxidized in the conventional manner to sulfoxides and sulfones.

The pyridylalkyl sulfides reported since the last compilation (1st Ed., Part IV, p. 402) are listed in Table XV-39.

V. Mercaptopyridoxal Derivatives

A large number of mercaptopyridoxal derivatives have been described and are listed in Tables XV-39, XV-41, and XV-46. For the synthesis of these compounds use has been made of methods discussed in other sections of this chapter, that is, procedures for the preparation of pyridinethiones, pyridylalkyl-thiols, pyridylalkyl sulfides, and pyridylalkyl disulfides; consequently, there is no need to discuss these further.

VI. Pyridyl Sulfoxides and Pyridyl Sulfones

The pyridyl sulfoxides are prepared by oxidation of the corresponding sulfides with hydrogen peroxide in methanol (84) or in acetic acid (280), or with nitric acid (162). The new derivatives are listed in Tables XV-15 to XV-18.

A greater variety of methods are available for the oxidation of sulfides to sulfones; these have included hydrogen peroxide, alone, in formic acid (2, 273, 455), or in acetic acid (1, 87, 162, 267, 273), nitric acid (91), molecular oxygen in polyphosphoric acid (39), and potassium permanganate in aqueous acetic acid

(2, 84). When a large excess of hydrogen peroxide is employed, the 1-oxide of the sulfone may be the product (299), and, occasionally, the sulfonic acid is a by-product (267). Activated halogens in halopyridines, for example, 2-chloro-5-nitropyridine, undergo displacement with sodium p-toluenesulfinate in boiling ethanol to give 5-nitro-2-pyridyl p-tolyl sulfone (74, 371).

A novel synthesis of a 2-pyridyl sulfone also demonstrates the reactivity of the 2-pyridinethiosulfonate group toward strong nucleophiles (XV-75a) (438).

XV-75a

A synthesis involving pyridine ring formation has led to a series of 2-pyridylsulfones (XV-75b) (503).

XV-75b

Alkyl and aryl 2-, 3-, and 4-pyridyl sulfones, and their 1-oxides are unstable in ethanolic or aqueous sodium hydroxide solution under reflux. The reactions involve a rather unusual cleavage of the sulfur-pyridine linkage, and the products are the corresponding pyridyl ethyl ether (546) or pyridinol (483); the 4-methylsulfonyl group attached to a tri- or tetrachlorinated pyridine is also displaced by the cyanide, pyrrolidinyl, or alkylthiaxanthyl nucleophiles (483, 519).

With one molar equivalent of a variety of nucleophiles, Michael addition to the vinyl group of 2,3,5,6-tetrachloropyridyl vinyl sulfone was the exclusive reaction to give XV-75c; a second mole of the nucleophile attacked position-2 to give XV-75d (579).

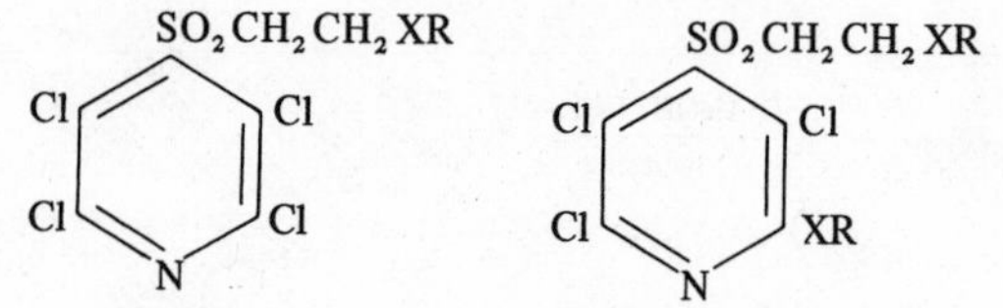

XV-75c   **XV-75d**

R XH = ammonia, aniline, benzylamine, morpholine, piperidine, phenol, ethanol, 2,3,5,6-tetrachloro-4-pyridinethiol, pentachlorothiophenol.

The hydrogenation of phenyl 4-pyridyl sulfone-1-oxide over Raney nickel yields phenyl 4-pyridyl sulfone (140). Iron and ammonium chloride in aqueous ethanol reduces 5-nitro-2-pyridyl phenyl sulfone to 5-amino-2-pyridyl phenyl sulfone (1). As illustrated in **XV-76**, 3-acetamidopyridyl sulfones can undergo the Smiles rearrangement in the presence of base; in methanolic base, the rearranged amine forms the stable methyl sulfinate, and under these conditions, no cyclization to a tricyclic heterocycle *via* the nitro group occurs (216). Rearrangement also occurs under acidic conditions, but the intermediate sulfinic acid derivative is unstable and may degrade partially or completely by loss of SO_2 **(XV-76)** (218).

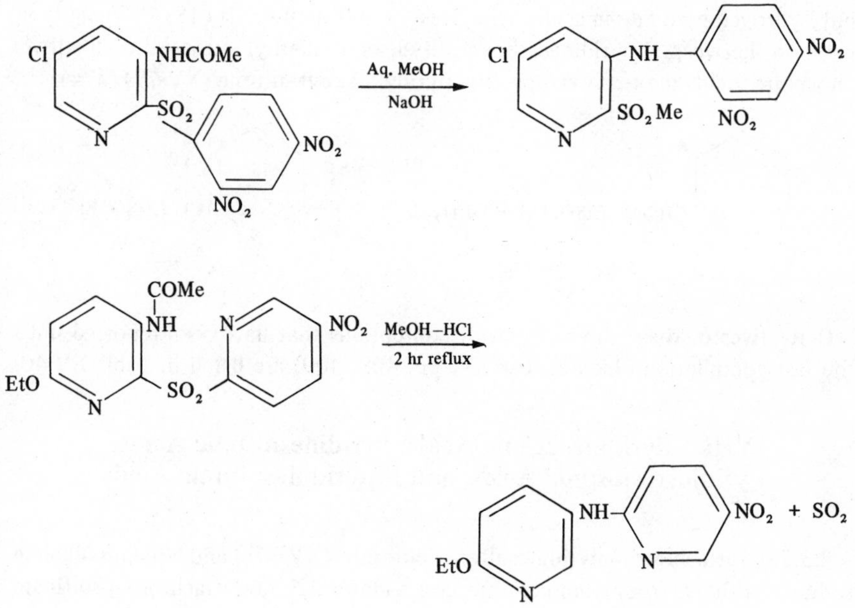

XV-76

VII. Pyridylalkyl Sulfoxides, Pyridylalkenyl Sulfones, and Pyridylalkyl Sulfones

Hydrogen peroxide in neutral solvents converts pyridylalkyl sulfides to the sulfoxides while potassium permanganate gives the sulfones (45). Hydrogen peroxide, with a catalytic amount of hydrochloric acid, yields the sulfone (380), but hydrogen peroxide in acetic acid gives the sulfone-1-oxide (45). A Wittig-type reaction between 2-pyridinecarboxaldehyde and diethyl ethylsulfonylmethyl-phosphite yields the first example of a pyridylalkenyl sulfone (**XV-77**) (374).

Derivatives of these three classes of compounds that have been reported since the last compilation (1st Ed., Part IV, pp. 408, 409) are listed in Table XV-40.

VIII. Pyridinesulfinic Acids, Pyridinesulfonic Acids, Pyridinedisulfonic Acids, and Bipyridinesulfonic Acids

2,3,5,6-Tetrachloro-4-pyridinesulfonyl chloride (**XV-77a**) and aqueous alkaline sodium sulfite at room temperature give sodium 2,3,5,6-tetrachloro-4-sulfinate (**XV-77b**); the addition of **XV-77b** to cold concentrated aqueous hydrochloric acid precipitates the sulfinic acid **(XV-77c)**. The latter melts and decomposes at 90° to give 2,3,5,6-tetrachloropyridine and sulfur dioxide (460).

Hydrolysis of **XV-77d** (R = H or NO$_2$) with aqueous sodium bicarbonate-tetra-hydrofuran gives the corresponding 2-pyridinesulfonate. The 2-pyridinesulfinates are obtained in similar fashion as oils that decompose with loss of sulfur dioxide; however, solutions of the sodium 2-pyridinesulfinates are stable for several days (438, 439).

Pyridine, pyridine-1-oxide, and 2-picoline have been converted in good yield to the corresponding 3-sulfonic acid by passing a stream of air containing 15 to 20% of sulfur trioxide through each compound, containing a catalytic amount of mercury, at 170 to 180°. It has been suggested that the mechanism involves the sequence **(XV-78)** (118, 196). 3-Pyridinesulfonic acid and sulfur trioxide, again with a mercury catalyst, give 3,5-pyridinedisulfonic acid (95). When 2,6-di-*t*-butylpyridine and sulfur trioxide are heated at 240 to 250° in a sealed vessel, two products are obtained: the 3-sulfonic acid in 30 to 35% yield and the bicyclic heterocycle **XV-79** in 15 to 20% yield.

The 2,6-di-*t*-butylpyridines are especially reactive toward sulfur trioxide in liquid sulfur dioxide (1st Ed., Part IV, p. 368). In an interesting extension of this reaction, the unusually stable 4-chloro-2,6-di-(*t*-butyl)pyridine **(XV-80)** has been sulfonated under these conditions to 4-chloro-2,6-di-(*t*-butyl)-3-pyridinesulfonic acid. The yield, however, is only 10 to 15%. Thus the shielding of the pyridine nitrogen lone pair by the two bulky *t*-butyl groups prevents coordination with the sulfur trioxide and the consequent deactivation of the nucleus. The 3-position is deactivated by the chlorine atom at C-4, so that the rate at which **XV-80** is sulfonated in significantly lower than that observed with 2,6-di-(*t*-butyl)pyridine. 2,6-Di-(*t*-butyl)-4-ethoxypyridine is far more reactive than **XV-80** and gives the 3-sulfonic acid in 40 to 45% yield (267).

6-Hydroxy-2-pyridone and chlorosulfonic acid or oleum at room temperature give **XV-80a** (521).

XV-80   XV-80a

A more generally applicable preparation of 2- and 4-pyridinesulfonic acids involves the oxidation of the corresponding pyridinethione. One of the oxidizing agents used is alkaline permanganate. The only limitation as to other substitution in the pyridine ring is the obvious one: methyl groups are converted simultaneously to carboxylic acid groups. Apparently, no oxidation of the pyridine nitrogen atom occurs (77, 79, 81). Other oxidizing agents, for example, nitric acid at 100° (3, 327), performic acid (455), or perbenzoic acid (353) have

XV-81

also been employed to oxidize 1-hydroxy-2(1*H*)-pyridinethione to the corresponding sulfonic acid. 1-Alkyl-4-pyridinethiones have been converted with peracetic acid to sulfobetaines **(XV-81)** (97).

Diazotized 4-amino-2-picoline-1-oxide reacts with sulfur dioxide to give 2-picoline-4-sulfonic acid-1-oxide (78).

The displacement of a suitably activated halogen in a halopyridine by sulfite ion has been carried out under a variety of conditions and several examples of this reaction are summarized in **XV-82** (3, 121, 327, 332, 358, 455). In a related reaction, 1-(4-pyridyl)pyridinium chloride and sodium sulfite also give 4-pyridinesulfonic acid (103).

XV-82

At 225 to 235°, in 20% oleum, and with mercuric sulfate as the catalyst, 2-picoline gave 6-methyl-3-pyridinesulfonic acid in 62% yield, 3-picoline gave a 61% yield of 5-methyl-3-pyridinesulfonic acid, and 4-picoline gave a 60% yield of 4-methyl-3-pyridinesulfonic acid; while precise details are lacking, the implication appears to be that the required time for these reactions are 3, 16, and 5 hr, respectively (77). Hexamethylene-2,2'-dipyridine 1,1'-dioxide was not sulfonated under these conditions in contrast to 2-picoline 1-oxide, which gave the 3-sulfonic acid (406). 5-Nitro-2-pyridone and chlorosulfonic acid at 145° gave 5-nitro-2-pyridone-3-sulfonic acid (100). Sulfonic acids have also been formed as by-products of the reduction of 4-nitro-2,6-lutidine-1-oxide with sodium hydrosulfite (101) and by a nucleophilic displacement of a 4-nitro group in a pyridine-1-oxide by sulfite ion (XV-83) (275). In an unusual reaction, 2-dibromoacetamidopyridine and sulfuryl chloride in acetic acid are reported to give a 2-dichloroacetamido-x-pyridinesulfonic acid (367).

In general, the pyridine ring is far more difficult to sulfonate than is benzene. For example, 4-phenylthiopyridine and chlorosulfonic acid at 50° give only p-4-pyridylthiobenzenesulfonyl chloride (305).

Finally, a copolymer of acrylic acid and vinylpyridine has been converted to a poly(acrylic acid-vinylpyridine)sulfonic acid by reaction with concentrated sulfuric acid, employing cadmium chloride as the catalyst (178).

The dissociation constants of a number of pyridinesulfonic acids have been determined spectroscopically in water and the relevant data are summarized below (102). It is apparent that the inductive effect of the electron attracting sulfonic acid group is greater, the closer the group is to the pyridine nitrogen atom. The ultraviolet spectra of these compounds indicate that in aqueous

Sulfonic Acid	pK_a (25°C)
2-Pyridine-	1.75
3-Pyridine-	3.22
4-Pyridine-	3.44
2,6-Lutidine-3-	4.89
2,6-Lutidine-4-	5.09
2,6-Di-t-butyl-3	4.12
(Reference: Pyridine)	5.17

solution they form the zwitterionic species **XV-84**; their infrared spectra in the solid state, however, show the characteristic sulfonic acid group absorption at 8.0–8.5 and 9.26–9.90 μ (102).

XV-84

The isomeric 2,2′-, 3,3′-, and 3,4′-bipyridyls have been treated with concentrated sulfuric acid at 260 to 315° apparently in the absence of a catalyst to yield mono-, or mixtures of mono- and di-sulfonic acids (250–253, 370). The entering group always enters *beta*- to the pyridine nitrogen atom. Some preference is shown, however, as summarized in **XV-85**. Structures in each instance, have been established by conversion to known pyridinedicarboxylic acids **(XV-86)**.

The reactions of the pyridinesulfonic acids are diverse. 3-Pyridinesulfonic acid has been converted to 3-pyridinesulfonyl chloride in 79% yield with phosphorus pentachloride; yields of 20 to 26% of the same compound are obtained (*a*) from the calcium salt of the acid and thionyl chloride in *N,N*-dimethylformamide or (*b*) from the potassium salt and phosphorus trichloride in the same solvent (118, 119); and sodium 2,3,5,6-tetrafluoro-4-pyridinesulfonate and sodium cyanide in that solvent gave 2,3,5,6-tetrafluoroisonicotinonitrile (455). The hydrolysis of 3-pyridinesulfonic acid to 3-pyridinol by means of aqueous alkali at 240 to 260° has been described (94), as well as the hydrolysis of 3-nitramino-2-pyridinesulfonic acid to 3-hydroxy-2-pyridinesulfonic acid with concentrated sulfuric acid at 60 to 80° (68, 69). The ammonolysis of (*a*) 2,6-di-(*t*-butyl)-4-pyridinesulfonic acid to 4-amino-2,6-di-(*t*-butyl)pyridine by means of 25% aqueous ammonia at 180° (267) and (*b*) 4-pyridinesulfonic acid to 4-aminopyridine with 28% aqueous ammonia containing zinc as catalyst, at 150 to 160° (328) have

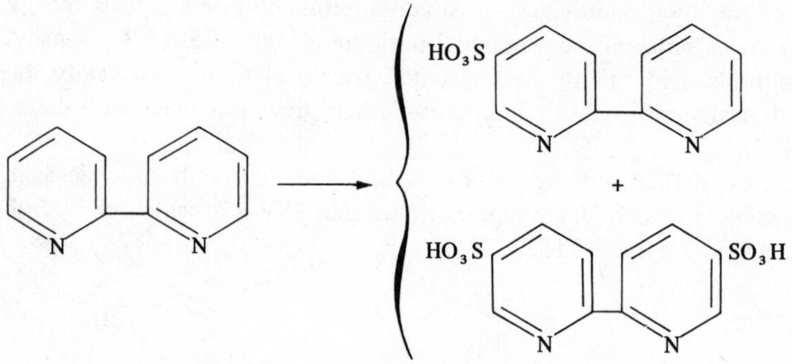

XV-85

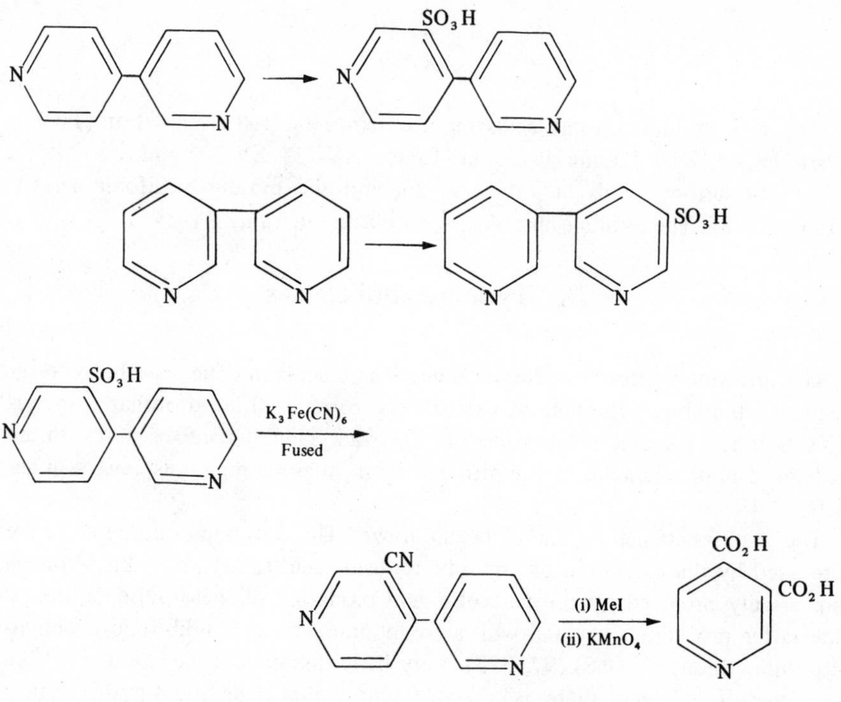

XV-86

been successful. Cleavage of the ether linkage in 2,6-di-(t-butyl)-4-ethoxy-3-pyridinesulfonic acid can be effected with aluminum chloride in tetrachloroethane at 120° and gives 2,6-di-(t-butyl)-4-hydroxy-3-pyridinesulfonic acid (267). Reductive dechlorination of 2,6-di-(t-butyl)-4-chloro-3-pyridinesulfonic acid to 2,6-di-(t-butyl)-3-pyridinesulfonic acid occurs with Raney nickel in aqueous ethanolic sodium hydroxide (267). 6-Amino-3-pyridinesulfonic acid is readily acylated to give a series of 6-acylamido-3-pyridinesulfonic acids (126). Finally, 3-pyridinesulfonic acid can be hydrogenated over a platinum catalyst to the piperidine derivative (115). 3-Pyridinesulfonic acid does not react with diazomethane (405).

Furfural, treated first with hypochlorous acid and then with sulfamic acid, gives 3-hydroxy-2-imino-1(2H)-pyridinesulfonic acid (XV-87), which on hydrolysis, gives 2-amino-3-pyridinol (125, 391).

XV-87

The new pyridinesulfonic acids reported since the last compilation (1st Ed., Part IV, p. 410 ff) are listed in Tables XV-21, XV-22, and XV-23; the bipyridinesulfonic acids in Table XV-26; and the pyridinedisulfonic acids in Table XV-35. The pyridinesulfonic acids are listed in Table XV-24.

IX. Pyridinesulfobetaines

The ultraviolet spectra of the pyridinesulfonic acids in aqueous solution offer evidence that these compounds exist in the form of the 1-protonated species (XV-84). The pyridinesulfobetaines are the N-alkylated derivatives, and with the enhanced basic character of the nitrogen atom, these compounds exist as inner salts.

The 2-sulfobetaines appear to be unknown. The 3-pyridinesulfobetaines are prepared by the alkylation of 3-pyridinesulfonic acid (283) while the 4-isomers are usually prepared by the peracetic acid oxidation of the 4-pyridinethiones; the latter procedure may fail with a compound such as 3,5-dibromo-1-methyl-4-pyridinethione (XV-88) (97, 397). Very little is known of the reactions of the pyridinesulfobetaines; there is a single report that 1-methyl-4-pyridinesulfobetaine could not be brominated (97).

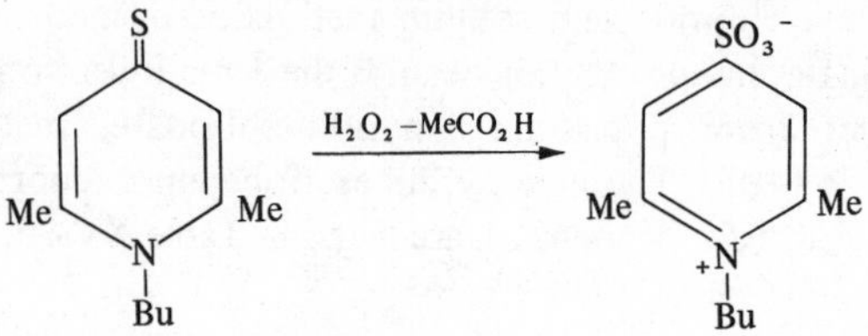

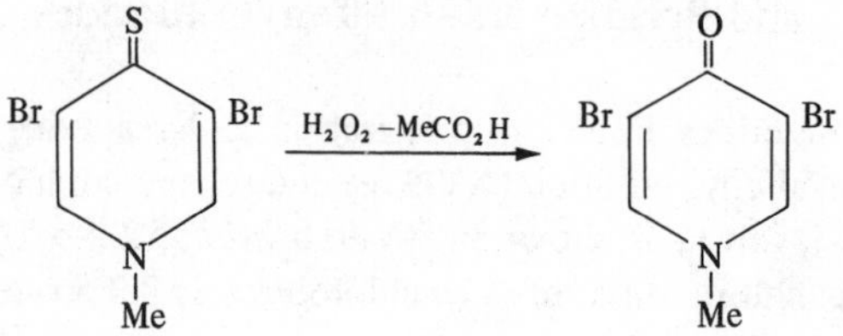

XV-88

One of the mechanisms suggested for the sulfonation of pyridine involves the sequence **XV-89** and assigns the structure **XV-90** to the pyridine sulfur trioxide adduct (102, 196). Several derivatives of the species **XV-90** have recently been

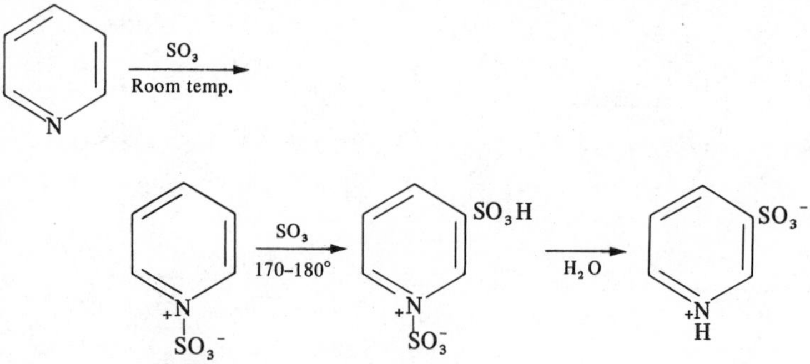

XV-90

XV-89

described **(XV-91)**; these have been prepared by a ring opening with hypochlorous acid of furfural followed by cyclization with sulfamic acid (125–128).

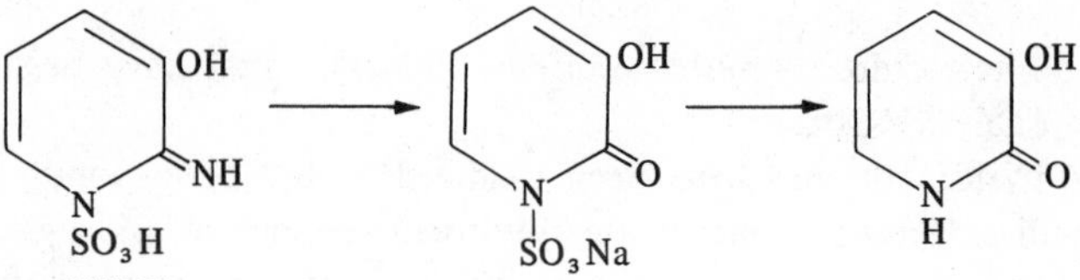

XV-87

XV-91

3-Pyridinesulfonyl chloride and sodium methoxide in methanol do not form
the methyl sulfonate; the product, instead, is the 1-methylsulfobetaine, identical
with the product from potassium 3-pyridinesulfonate, methyl iodide and
aqueous hydroxide (405). The new pyridinesulfobetaines reported since the last
compilation (1st Ed., Part IV, p. 412), are listed in Table XV-25.

X. Pyridinesulfenyl Chlorides, Pyridinesulfonyl Chlorides and Pyridine-3,5-disulfonyl Chlorides

Recent work describes both the preparation and reactions of 2,3,5,6-tetra-
chloro-4-pyridinesulfenyl chloride (**XV-91a**) and related compounds; a summary
of the reactions involved is shown in **XV-91b** (460, 522, 570); **XV-91a** is also
prepared by the chlorination of 4-(p-chlorobenzyl)-2,3,5,6-tetrachloropyridyl
sulfide (529).

The 2,3,5,6-tetrachloro-4-pyridinesulfenyl chlorides that have been reported
are listed in Table XV-26a.

2- And 4-pyridinethiones have been oxidized to the corresponding sulfonyl
chlorides with chlorine in concentrated hydrochloric acid or 50% aqueous acetic
acid at −5 to 0°C. At room temperature, these derivatives evolve sulfur dioxide

and yield the 2- or 4-chloropyridines. Presumably because of the more positive character of the pyridine nitrogen atom, the 1-oxides of 2- and 4-sulfonyl chlorides are somewhat more stable but still require storage and reaction at 0 to 5°C (21–23). 3-Pyridinesulfonyl chloride can similarly be prepared from the thiol and is isolable (24, 207). Sodium chlorate may be used in place of chlorine (109). Other substituents in the pyridine ring, for example, Br, Cl, Me, MeCONH, SO_2NH_2, or 1-oxide, do not interfere or become involved during these oxidations (4, 230, 249, 376, 384).

3-Pyridinesulfonyl chloride has been prepared in 79% yield from 3-pyridinesulfonic acid and phosphorus pentachloride; yields of 20 and 26%, respectively, are obtained from the calcium salt and thionyl chloride in N,N-dimethylformamide or from the potassium salt and phosphorus trichloride in the same solvent as mentioned earlier (118, 119). 4-Pyridinesulfonic acid-1-oxide and phosphorus pentachloride give 4-chloropyridine-1-oxide (125). 2,3,5,6-Tetrachloro-4-pyridinesulfonyl chloride and phosphorus pentachloride at 170° gave 2,3,4,5,6-pentachloropyridine (460).

3-Pyridinesulfonyl chloride and 4-pyridinesulfonyl chlorides and hydrazine do not yield the corresponding hydrazides; instead complex reactions occur and the only isolable products are bis(3-pyridyl) disulfide and bis(4-pyridyl) disulfide, respectively (203); surprisingly, 4-pyridinesulfonyl chloride 1-oxide reacts smoothly to give 4-pyridinesulfonic acid hydrazide-1-oxide (4). 3-Pyridinesulfonyl chloride and sodium methoxide in methanol forms the 1-methylsulfobetaine and not the sulfonate ester (see p. 234) (405).

XV-92

Essentially all of the effort devoted to the synthesis of the 3,5-pyridine-disulfonyl chlorides has been concerned with the ultimate syntheses of pyrido [2,3-e]-1,2,4-thiadiazine-1,1-dioxides and 3,4-dihydro-2H-pyrido [2,3-e]-1,2,4-thiadizine-1,1-dioxides (XV-92) (22, 54, 66, 67, 308, 325, 356, 363, 364). The remarkable stability of 2-amino-3,5-pyridinedisulfonyl chloride is best demonstrated by its synthesis in 79% yield from 2-aminopyridine and 10 molar equivalents of chlorosulfonic acid at 150° for 116 hr (67). It should be noted that in these chlorosulfonations, the sulfonic acid is formed initially, and this reacts with excess chlorosulfonic acid at 150° to give the sulfonyl chloride.

The pyridinesulfonyl chlorides and pyridine disulfonyl chlorides reported since the last compilation (1st Ed., Part IV, p. 413) are listed in Tables XV-27 and XV-36, respectively.

XI. Pyridinesulfonamides and Pyridinedisulfonamides

The crude unstable 2- and 4-pyridinesulfonyl chlorides are treated with ammonia or amines at moderate temperatures to give amides; with the amines, pyridine is a useful solvent (384). 3-Pyridinesulfonyl chloride is far more stable and has been converted to sulfonamides with ammonia and a large variety of primary and secondary aliphatic and aromatic amines (3, 4, 10, 26, 100, 139, 191, 230, 249, 290, 305, 376, 384).

The sulfonamide group undergoes the usual acylation reactions with acid halides (89); with n-butyl isocyanate and base in acetone, 2-pyridinesulfonamide (XV-93) yields 3-n-butyl-1-(2-pyridyl)sulfonylurea (384); 2-pyridinesulfona-mide-1-oxide (XV-94) is prepared by heating XV-93 with peracetic acid at 70 to 80° for 6 hr (376). The reaction products of XV-93 and XV-94 with ketene dimer undergo several interesting reactions, including a rearrangement followed by loss of SO_2 (89, 90, 376). Similar reactions with two related compounds, XV-97 and XV-98, have indicated that with the 1-oxides the loss of SO_2 is generally observed (XV-99) (237).

XV-93

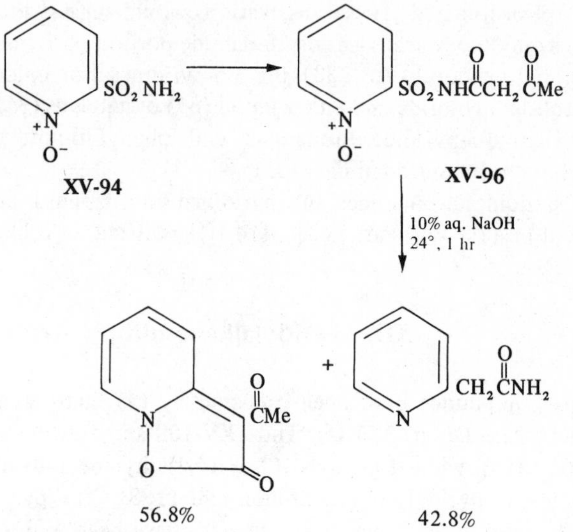

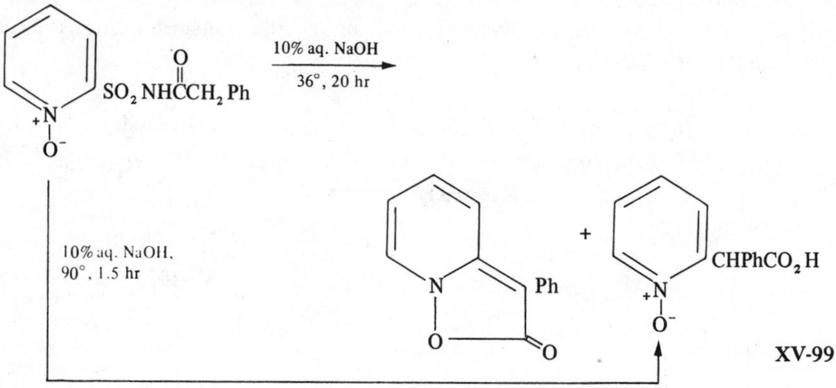

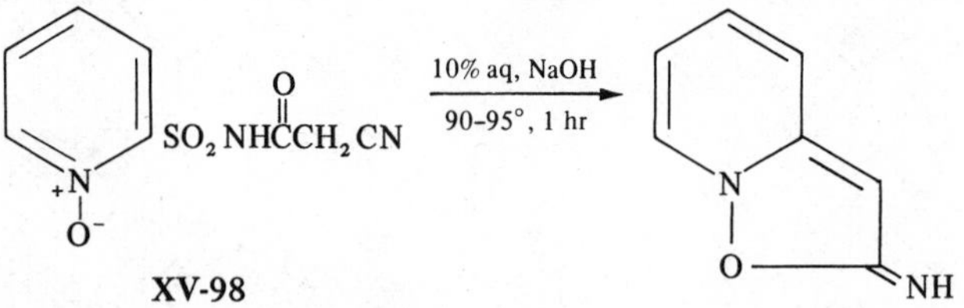

XV-94 → XV-96

10% aq. NaOH
24°, 1 hr

56.8%   +   42.8%

10% aq. NaOH
36°, 20 hr

10% aq. NaOH,
90°, 1.5 hr

+   XV-99

XV-98

10% aq, NaOH
90-95°, 1 hr

A reaction mechanism for these rearrangements has not been proposed. The 1-oxide substituent plays an important role in the base cleavage of the carbon–sulfur linkage, but that linkage is known to be stable to base in **XV-94**, and to peracetic acid. These observations would suggest that the rearrangement involves only the *N*-acetoacetylsulfonamide portion of the molecule.

As noted previously (p. 238), the 3,5-pyridinesulfonamides are prepared from the disulfonyl chloride and either liquid (67) or aqueous (363) ammonia.

N,N-Diethyl-3-pyridinesulfonamide and phenyllithium yield *N,N*-diethyl-2-phenyl-3-pyridinesulfonamide (405).

The pyridinesulfonamides and pyridinedisulfonamides reported since the last compilation (1st Ed., Part IV, p. 414 ff) are listed in Tables XV-28–XV-30 and XV-37.

XII.   Pyridylalkanesulfonic Acids

These compounds have been prepared by the methods previously described (1st Ed., Part IV, p. 375 ff). Thus **XV-100** and aqueous sodium bisulfite give **XV-101** (313), while 1-phenyl-1-(2-pyridyl)ethylene and the same reagent yield 1-phenyl-1-(2-pyridyl)ethanesulfonic acid (188). The pyridine heterocycles of both 2- and 4-pyridylethanesulfonic acids have been hydrogenated catalytically to the corresponding piperidines (56, 115). α-Hydroxy-3-pyridinemethanesulfonic acid could not be hydrogenated under the same conditions used successfully with **XV-102**.

XIII. 2- and 4-Pyridinesulfenamides

The interaction of the sodium salt of 2-pyridinethione or its 1-oxide with chloramine in water yields the stable 2-pyridinesulfenamides; reaction of these with pyridinecarboxaldehydes and pyridyl ketones gives 2-pyridinesulfenimines **(XV-103)**, which with various transition metal salts form stable ligands (184, 286).

XV-103

When **XV-103a** was treated with chlorine under anhydrous conditions, the sulfenyl chloride **(XV-103b)** was obtained as an orange-colored oil. With piperidine, **XV-103b** gave the sulfenamide **(XV-103c)**.

XV-103a   **XV-103b**   **XV-103c**

The 2- and 4-pyridinesulfenamides are listed in Table XV-33.

XIV. 2- and 3-Pyridinesulfonylureas

2- And 3-pyridinesulfonamide react with *n*-butylisocyanate in aqueous acetone in the presence of base to give *N*-(2-pyridinesulfonyl)-*N'*-*n*-butylurea and *N*-(3-pyridinesulfonyl)-*N'*-*n*-butylurea (35, 384). These compounds are listed in Table XV-56.

XV. 4-Pyridinesulfonic Acid Hydrazide-1-oxide

In general, the reaction of a pyridinesulfonyl chloride with hydrazine involves a complex series of reactions and yields only the disulfide (203). One successful synthesis, however, has been reported: 4-pyridinesulfonyl chloride-1-oxide and hydrazine hydrate in chloroform give 4-pyridinesulfonic acid hydrazide-1-oxide. This derivative reacts normally with aldehydes and ketones to give hydrazones (4).

These compounds are found in Table XV-30.

XVI. 3-Pyridinesulfonylazide

This compound has been mentioned in a patent abstract as capable of modifying polyolefin structure; the synthetic method has not been disclosed (146).

XVII. Pyridinethiosulfinates, Pyridinesulfonates, and Pyridinethiolsulfonates

These esters are prepared from the pyridinesulfonyl chloride and an alkali metal alkoxide or alkylmercaptide. As anticipated, the methylsulfonate group in the 2- or 4-positions is a good leaving group and is readily displaced by various nucleophiles (**XV-104**) (12, 337).

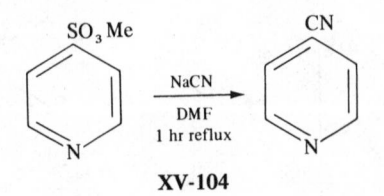

XV-104

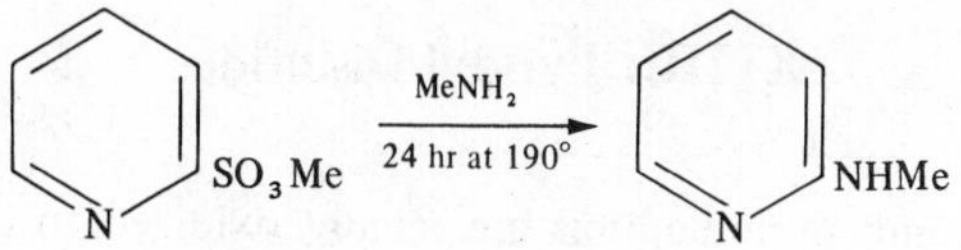

A recently described procedure (438, 439) employs a pyridinethione, an aryl- or alkylthiol, and N-chlorosuccinimide to prepare unsymmetrical disulfides (XV-104a), presumably *via* the intermediate sulfenyl chloride. Perbenzoic acid oxidation of the disulfides is remarkably selective in that at 20° the reaction involves only the sulfur atom attached to pyridine to give XV-104c; at 0°, employing 1 mole of perbenzoic acid, sequentially, XV-104b can first be isolated, and then be converted to XV-104c. Potassium iodide reduces XV-104b to XV-104a.

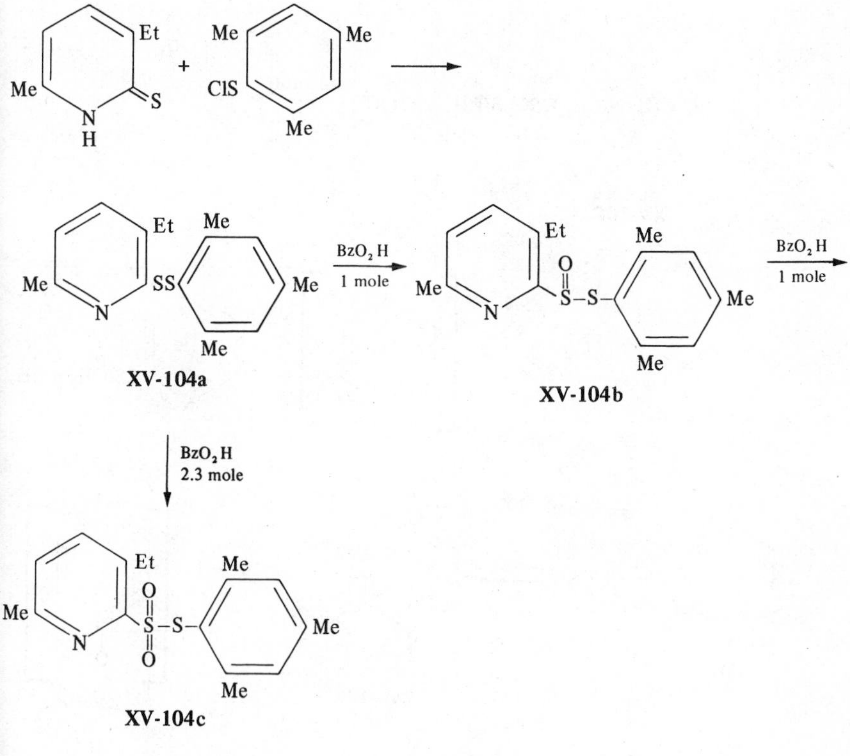

XV-104a

XV-104b

XV-104c

Table XV-31 lists the known derivatives of these classes of pyridine derivatives.

XVIII.　Pyridyl Disulfides

Pyridinethiones and pyridinethiols are readily oxidized to the disulfides and this behavior has been discussed in the previous compilation (1st Ed., Part IV, p. 363). Additional examples of the use of hydrogen peroxide (21, 201), potassium ferricyanide (273), and bromine (455) to affect this oxidation are to be found in the more recent literature. In addition, several novel procedures have been reported: 2- and 4-pyridylthiocyanates, aqueous sodium hydroxide, hydrogen peroxide, and iodine give 2- and 4- pyridyl disulfide, respectively (117), while 2-pyridinethione and the bisquaternary salt **XV-105** give both the symmetrical and unsymmetrical disulfides (110). The mixed disulfide **XV-106** is readily prepared from 1-hydroxy-2(1*H*)-pyridinethione and trichloromethanethiol (248). Picolinic acid-1-oxide and sulfur chloride are reported to yield **XV-106a** (489).

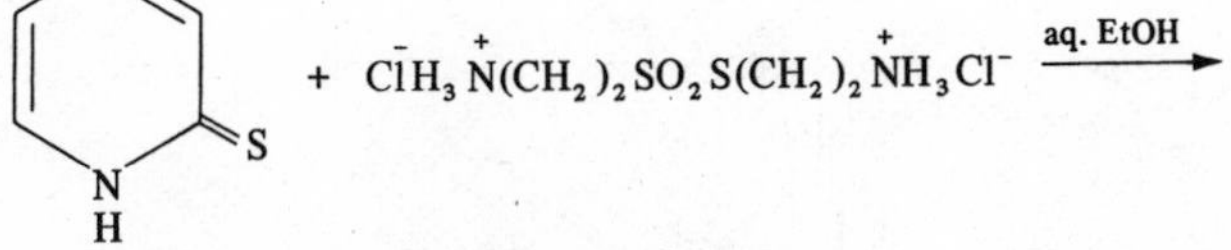

XV-105

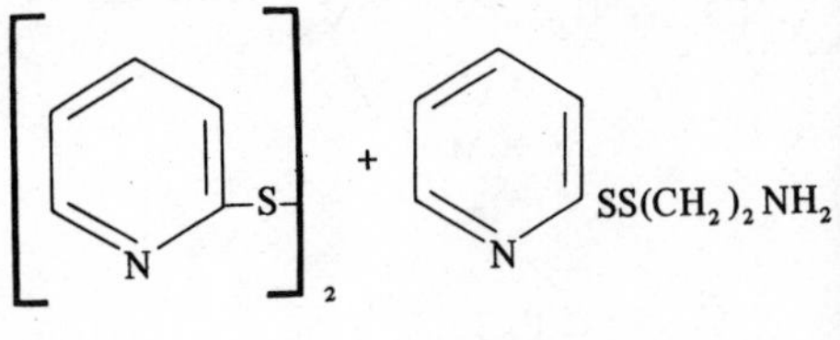

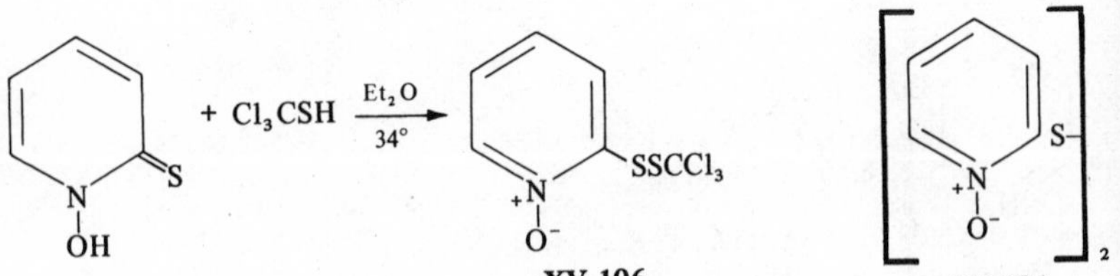

XV-106　　　　**XV-106a**

The pyridyl disulfides reported since the last compilation (1st Ed., Part IV, p. 403) are listed in Table XV-41.

XIX. Pyridylalkyl Disulfides

A major portion of the synthetic effort devoted to pyridylalkyl disulfides has been motivated by the interest in mercaptopyridoxal derivatives. The structures (XV-107) summarize the reactions that have been employed (157, 206a, b, 330, 431). By-product disulfide formation has also been observed in the reaction of 2-pyridylmethyl chloride-1-oxide with a hydrogen sulfide-saturated solution of

XV-107

sodium ethoxide in ethanol, but 2-pyridylmethanethiol is the major product; disulfide is the only product when the same chloride is treated in water with sodium polysulfide. Aeration of the methanethiol also gives the disulfide (172). The new pyridylalkyl disulfides are listed in Table XV-41.

XX. Pyridylalkyl Thioketones

The side-chain metalation of 2-picoline with phenyllithium gives 2-picolyl-lithium, which with thio- or dithio esters has yielded the novel pyridylalkyl thioketones (XV-108). As would be anticipated, these thioketones form chelates with heavy metal salts (339).

The pyridylalkyl thioketones are listed in Table XV-55.

XXI. 2- and 4-Pyridylmercaptoketones and 2-Pyridylmercaptoacetals

α-Haloketones and β-haloacetals react at room temperature with 2-pyridine-thione in ethanolic sodium ethoxide to give α-(2-pyridylmercapto)ketones and β-[2-(pyridylmercapto)]acetals. Concentrated sulfuric acid or 48% hydrobromic acid induces cyclodehydration of these derivatives and yields the polycyclic system XV-109; the latter are usually isolated as their stable perchlorates (6, 37–41).

The 2-pyridylmercaptoketones are listed in Table XV-53 and the 2-pyridyl-mercaptoacetals in Table XV-54.

XXII. Pyridylthiocyanates, Pyridylalkylthiocyanates, and Pyridylisothiocyanates

The lead salts of 2- and 4-pyridinethione and cyanogen iodide in ethanol give 2- and 4-pyridylthiocyanates, respectively (117). The halogen atoms in dimethyl 4-chloro-2,6-pyridinedicarboxylate (221), 4-chloro-3-nitropyridine-1-oxide (329, 373), and 3-fluoro-4-nitropyridine-1-oxide (332) react readily with potassium or ammonium thiocyanate in boiling ethanol to yield the corresponding pyridyl-thiocyanates; from 3-bromo-4-chloropyridine-1-oxide and potassium thiocyanate under the same conditions, only 3-bromo-4-pyridylthiocyanate is obtained (329, 373). Both 3- and 4-pyridinediazonium sulfate undergo the Sandmeyer reaction with potassium thiocyanate in the presence of CuSCN to give 3- and 4-pyridylthiocyanates, respectively (117). 2-Aminopyridine and cyanogen bromide give 2-amino-5-pyridylthiocyanate; however, contrary to an earlier report (1st Ed., Part IV, p. 377), 2,6-diaminopyridine and the same reagent give first the 3-thiocyanate and then the 3,5-dithiocyanate, and these derivatives can be isolated without their undergoing cyclization to the pyrido[2,3-d]thiazole (117).

2-Pyridinethione, sodium thiocyanate, and bromine in potassium bromide-saturated methanol have yielded the novel 2-pyridylsulfenyl thiocyanate **(XV-110)** (117).

| XV-110 | XV-111 | XV-112 |

The oxidation of 2-pyridylthiocyanate with hydrogen peroxide and iodine gives di-(2-pyridyl)disulfide (117).

2-Pyridylmethyl chloride reacts with barium thiocyanate to give 2-pyridylmeth-ylthiocyanate; the latter readily undergo hydrolysis to **XV-111** (83). 3-(Benzoyl-oxymethyl)-5-hydroxy-6-methyl-4-pyridylmethyl chloride and sodium thio-cyanate give the corresponding 4-pyridylmethylthiocyanate (313).

Several structures have been proposed for the red colored 2-pyridyl isothio-cyanate dimer; recently, structure **XV-112** has been proposed on the basis of its PMR spectrum (30).

The pyridylthiocyanates and pyridylalkylthiocyanates prepared since the last compilation (1st Ed., Part IV, p. 422) are listed in Table XV-44.

XXIII. Pyridylxanthates and Pyridylalkylthio- and dithiocarbamates

4-Pyridylxanthates have been prepared by the reaction of an alkali metal xanthate with either pentachloropyridine or 2,3,5-trichloro- or 2,3,5,6-tetrachloropyridyl 4-methylsulfone (519).

2-Pyridylalkanethiols and isocyanates give thiocarbamates while isothiocyanates yield dithiocarbamates. The latter class of compounds was also obtained from 2-vinylpyridine, an amine, and carbon disulfide (411).

XXIV. Sugar Derivatives of Sulfur Compounds of Pyridine

A substantial effort has been devoted to the synthesis of (β-D-glucopyranosyl)-thiopyridines, (β-D-glucopyranosyl)sulfonyl pyridines, and their acetyl derivatives. The interest in these derivatives is based on their use as model compounds for the study of the mechanism of methanolysis and enzymolysis of glucosides. In one typical approach to these compounds, 2-pyridinethione, tetra-O-acetyl-α-D-glucopyranosyl bromide, and potassium hydroxide in aqueous acetone are shaken at room temperature to give 2-[tetra-O-acetyl-(β-D-glucosyl)thio]-pyridine (**XV-113**); when the latter is heated for 1 min under reflux with methanolic sodium methoxide, 2-(β-D-glucosyl)thiopyridine (**XV-114**) is

XV-113; R = Ac
XV-114; R = H

obtained. Other pyridinethiones and pyridinethiols, for example, nitro- and 1-oxides, react similarly. The thioglucoside linkage is reasonably stable, so that nitro derivatives can be hydrogenated catalytically to the amine while 1-oxides are deoxygenated by catalytic reduction. A direct alternate synthesis of **XV-114** is from 2-bromopyridine-1-oxide and sodium 1-thio-D-glucose in aqueous acetone. The thio derivatives are oxidized to the sulfones in the conventional manner with perbenzoic acid in chloroform.

1-(Tetra-O-acetyl-β-D-glucosyl)-2-pyridinethione is prepared from the corresponding 2-pyridone and phosphorus pentasulfide in pyridine at 130 to 160° (349–355).

Cleavage of these derivatives by methanol or almond emulsin always occurs between the pyridine ring and the thio sugar. The 2-thio derivatives react more rapidly than the 4-; the 3-isomers are not cleaved. The same order of reactivity is observed with the corresponding 1-oxides.

The known derivatives of this novel class of compounds are listed in Table XV-52.

XXV. Pyridylthiophosphates and Pyridylalkylthiophosphates

These compounds have been prepared by the reaction of a halopyridine or a pyridylalkyl halide with a dialkyl phosphorothioic acid or a phosphorodithioic acid or a salt of one of these acids (99, 104, 213, 375). An alternate synthesis useful only for the 2-pyridylethyl derivatives involves the addition of the thioic acid to 2 vinylpyridine (7) or to β-nitrovinylpyridine (357) **(XV-115).**

These derivatives are listed in Tables XV-57 and XV-58.

XV-115

XXVI. Dihydropyridine Derivatives

The reduction of pyridinium salts (**XV-116**, R = benzyl, *o*-chlorobenzyl) to 1,4-dihydropyridines (**XV-117**) with sodium dithionite, may involve intermediate addition compounds such as **XV-118**; NMR and UV spectral data have been offered in support of this hypothesis (27).

The addition of ethyl chloroformate to 4-pyridinethione gives a red solution from which **XV-119** can be isolated as a high melting solid; the free base **XV-120** is a yellow oil showing no absorption in the UV at *ca* 370 mμ. On keeping under vacuum, **XV-120** slowly changes to an unidentified red solid (**XV-121**); recrystallization of **XV-121** gives an orange solid (**XV-122**). The pyridinethione structure assigned to **XV-122** is suggested by the similarity of its UV spectrum (maxima at 245 and 370 mμ) with that of 4-pyridinethione which shows maxima at 230 and 340 mμ; the bathochromic shift is attributed to the contribution of the quinoïd structure **XV-123**. The suggestion has been made that the hydrochloride **(XV-119)** is stable, but that the base **(XV-120)** readily rearranges to **XV-121**. Thus, when 4-pyridinethione and ethyl chloroformate undergo reaction in aqueous sodium bicarbonate, only **XV-122** is obtained. Interestingly, **XV-122** is also unstable and, after several days, it loses its orange color to give colorless di-(4-pyridyl)sulfide **(XV-124)** in 90% yield. The conversion of **XV-122** to **XV-124** is accelerated by sunlight, ultraviolet light, and heat, and yields, in addition to **XV-124**, carbon dioxide, carbon oxysulfide, ethyl carbonate, and ethyl thiocarbonate. The mechanism proposed for the formation of **XV-124** involves a dihydro- intermediate and is summarized in **XV-125**. It is noteworthy, too, that **XV-120** and hydrazine give only 4-pyridinethione (61).

XV-125

XV-124

XXVII. Tetrahydropyridine Derivatives

1,2,3,6-Tetrahydropyridine reacts with *p*-toluenesulfonyl chloride to give the *N*-tosyl derivative **(XV-126)**; the latter adds bromine to give a 4,5-dibromide, while oxidation with peracetic followed by hydrolysis gives the 4,5-diol (142). Heating a solution of benzenesulfonamide and 2-methoxy-3,4-dihydro-2*H*-pyran in acetonitrile under reflux gives glutaraldehyde hemiacetal **(XV-127)** and

N-benzenesulfonyl-2-methoxy-1,2,3,4-tetrahydropyridine **(XV-128)** as the major products, along with a lesser amount of *N*-benzenesulfonyl-2-benzenesulfona-mido-1,2,3,4-tetrahydropyridine **(XV-129)**. A mechanism proposed for the formation of these products has been reported and involves opening of the pyran ring, a process facilitated by the methoxyl group at position 2, since dihydropyran and benzenesulfonamide form only the adduct **(XV-130)** (113). Alternatively, elimination of methanol from the intermediate piperidine would lead to the same product **(XV-129)**.

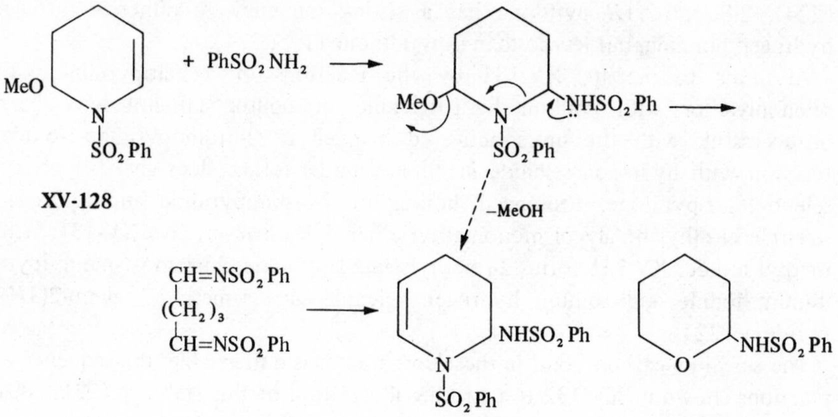

XV-128   + PhSO₂NH₂ ⟶

—MeOH

XV-129   **XV-130**

The new tetrahydropyridine derivatives are found in Table XV-59.

XXVIII. Selenium Compounds of Pyridine

An investigation paralleling the effort devoted to the preparation of (β-D-glucopyranosyl)thiopyridines and their derivatives (p. 248) has concerned itself with the corresponding selenium compounds. Again, the purpose of the study has been to employ the selenium derivatives as model compounds in methanolysis and enzymolysis reactions to elucidate cleavage mechanisms. The chemistry involved differs little from that already discussed with the sulfur derivatives and is not elaborated further at this time; the rates of methanolysis and enzymolysis for the isomeric selenium derivatives are very similar to those observed with the sulfur compounds (348–355).

The UV spectrum of 2-seleno-2(1*H*)-pyridone (**XV-131**) resembles that of the

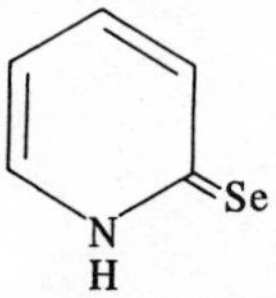

XV-131

N-methyl derivative and is somewhat different from that of 2-(methylselenyl)-pyridine; hence, **XV-131** exists predominantly as the selenone. As anticipated, **XV-131** is a stronger acid than 2-pyridinethione and far stronger than 2-pyridone

(224). 2-Seleno-2(1*H*)-pyridone has a strong tendency to dimerize through hydrogen bonding, but less so than 2-pyridinone (193, 224).

Attempts to prepare **XV-131** by the reactions of a halopyridine with selenourea or with sodium hydroselenide in boiling ethanol have been unsuccessful; with the more activated halogen in 2-bromopyridine-1-oxide, reaction with hydrogen selenide in ethanol under reflux, does give 1-hydroxy-seleno-2(1*H*)pyridone. Prolonged boiling of 2-bromopyridine and hydrogen selenide in ethylene glycol monomethyl ether did, however, give **XV-131**. With methyl iodide, **XV-131** forms 2-(methylselenyl)pyridine; 2-bromo-1-methylpyridinium iodide and sodium hydrogen selenide give 1-methyl-2-seleno-2(1*H*)-pyridone (224).

The selenium-carbon bond in these compounds is quite stable; the sequence of reactions shown in **XV-132** is a striking illustration of this stability (221, 380).

The selenium derivatives of pyridine have been utilized as intermediates for the preparation of the selenium heterocycles, **XV-133** and **XV-134** (569).

The selenium compounds reported since the last compilation (1st Ed., Part IV, p. 425 ff) are listed in Table XV-61.

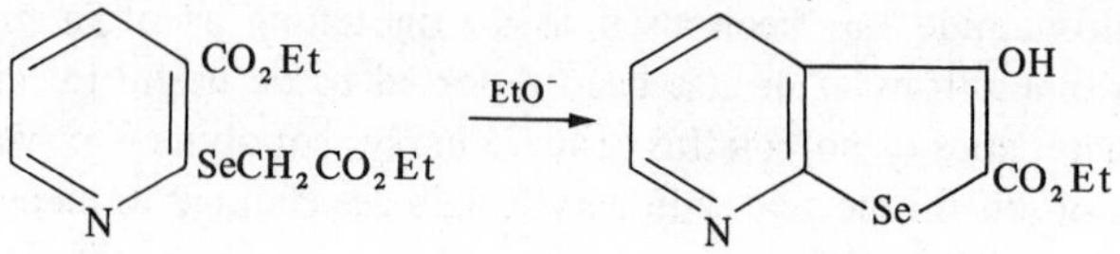

XV-133

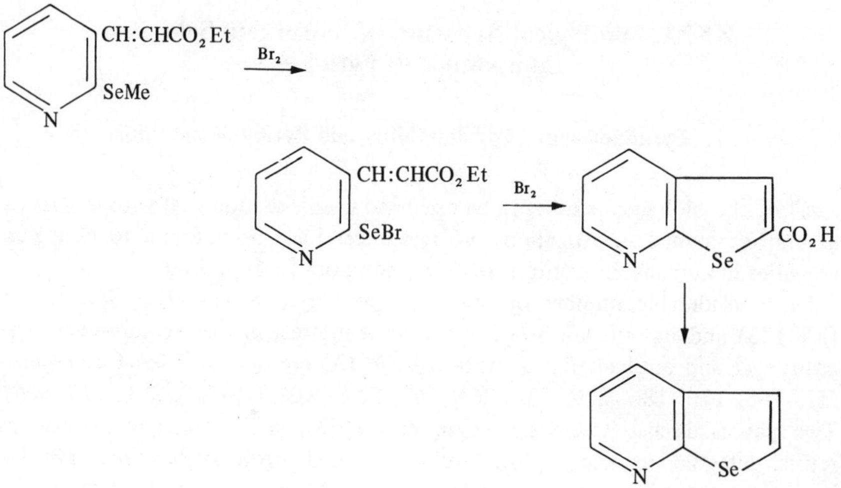

XV-134

XXIX. 1-Hydroxy-2(1*H*)-pyridinethione in Analytical Procedures

1-Hydroxy-2(1*H*)-pyridinethione has been suggested as a reagent for the determination of trivalent iron (70, 71). A procedure has been proposed for the polarographic assay of metal derivatives of 1-hydroxy-2-pyridinethione (200); another procedure makes use of the Autoanalyzer and a turbidometric microbiological assay for the sodium salt (268). 2,2'-Dithiobis-pyridine, 2,2'-di-thiobis(5-nitropyridine) (466, 482, 484), and 6,6'-dithiodinicotinic acid (467) have been used for the determination of thiol groups.

XXX. Nonbiological Applications of Sulfur Compounds of Pyridine

2-(1*H*)-Pyridinethione has been mentioned as an activator for color film (318). *N,N*-Dimethyl-4-pyridyldithiocarbamate has been suggested as a sensitizer for photographic emulsions (177). 2-Pyridinesulfenamide and 1-morpholinomethyl-2-pyridinethione have been employed as vulcanization accelerators (234).

3-Pyridinesulfonamide has been used as a brightening agent in nickel plating (139). 3-Pyridinesulfonylazide has been reported to be useful in improving the dyeing characteristics of polyolefins (146). Finally, copolymers of vinylpyridinesulfonic acid or -disulfonic acid with acrylic acid are claimed to be coagulants for photographic emulsions (178).

XXXI. Biological Activities of Sulfur and Selenium Compounds of Pyridine

1. Pyridinethiones, Pyridinethiols, and Pyridylalkanethiols

All of the biological studies to be discussed in this section deal with *in vitro* or laboratory animal investigations; no references have been found to biological activities in humans for sulfur or selenium compounds of pyridine.

The considerable number of references to 1-hydroxy-2-(1*H*)-pyridinethione **(XV-133)** and its salts are proof of the great interest in these compounds. The antifungal and antibacterial activities of **XV-133** are discussed in 16 references (13, 165, 170, 186, 209, 223, 261, 263, 277, 303, 310–312, 321, 322, 336). Two papers discuss its toxicity: cytoxicity (65) and its tendency to produce retinal detachment in dogs (76). Antibacterial and antifungal activities have also been reported for a variety of salts and complexes of **XV-133**: sodium salts (143–145, 167, 240), copper salt (51), zinc, cadmium, tin, zirconium, and manganese salts and complexes (171, 176, 179, 204, 208, 314, 319, 320, 342, 343, 359), phenylmercury salt (149); triphenyltin salt (32), and triphenylborane complex (341). The copper salt is said to be an animal repellent (93), while the administration of **XV-133** itself is claimed to aid in the elimination of ionic mercury from the kidneys (241).

2,3,5,6-Tetrachloro-4-pyridinethione is stated to have anthelmintic and herbicidal activities (150), while 2-mercaptoisonicotinic acid hydrazide has demonstrated antituberculous activity (243, 256).

The mercaptopyridoxines have been found to afford protection against ionizing radiation (33, 42, 111, 184, 185, 194, 369), to increase glucose permeability in the brain (276), to restore normal amplitude to the heart after cadmium poisoning (34), as an antidote for methyl mercury poisoning (574) and to inhibit tumor growth (132–134).

2. Pyridyl- and Pyridylalkyl Sulfides, Disulfides, and Sulfones

Antibacterial, herbicidal, and fungicidal activity have been reported for a variety of pyridyl sulfides (1, 31, 62, 63, 75, 132, 265, 269, 271, 323, 488,

518); di-(2-pyridylmethyl) sulfide has been recommended as a flavoring agent for food and beverages (244, 245). 4-(p-Aminophenylthio)pyridine inhibits the biosynthesis of testicular steroids (239). 2-Aminoethyl 2-pyridyl disulfide has been evaluated as a protective agent against ionizing radiation (110). 4-(2-Hydroxyethylthio)-3-pyridinecarboxamide has been shown to have activity against the Walker carcinoma and the L 1210 strain of leukemia (291). Di-(2-pyridyl)disulfide-1,1'-dioxide has also shown antibacterial, antifungal, and antitumor activity (21, 48). Acaricidal activity has been seen with 2-(p-chlorophenylsulfonyl)pyridine (371, 372). 5-Amino-2-(phenylsulfonyl)pyridine has shown antibacterial activity (1).

3. Pyridinesulfonic Acids

3-Pyridinesulfonic acid, considered an isotere of nicotinic acid, has been evaluated in a large variety of screening tests to detect competitive inhibition of that biologically important vitamin (18, 52, 92, 96, 124, 126, 168, 189, 190, 197–199, 210, 238, 242, 296). An unusual study has demonstrated retention of rare earth metals in the liver and spleen following the ingestion of rare earth salts of sulfoisonicotinic acid (344).

4. Pyridinesulfonamides

The diuretic activity seen with substituted benzenesulfonamides has stimulated the search for the same activity in the pyridinesulfonamides (19, 22, 26, 36, 230, 255, 305). Other possible uses for this class of compounds are suggested by their anticonvulsive (19, 109), hypoglycemic (35), analgesic (109), antibacterial (10, 290), and antituberculous (3, 4) activities.

5. Pyridinealkanethiophosphates

Several studies of this class of compounds refer to their application as insecticides (7, 104, 295).

6. Miscellaneous Structural Types

Antibacterial activity is reported for allyl 3-pyridinethiolsulfonate (337). Antiprotozoal activity is attributed to 2,6-diamino-4-pyridinethione and 2,6-diamino-4-pyridineselenone (380), and parasiticidal properties are claimed for diethyl 6-methanesulfonyl-2-pyridylphosphate (385).

XXXII. Tables

TABLE XV-1. Reaction of Alkoxy- and Acycloxypyridinium Salts with Thiols (14-17)

H or Me in pyridine ring	EX	R in SR	Total sulfides (%)	Distribution of isomeric sulfides (%)		
				2-	3-	4-
H	$(EtO)_2 SO_2$	n-Pr	30	0	86	14
H	$(EtO)_2 SO_2$	n-Octyl	6.7	0	(Not sepd.)	
H	$PhSO_2 Cl$	n-Pr	32	50	50	0
H	$(MeCO)_2 O$	n-Pr	67	61	39	0
2-Me	$PhSO_2 Cl$	n-Pr	9.4	52	42	6
3-Me	$PhSO_2 Cl$	n-Pr	7.2	67^a	33	0
4-Me	$(EtO)_2 SO_2$	n-Pr	31	9	22	−
4-Me	$PhSO_2 Cl$	n-Pr	26	77	23	−
4-Me	$(MeCO)_2 O$	n-Pr	31	50	50	−
H	$(EtO)_2 SO_2$	n-Bu	15	16	60	24
H	$p\text{-}MeC_6 H_4 SO_3 Et$	n-Bu	11	11	74	15
H	$(MeCO)_2 O$	n-Bu	67	61	39	0
H	MeCOCl	n-Bu	10	89	9	2
H	PhCOCl	n-Bu	19	81	18	1
H	PhCOCl	n-Bu	16	81	15	4
H	$PhSO_2 Cl$	n-Bu	32	50	50	0
H	$p\text{-}MeC_6 H_4 SO_3 Me$	$PhCH_2$	0^b	−	−	−

^{a}Ratio of 2- to 6- isomers is 3: 1.
^{b}N. A. Coates and A. R. Katritzky, *J. Org. Chem.*, **24**, 1836 (1959).

TABLE XV-2. 2(1*H*)-Pyridinethiones

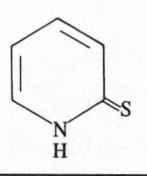

Substitutents	M.p. (°C)	Ref.
None	126–129 (Thiolacetate, 88–88.5°, b.p. 96–99°/4 mm; thiolpropionate, b.p. 100°/0.7 mm; thiolbutyrate, b.p. 113–114°/1 mm; thiolhexanoate, b.p. 144–144°/2 mm; thioldecanoate, b.p. 145–149°/0.3 mm; thiolbenzoate, m.p. 44–45°, b.p. 171–172°/3 mm)	25, 26, 35, 117, 132, 175, 234, 246, 318, 338, 387, 408, 409 425, 489, 495, 537
3-Amino	131–133	288, 573
3-Amino-5-chloro	204–205	216
3-Amino-6-methoxy	–	214
3-[3-Amino-2-(pyridylamino)]	204–208 (decomp.)	289
5-Acetamido	–	249
4-*t*-Butyl-3-dimethylamino	–	491, 492
3-Carboxy	–	81
6-Carboxy	196–197	79
3-Carboxy-6-methyl	246–247	366
4-Carbohydrazide	247	243, 256
3-Chloro	197–206	37
5-Chloro	–	37
5-Chloro-3-nitro	–	216
3-Cyano-6-methyl	235–236	366
3-(2-*p*-Chlorophenethyl)	185–186	377
3,5-Dinitro	182	331
4,6-Diphenyl	197	201
3-Ethyl-6-methyl	147–149	439
3-(2-Hydroxy-2,2-diphenylethyl)	240–241	377
6-(2-Hydroxy-2,2-diphenylethyl)	231–232	377
6-Mercapto	148–152	284
3-Methyl	–	408
4-Methyl	179–180	79, 6, 376, 408, 409
5-Methyl	–	408
6-Methyl	154–156	79, 408, 409
3-[3-Methyl-5-nitro-2-(pyridylamino)]	274–277	288
3-[5-Methyl-3-nitro-2-(pyridylamino)]	257–257.5 (decomp.)	289
3-Nitro	172–174	37
5-Nitro	–	37, 105, 132
3-[5-Nitro-2-(pyridylamino)]	262–263 (decomp.)	288

TABLE XV-2. 2(1H)-Pyridinethiones (*Continued*)

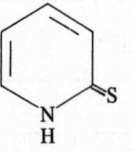

Substitutents	M.p. (°C)	Ref.
3-[3-Nitro-2-(pyridylamino)]	242–244 (decomp.) *N*-acetyl, m.p. 170– 172 (decomp.)	288, 289
3-[3-Nitro-5-methyl-2-(pyridylamino)]	255–257	288
3-(2-Phenethyl)	177–179	377
6-(2-Phenethyl)	145–159	377
3-Phenyl	229–237	535
3-SO$_3$H	235	81
5-Stearamido	123	126

NC — Et — Me — O — N(H) — SH

	295–297	130
3-Methyl-5-nitro	195	443
4-Methyl-3-nitro	185	443
4-Methyl-5-nitro	194	443
5-Methyl-3-nitro	207	443
6-Methyl-3-nitro	186	443
6-Methyl-5-nitro	205	443

TABLE XV-3. 1-Substituted-2-Pyridinethiones

Substituents	R	M.p. (°C)	Ref.
None	Et	B.p. 182°/15 mm	409
None	[1,1-^{2}H$_2$]-Et	—	409
None	[2,2,2-^{2}H$_3$]-Et	—	409
None	OH	70–73	23, 65, 70, 71, 76, 138, 144, 147, 165, 167, 170, 179, 186, 187, 193, 200, 209, 223, 226, 227, 241, 261, 263, 277, 303, 304, 310–312, 316, 324, 341, 343, 344, 433, 471, 487, 537, 556, 43, 72, 138, 154
		Phenylthiourethane, oil; phenyldithiourethane, oil; octyldithiourethane, oil; ethyldithiourethane, oil; p-chlorophenyldithiourethane, m.p. 76–78°; thiolbenzoate, 3,5-dinitrothiolbenzoate, 2-furyl-5-nitrothiolacrylate, m.p. 165–166° (decomp.)	
		Copper salt	93, 176
		Zinc salt	13, 51, 116, 204, 314, 359, 407, 506, 550

TABLE XV-3. 1-Substituted-2-Pyridinethiones (Continued)

Substituents			
	R	M.p. (°C)	Ref.
None		Sodium salt	143, 145, 171, 240, 268, 511
		Manganese salt	204, 285, 286, 322, 342
		Thiourea salt	204
		Aluminum salt, m.p. 254.5–257°	254
		Molybdenum salt	319
		Et₃SnOH salt,	179
		SnCl₂ complex,	
		m.p. 313° (decomp.);	320
		morpholine salt, m.p. 147–148°;	336
		diethylamine salt, m.p. 130–132°;	336
		piperazine carboxaldehyde salt,	
		m.p. 112–115°	336
		Tetraethylammonium salt,	
		m.p. 145.0–146.5°	572
		Pyridine salt, —	489
		Tributylamine salt, —	489
		Phenylhydrazine salt, m.p. 72–74°	585
		methylhomopiperazine salt,	
		m.p. 123–124°; cetyl-	
		trimethylammonium salt,	23
		m.p. 152–158° (decomp.)	
Me		—	187, 409

None	CH₂:CH	65.5	409
None	Pr	n_D^{25} 1.655	409
None	HO₂CCH₂CH₂	—	107, 187
None	CNCH₂CH₂	B.p. 133–135°/2 mm	187, 346
None	MeO₂CCHMeCH₂	—	187
None	N-morpholinomethyl	52–53	234
None	Tetra-O-acetyl-β-D-glucosyl	92–93	352
None	β-D-glucosyl	$[\alpha]_D^{19}$ 213.20 (c, 2.5, CHCl₃) 114–115 $[\alpha]_D^{20}$ 125.5° (c, 2.5, H₂O)	353
None	CNCH₂CH₂	92–93	107, 108
None	EtO₂CCH₂CH₂	158–160	107
None	p-MeC₆H₄SCH₂CH₂	76	11
4-Bromo	OH	—	564
5-Bromo	OH	Acetate, —	527, 552
4-Chloro	OH	Propionate, —	507, 527
4-Chloro-5-methyl	OH	Unstable	433
3,5-Dicyano-6-mercapto-4-methyl	H	—	44
		(Potassium salt, m.p. >350°; ammonium salt, m.p. 325°; piperidine salt, m.p. 249°)	
3,5-Dicyano-4-ethyl-6-mercapto	H	178	44
		(Ammonium salt, m.p. 320°; sodium salt, m.p. 300°; pyridine salt, m.p. 212°)	
3,5-Dicyano-6-mercapto-4-propyl	H	—	44
		(Ammonium salt, m.p. 188°; pyridine salt, m.p. 212°; piperidine salt, m.p. 254°)	

TABLE XV-3. 1-Substituted-2-Pyridinethiones *(Continued)*

Substituents	R	M.p. (°C)	Ref.
3,5-Dicyano-4-hexyl-6-mercapto	H	Ammonium salt, m.p. 236°; potassium salt, m.p. 281°; pyridine salt, m.p. 208°; piperidine salt, m.p. 179°	44
4-Ethoxy	OH	—	507
4-Methoxy	OH	Acetate, m.p. 98° (benzoate, — pivaloate, — stearate, —)	507, 527, 552
4-Methyl	Et	58	409
6-Methyl	Et	88	409
3-Methyl	Me	120	409
3-Methyl	OH	73.5–75	438
4-Methyl	Me	138	409

4-Methyl	OH	59	433
5-Methyl	Me	119	409
6-Methyl	Me	104	409
5-Nitro	HO$_2$CCH$_2$CH$_2$	133–134	346
5-Nitro	NCCH$_2$CH$_2$	188–189	346
4,5,6-Triphenyl	H$_2$N	195	98
3,4-Dimethyl	OH	128–129	433
4,5-Dimethyl	OH	121–122	433
4,6-Dimethyl	Me	127	409
		120–122 (2,4-Dinitrophenylsulfenyl deriv., m.p. 175–177°; Zn salt, m.p. > 300°)	433

TABLE XV-4. 3-Pyridinethiols

Substituents	M.p. ($^\circ$C)	Ref.
None	74–78 (hydrochloride)	118, 182
2-Carboxy	–	525
2-Carbomethoxy	S-benzoate, –	525
4-Carboxy	259–260	80
5-Carboxy	162–165 (decomp.)	458
6-Hydroxy	Oil	280
	Hydrochloride, m.p. 201–203 (decomp.)	133
	Hydrochloride, m.p. 189–191 (decomp.)	133, 526
	[3-(2,2-Dimethylpropionate), no m.p.]	454
	100	518
	–	526

TABLE XV-4. 3-Pyridinethiols (*Continued*)

Substituents	M.p. (°C)	Ref.
	–	526
	–	526
		526

TABLE XV-5. 4-(1H)-Pyridinethiones

Substituents	M.p. (°C)	Ref.
None	177	28, 61, 106, 159, 187, 317
	(acetate, m.p. 89°,	
	benzoate, m.p. 76°)	
3-Carbamoyl	182	358
2-Carbamoyl-3,5,6-trichloro	220–250 (decomp.)	518, 583
2-Carboxy	188–190	80
3-Carboxy	234–237	80
2-Carboxy-3,5,6-trichloro	186	583
2-Chloro-3,5-dibromo-6-methyl	126–128	544
	(Na salt, m.p. 250°	
	acetate, m.p. 68–70°	
	3-chloropropionate, m.p. 96–97°	
	2,2-dimethylpropionate, m.p. 104–106°	
	butyrate, m.p. 84–85°	
	cyclopropylcarboxylate, m.p. 86–90°	
	cyclohexanoate, m.p. 93–96°	
	2-ethylhexanoate, oil	
	oleate, oil	
	stearate, m.p. 40–45°	
	p-Toluenethiosulfonate, m.p. 118–119°)	
5-Chloro-2,3,6-trifluoro	128–130	151
2-Cyano-3,5,6-trichloro	136	518, 545, 578, 583, 584
3-Cyano-2,5,6-trichloro	Ammonium salt	518, 578, 583
2,6-Dicarboethoxy		221

2,6-Dichloro	—	493
3,5-Dichloro-2,6-dicyano (mixture with 4,5-dichloro-2,6-dicyano-3-pyridinethiol)	—	518, 578
3,5-Dichloro-2,6-difluoro	Sodium salt, 46–47 (sublimes, 35°/0.1 mm); thiolacetate, m.p. 69–70° or 11.5–112.5°; thiolpropionate, m.p. 47–48°	150, 151
3,5-Dichloro-6-methylsulfonyl	—	545
2-Methyl	147	273
2-Methyl-3,5,6-trichloro	102–104	544
3-Methyl	159–160	159
2,3,5,6-Tetrachloro	165–166	150, 378, 434, 460
2,3,5-Tribromo-6-methylsulfonyl	—	545
2,3,5,6-Tetrafluoro	26.5; b.p. 156–157	394, 455
2,3,5-Trichloro	67 (sodium salt, m.p. > 250°)	378, 434, 584
2,3,6-Trichloro	80; 103–103.5	378, 434
3,5,6-Trichloro	—	280
2,3,5-Trichloro-6-(trifluoromethyl)	77–78	407, 539, 540, 571
2,5,6-Trifluoro	B.p. 59/3 mm	150

TABLE XV-6. 1-Substituted-4-Pyridinethiones

Substituents	R	M.p. (°C)	Ref.
None	HO	42–43 (thiolacetate, m.p. 88–90°)	247, 329, 37
None	NCCH₂CH₂	117–118	106, 187
None	MeO₂CCH₂CH₂	123–124	106
None	MeO₂CCHMeCH₂	136–138	106
None	EtO₂CCH₂CH₂	185–187	107
None	N-Morpholinomethyl	107–108	234
None	2,3,4,6-Tetra-O-acetyl-β-D-glucopyranosyl	198.5–199.5, $[\alpha]_D^{21}$ 84.6° (c, 5.0, CHCl₃)	353
None	β-D-glucosyl	251–253 (decomp.), $[\alpha]_D^{19}$ 34.5° (c, 5.0, H₂O)	353
None	MeO₂C	68–69	61
None	EtO₂C	90	61
None	ClCH₂CH₂O₂C	92	61
None	CH₂:CHCH₂O₂C	65	61
None	PhCH₂O₂C	68	61
None	MeCO	89	61
None	Me	150–152 (dodecylammonium bromide, methylene di(sulfonium bromide), m.p. 78–80°)	31, 187, 281
2-Carboxy-1-hydroxy		155	275
3,5-Dibromo-2,6-dimethyl	Me	300	97
3,5-Dibromo-2,6-dimethyl	Et	262	97
3,5-Dibromo-2,6-diphenyl	Me	244	97
3,5-Dibromo-2,6-diphenyl	Et	276	97

2,6-Dimethyl	Me	268	397
2,6-Dimethyl	Bu	206	97
2,6-Dimethyl	PhCH$_2$	208	97
2,6-Diphenyl	Me	248	397
2,6-Diphenyl	Bu	221	97
2,6-Diphenyl	PhCH$_2$	161	97
2-Ph-6-(p-MeOC$_6$H$_4$)	Me	200	397
2,6-Di-(p-MeOC$_6$H$_4$)$_2$	Me	225	397
2,6-Dimethyl	Et	248	397
2,6-Diphenyl	Et	210	397
2-Ph-6-(p-MeOC$_6$H$_4$)	Et	162	397
2,6-Di-(p-MeOC$_6$H$_4$)	Et	220	397

TABLE XV-7. 2-Pyridylsulfides

Substituents	R	M.p. (°C)	Ref.
None	Me	B.p. 192°/750 mm	187, 409
None	[^{2}H$_3$]-Me	—	409
None	Et	B.p. 205°/750 mm	409
None	[1,1-^{2}H$_2$]-Et	—	409
None	[2,2,2-^{2}H$_3$]-Et	—	409
None	Pr	B.p. 108–110°/18 mm (picrate, m.p. 122–123°)	15
None	C$_8$H$_{17}$	B.p. 113–115°/2.0 mm (picrate, 77–78)	15
None	NCCH$_2$	B.p. 97°/3 mm (d_{20} 1.2180; n_D^{20} 1.5873; hydrochloride, m.p. 147–148°)	107
None	NCCH$_2$CH$_2$	B.p. 134–135/2 mm (d_{20} 1.1645; n_D^{20} 1.5710; hydrochloride, m.p. 149–150)	107, 108, 187, 346
None	HOCH$_2$CH$_2$	B.p. 108°/0.8 mm (n_D^{24} 1.5954)	409
None	EtO$_2$CCH$_2$CH$_2$	73–75 (picrate, m.p. 124–126°)	107

None	$MeO_2CCH_2CH_2$	B.p. 133–134°/4 mm [d_{20} 1.1568; n_D^{20} 1.5534; picrate, m.p. 101–102° (decomp.)]	108
None	$H_2NCH_2CH_2$	Dihydrochloride, m.p. 201°	105a
None	$MeCONHCH_2CH_2$	70	105a
None	$(MeCO)_2NCH_2CH_2$	102	105a
None	$HO_2CCH_2CH_2$	—	105a
None	HO_2CCH_2	136–137	187
None	EtO_2CCH_2	133–134	175
None	$(MeO)_2CHCH_2$	—	257
None	$Me_2NCH_2CH_2$	B.p. 91–93°/3 mm [d_{20} 951; n_D^{20} 1.560 hydrochloride, m.p. 194–196°; picrate, m.p. 130–132°; methiodide, m.p. 183–184° (decomp.)]	347
None	$MeO_2CCHMeCH_2$	B.p. 142–143°/6 mm (d_{20} 1.1711; n_D^{20} 1.5596)	108
None	Ph	B.p. 107–110°/0.3 mm	41, 47, 48, 132, 235, 473
None	$o\text{-}MeC_6H_4$	B.p. 165–175°/6 mm (acetonyl iodide, m.p. 159–160°)	41

TABLE XV-7. 2-Pyridylsulfides (Continued)

Substituents	R	M.p. (° C)	Ref.
None	m-MeC_6H_4	B.p. 174-180°/3 mm (acetonyl iodide, 126-127°)	41
None	p-MeC_6H_4	B.p. 175-180°/6 mm (acetonyl iodide, 163-164°)	41
None	p-ClC_6H_4	B.p. 185-190°/10 mm (acetonyl iodide, m.p. 167-168° hydrochloride, m.p. 193-196°)	41, 265, 371, 372
None	2,4,5-Cl_3C_6H_2	50.5-51.5 [hydrochloride, m.p. 170-173° (decomp.)]	265, 371, 372
None	2,4,6-Cl_3C_6H_2	—	163
None	p-Me_3CC_6H_4	46-49 (b.p. 205-210°/8 mm; acetonyl iodide, m.p. 168-169°)	39
None	m-CF_3SO_2NHC_6H_4	119.5-121.0	517
None	CH(CO_2Et)COPh		547
None	2-C_{10}H_7	B.p. 197-200°/1 mm (picrate, m.p. 180-181°; acetonyl iodide, m.p. 157-158°)	40

274

None	1-C$_{10}$H$_7$	B.p. 185–190°/1 mm (acetonyl iodide, m.p. 188–189°)	40
None	Me$_2$NCH$_2$CH$_2$	—	436
None	Et$_2$NCH$_2$CH$_2$	B.p. 130–132°/4 mm (d_{20} 1.018; n^{20}_D 1.5400; picrate, m.p. 103–105°)	347, 436
None	NCH$_2$CH(OH)CH$_2$ (Me$_2$ ... Me$_2$)	(Dihydrochloride, m.p. 209–212°)	453
5-Acetamido	Et$_2$NCH$_2$CH$_2$	(Dihydrochloride, m.p. 181–182°; dimethiodide, m.p. 140°)	105
3-Acetamido-5-chloro	2,4-(O$_2$N)$_2$C$_6$H$_3$	162–164	216
3-Acetamido-6-ethoxy	2,4-(O$_2$N)$_2$C$_6$H$_3$	205–206	214
5-Amino	2,4-Cl$_2$C$_6$H$_3$	84–85 (acetate, m.p. 125–126°)	1
5-Amino	3,5-Cl$_2$C$_6$H$_3$	87–88	1
5-Amino	p-HOC$_6$H$_4$	169–170 (acetate, m.p. 168–169°)	1
5-Amino	2,4-Cl(HO)C$_6$H$_3$	205–208 (acetate, m.p. 159–160°)	1
5-Amino	3,4-Cl(HO)C$_6$H$_3$	148–149 (acetate, m.p. 169–170°)	1
5-Amino	2,5,4-Cl$_2$(HO)C$_6$H$_2$	213–214	1

TABLE XV-7. 2-Pyridylsulfides (*Continued*)

Substituents	R	M.p. (°C)	Ref.
5-Amino	$3,5,4\text{-}Cl_2(HO)C_6H_2$	222–223 (acetate, m.p. 172–173°)	1
5-Amino	$Et_2NCH_2CH_2$	Trihydrochloride, m.p. 130°	105
6-Amino	Ph	117–117.5	161
3-Amino-5-chloro	Me	65	216
3-Amino-5-chloro	$2,4\text{-}(O_2N)_2C_6H_3$	183–184	216
5-Amino-3-chloro	$2,4\text{-}Cl_2C_6H_3$	114–115 (acetate, m.p. 121–122°)	1
5-Amino-3-chloro	$3,4\text{-}Cl_2C_6H_3$	103–104 (acetate, m.p. 118–119°)	1
5-Amino-3-chloro	$3,5\text{-}Cl_2C_6H_3$	129–130 (acetate, m.p. 148–149°)	1
5-Amino-3-chloro	$p\text{-}HOC_6H_4$	217–218 (acetate, m.p. 192–193°)	1
5-Amino-3-chloro	$2,4\text{-}Cl(HO)C_6H_3$	172–174 (acetate, m.p. 217–218°)	1
5-Amino-3-chloro	$3,4\text{-}Cl(HO)C_6H_3$	171–172 (acetate, m.p. 137–138°)	1
5-Amino-3-chloro	$2,5,4\text{-}Cl_2(HO)C_6H_2$	178–179 (acetate, m.p. 137–138°)	1

5-Amino-3-chloro	3,5,4-Cl$_2$(HO)C$_6$H$_2$	223–224 (acetate, m.p. 167–168°)	1
3-Amino-6-methyl	Ph	—	541
3-Amino-6-ethyl	Me	124	214
3-Amino-6-ethoxy	2,4-(O$_2$N)$_2$C$_6$H$_3$	174–176	214
3-Amino-6-methoxy	Me	120	214
5-Bromo	Ph	Picrate, m.p. 129–130°	448
3-Bromo	Ph	Picrate, m.p. 111°	448
3,5-Dibromo-1-oxide		Hydrobromide	62
3-Carboethoxy	Et	B.p. 101°/0.05 mm (n_D^{12} 1.5652)	430
3-Carboxy	5,2-Cl(Me)C$_6$H$_3$	200–202	29
3-Carboethoxy-6-methyl	Et	—	542
3-[2-(Carboethoxy)aceto]-6-methyl	Et	—	542
3-(1,3-Dihydroxypropyl)-6-methyl	Et	—	542
3-Carboxy-6-methyl	Ph	180–181	294

TABLE XV-7. 2-Pyridylsulfides (*Continued*)

Substituents	R	M.p. (°C)	Ref.
3-Carboxy-4,6-dimethyl	Ph	193-195	294
3-Carboxy-6-methyl	HO_2CCH_2	194-195	366
5-Chloro	p-ClC$_6$H$_4$	60-61 [hydrochloride, m.p. 170-173° (decomp.)]	265, 371, 372
5-Chloro	2,4,5-Cl$_3$C$_6$H$_2$	120-122	265, 371, 372
6-Chloro	m-CF$_3$SO$_2$NHC$_6$H$_4$	119.0-120.5	517
3-Chloro-5-nitro	2,4-Cl$_2$C$_6$H$_3$	135-136	1
3-Chloro-5-nitro	3,4-Cl$_2$C$_6$H$_3$	119-120	1
3-Chloro-5-nitro	3,5-Cl$_2$C$_6$H$_3$	138-139	1
3-Chloro-5-nitro	p-HOC$_6$H$_4$	188-189 (acetate, m.p. 136-137°)	1
3-Chloro-5-nitro	2,4-Cl(HO)C$_6$H$_3$	178-179 (acetate, m.p. 140-141°)	1
3-Chloro-5-nitro	3,4-Cl(HO)C$_6$H$_3$	139-140 (acetate, m.p. 144-145°)	1
3-Chloro-5-nitro	2,5,4-Cl$_2$(HO)C$_6$H$_2$	203-204 (acetate, m.p. 122-123°)	1

		m.p.	Ref.
3-Chloro-5-nitro	3,5,4′-Cl_2(HO)C_6H_2	186–188 (acetate, m.p. 108–109°)	1
5-Chloro-3-nitro	o-$H_2NC_6H_4$	178–179	219
5-Chloro-3-nitro	o-$MeCOHHC_6H_4$	—	217
6-Chloro-3-nitro	o-$H_2NC_6H_4$	(N-acetyl deriv.)	361, 362
5-Chloro-1-oxide	p-ClC_6H_4	156–157.5	265
5-Chloro-1-oxide	2,4,5-$Cl_3C_6H_2$	148.5–150	265
3-Chloro-6-trichloromethyl	$C_{14}H_{29}$	—	163
6-Chloro-4-trichloromethyl	C_6H_{13}	—	163
5-Chloro-4-trichloromethyl	$C_{12}H_{25}$	—	163
3-Cyano-6-methyl	Ph	80–81	294
3-Cyano-4,6-dimethyl	Ph	90–91	294
3,5-Diacetamido	Me	237–238	334
3,5-Dichloro	o-$H_2NC_6H_4$	131–132	219
3,5-Dichloro	o-$HCONHC_6H_4$	197–198	219
3,5-Dinitro			219
3,5-Dinitro	Me	123–124	334
3,5-Dinitro	Ph	122–123	331

TABLE XV-7. 2-Pyridylsulfides (Continued)

Substituents	R	M.p. (°C)	Ref.
(2,4-Dinitrophenylamino)-6-ethoxy	Me	194	216
3-(2,4-Dinitrophenylamino)	Me	223–224	216
3-Ethyl-1-oxide		Hydrofluoride	62
5-Fluoro-1-oxide		Hydrochloride	62
6-Fluoro-3,4,5-trichloro	HO_2CCH_2	148–149	152
6-Fluoro-3,4,5-trichloro	HO_2CCHMe	140–142	152
4-Heptyl	Ph	B.p. 176–180°/0.5 mm	2
6-Hydroxy	Et	150–156	280

5-Iodo-1-oxide		Hydrochloride	62
5-Linoleylamido	Me	48	126
5-Linoleylamido	$CH_2{:}CHCH_2$	30	126
5-Linoleylamido	Bu	25	126
3-Methyl	Pr	B.p. 60–61°/0.3 mm (n_D^{25} 1.5535)	17
3-Methyl	Ph	B.p. 130–132°/0.5 mm (n_D^{25} 1.6330, picrate, m.p. 150–152°; acetonyl iodide, m.p. 162–163°)	39, 448
4-Methyl	Me	(n_D^{20} 1.5650)	409
4-Methyl	Et	(n_D^{20} 1.5622)	409
4-Methyl	Pr	B.p. 67°/2 mm (picrate, m.p. 137–139°)	15
4-Methyl	C_8H_{17}	B.p. 135–138°/0.5 mm (n_D^{29} 1.5120)	15
4-Methyl	Ph	B.p. 132–136°/0.25 mm (n_D^{18} 1.6263, picrate, m.p. 159–160°; acetonyl iodide, m.p. 180–182°)	15
4-Methyl	$o\text{-MeC}_6H_4$	B.p. 120–122°/0.5 mm (n_D^{25} 1.6182, picrate, m.p. 127–129°; acetonyl iodide, m.p. 190–191°)	39

TABLE XV-7. 2-Pyridylsulfides (*Continued*)

Substituents	R	M.p. (°C)	Ref.
4-Methyl	m-MeC$_6$H$_4$	B.p. 148–151°/0.6 mm (n_D^{25} 1.6172, picrate, m.p. 134–136°; acetonyl iodide, m.p. 170–171°)	39
5-Methyl	Pr	B.p. 68–70°/1.3 mm (n_D^{25} 1.5505)	17
5-Methyl	1-C$_{10}$H$_7$	B.p. 199–200°/0.9 mm (picrate, m.p. 156–158°; acetonyl iodide, m.p. 172–174°)	40
4-Methyl	m-CF$_3$SO$_2$NHC$_6$H$_4$	113.0–115.5	517
5-Methyl	Ph	B.p. 130–132°/0.4 mm (n_D^{25} 1.6298, picrate, m.p. 165–167°; acetonyl iodide, m.p. 171–173°)	39, 448
5-Methyl	2-C$_{10}$H$_7$	65–70 (picrate, m.p. 177–179°; acetonyl iodide, m.p. 149–150°)	40
6-Methyl	Me	(n_D^{21} 1.5648)	409
6-Methyl	Et	(n_D^{21} 1.5600)	409
6-Methyl	[1,1-²H$_2$]-Et	—	409

		B.p.	
6-Methyl	Pr	B.p. 72-74°/0.5 mm (n_D^{25} 1.5435)	17
6-Methyl	Ph	B.p. 130-131°/0.5 mm (n_D^{25} 1.6240)	39
3-Nitro	CH_2CO_2H	161-163	484
3-Nitro	Ph	—	132
3-Nitro	$o\text{-}H_2NC_6H_4$	124-126 (N-acetyl deriv., m.p. 138-139°)	361, 362
5-Nitro	Ph	—	132
5-Nitro	$2,4\text{-}Cl_2C_6H_3$	112-113	1
5-Nitro	$3,4\text{-}Cl_2C_5H_3$	117-118	1
5-Nitro	$3,5\text{-}Cl_2C_6H_3$	125-126	1
5-Nitro	$o\text{-}H_2NC_6H_4$	111.5	215
5-Nitro	$o\text{-}CH_3CONHC_6H_4$	152	215
5-Nitro	$p\text{-}HOC_6H_4$	174-174 (acetate, m.p. 162-163°)	1
5-Nitro	$2,4\text{-}Cl(HO)C_6H_3$	175-176 (acetate, m.p. 183-185°)	1
5-Nitro	$3,4\text{-}Cl(HO)C_6H_3$	163-164	1
5-Nitro	$2,5,4\text{-}Cl_2(HO)C_6H_2$	194-195 (acetate, m.p. 142-143°)	1
5-Nitro	$3,5,4\text{-}Cl_2(HO)C_6H_2$	197-198 (acetate, m.p. 133-134°)	1

TABLE XV-7. 2-Pyridylsulfides *(Continued)*

Substituents	R	M.p. (°C)	Ref.
5-Nitro	$Et_2NCH_2CH_2$	50–52 (hydrochloride, m.p. 173–174°; dimethiodide, m.p. 169–170°)	105, 347, 436
5-Nitro	$NCCH_2CH_2$	91–92	346
5-Nitro	$Me_2NCH_2CH_2$	40–42 (picrate, m.p. 167–170°; hydrochloride, m.p. 194–196°)	347, 436
3-Nitro-6-methyl	Ph	—	541
3-(5-Nitro-2-pyridyl)	Me	142–143	288
3-(5-Nitro-2-pyridyl)amino-6-methoxy	Me	223–223	214
3-Trifluoromethylsulfonamido	Ph	153.5–155.0	517
6-Trifluoromethylsulfonamido	Ph	175.5–176.5	517
3-(5-Nitro-2-pyridyl)amino-6-ethoxy	Me	176–177	214
1-Oxide	Et	104–105 (picrate, m.p. 212–213°)	138
1-Oxide	$PhCH_2$	170–172	73
1-Oxide	$2,3,6\text{-}Cl_3C_6H_2CH_2$	237–238	73

1-Oxide	$2,6\text{-}Cl_2C_6H_3CH_2$	213–216	73
1-Oxide	$o\text{-}MeC_6H_4CH_2$	108–110	73
1-Oxide	(ring structure, N–H, C_6H_4)	Hydrobromide, m.p. 153–156.6° (decomp.)	62, 75, 181, 225, 321
1-Oxide	(ring structure, C_6H_{13})	Hydrobromide, m.p. 131.3–131.5° (decomp.)	62
1-Oxide	(ring structure, $(CH_2)_2Ph$)	Hydrobromide, m.p. 112.1–112.4°	62
1-Oxide	(ring structure, CH_2, N–H)	Hydrochloride, m.p. 192° (decomp.)	504

TABLE XV-7. 2-Pyridylsulfides (*Continued*)

Substituents	R	M.p. (°C)	Ref.
1-Oxide	PhCH	Hydrochloride, m.p. 175–177° (decomp.)	504
1-Oxide	2-Furfuryl	173.5–174 (decomp.)	63, 225
1-Oxide	5-Nitro-2-furfuryl	–	63, 225
1-Oxide	Ph	137–139°	138, 475
1-Oxide	PrOCH₂	Hydrochloride, m.p. 98.5–100.5°	266
1-Oxide	MeOCH₂	105–107	226
1-Oxide	CH₂:CCICH₂	114–115	278, 279
1-Oxide	CH₂:CBrCH₂	122–123	278, 279
1-Oxide	CHCl:CHCH₂	Oil	278, 279
1-Oxide	*trans*-CHCl:CClCH₂	Oil	278, 279

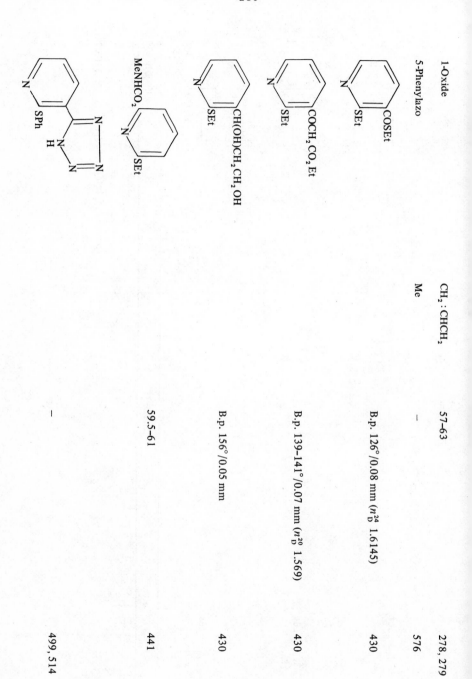

Structure	Substituent	B.p. / M.p.	Ref.
1-Oxide	CH₂:CHCH₂	57–63	278, 279
5-Phenylazo	Me	—	576
COSEt, SEt (pyridine)		B.p. 126°/0.08 mm (n_D^{24} 1.6145)	430
COCH₂CO₂Et, SEt (pyridine)		B.p. 139–141°/0.07 mm (n_D^{20} 1.569)	430
CH(OH)CH₂CH₂OH, SEt (pyridine)		B.p. 156°/0.05 mm	430
MeNHCO₂, SEt (pyridine)		59.5–61	441
SPh triazole structure		—	499, 514

TABLE XV-7. 2-Pyridylsulfides (*Continued*)

Substituents	R	M.p. (°C)	Ref.
CO$_2$H, Me Me		—	497
CONH$_2$, Me Me		—	497
CSNH$_2$, Me Me		—	497

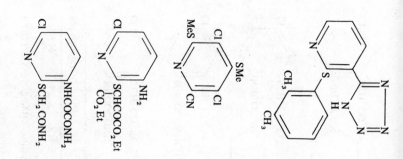

—   497

—   518

57°   501

—   501

—   501

TABLE XV-7. 2-Pyridylsulfides (*Continued*)

Substituents	R	M.p. (°C)	Ref.
Cl-pyridine, NHCOCONH₂, SCH₂CO₂Et	—	—	501
Cl-pyridine, NHCOCONHCH₂Ph, SCH₂CO₂Et	—	—	501
Cl-pyridine, NHCOCON(morpholine), SCH₂CO₂Et	—	—	501
Cl-pyridine, NHCOCON(piperidine), SCH₂CO₂Et	—	—	501

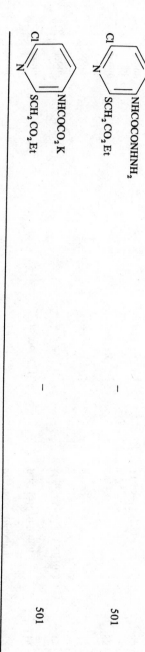

	501
	501

TABLE XV-8. 3-Pyridyl Sulfides

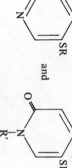

Substituents	R	M.p. (°C)	Ref.
None	Me	—	446
None	Pr	B.p. 124–128°/16 mm (n_D^{26} 1.5548, picrate, m.p. 113–115°)	15
None	Bu	B.p. 141/31 mm	14
None	C_8H_{17}	B.p. 120°/0.1 mm (picrate, m.p. 91°)	15
None	$C_{10}H_{21}$	B.p. 83–85°/0.19 mm	449
None	Ph	B.p. 130°/0.65 mm	449
None	m-MeOC$_6$H$_4$	B.p. 150°/0.2 mm	449
None	p-MeOC$_6$H$_4$	M.p. 45; b.p. 150°/0.2 mm	449
None	o-O$_2$NC$_6$H$_4$	93–95	55
None	o-H$_2$NC$_6$H$_4$	58–59	55
None	p-O$_2$NC$_6$H$_4$	112.5–114	55

None	p-$H_2NC_6H_4$	88–90	55
None	m-$CF_3SO_2NHC_6H_4$	123–125	517
None	2,4-$(O_2N)_2C_6H_3$	129–131	55
None	2,4-$(H_2N)_2C_6H_3$	129.5–130.5	55
None	4,2-$Cl(O_2N)C_6H_3$	95–96	55
None	4,2-$Cl(H_2N)C_6H_3$	115–117	55
None	2,4-$Cl(O_2N)C_6H_3$	111.5–112	55
None	2,4-$Cl(H_2N)C_6H_3$	94–94.5	55
None	4,2-$Me(O_2N)C_6H_3$	83–84	55
None	4,2-$Me(H_2N)C_6H_3$	102.5–103.5	55
None	p-$MeC_6H_4SO_2NHC_6H_4$	127–129	55
None	2,4,5-$Cl_3C_6H_2$CH:CH	118.5–120	338
6-Amino	Me	65.5–66.5	280
2-Carboxy	m-ClC_6H_4	168–169 (decomp.)	166
4-Carboxy	5,2-$Cl(Me)C_6H_3$	252–254	29
5-Chloro	Ph	B.p. 125–128°/0.5 mm	47
5-Chloro-2-hydroxy	Et	158–159	280

TABLE XV-8. 3-Pyridyl Sulfides (*Continued*)

Substituents	R	M.p. (°C)	Ref.
2,5-Dichloro-4,6-dicyano	Me	104–106	518, 576, 583
2,5-Dichloro-4,6-dicyano	Pr	48–50	518, 578, 583
4,5-Dichloro-2,6-dicyano	Me	63–66	518, 578, 583
3,5-Dichloro-2,6-difluoro	HO$_2$CCH$_2$	99–100	152
5-Ethylmercapto-4-hydroxy	Et	147–149	280
6-Hydroxy	2-Pr	53–54	280
6-Hydroxy	Me	75–76	280
6-Hydroxy	Et	79.5–80	280
6-Hydroxy	Ph	180–181.5	280
4,5-Bis(hydroxymethyl-2-methyl)	Et	Hydrochloride	194
2-Methyl	Pr	B.p. 70–71°/0.5 mm (n_D^{25} 1.5452)	17

4-Methyl	Pr	B.p. 139-140°/21 mm (n_D^{25} 1.5510, picrate, m.p. 140-143°)	16
4-Methyl	C_8H_{17}	B.p. 125-127°/0.25 mm (n_D^{34} 1.5188; picrate, m.p. 120-122°)	16
2,4-Dimethyl-6-hydroxy	Et	138.5-139.5	280
4-Nitro-1-oxide	Ph	148	332
5-Phenylthio	Ph	51-52	235
MeS—pyridine—OPS(OEt)₂		(n_D^{25} 1.5022)	385, 462, 465
EtS—pyridine—OPS(OEt)₂		(n_D^{25} 1.5486)	385, 444, 462, 465
EtS—pyridine—OPS(OEt)₂		(n_D^{25} 1.5232)	462, 465

TABLE XV-8. 3-Pyridyl Sulfides (*Continued*)

Substituents	R	M.p. (°C)	Ref.
EtS / SEt, OPS(OEt)$_2$		(n_D^{25} 1.5232)	385, 462, 465
2-PrS / OPS(OEt)$_2$		(n_D^{25} 1.5258)	385, 462, 465
PhS / OPS(OEt)$_2$		(n_D^{25} 1.5258)	385, 462, 465
EtS / Cl, OPS(OEt)$_2$		(n_D^{25} 1.5144)	385, 462, 465

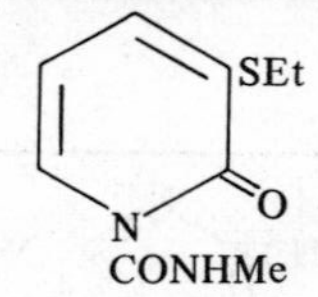

O
N—CONHMe
SEt
75–76   441

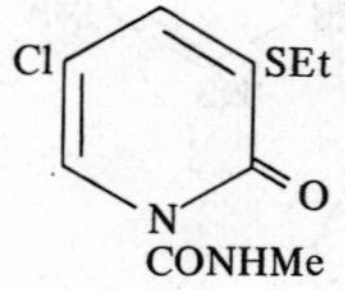

Cl
N—CONHMe
O
SEt
118–120   441

O
N—CONHMe
SMe
90.5–92   441

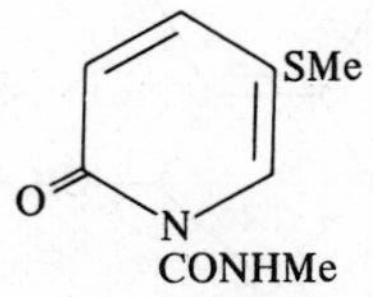

O
N—CONHMe
SPr-iso
66.5–68   441

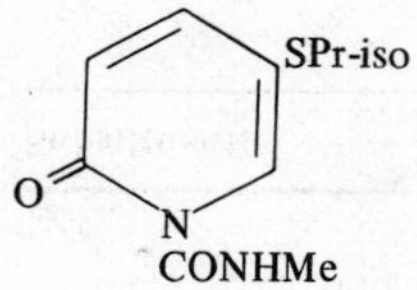

O
Me
N—CONHMe
Me
SMe
124–125   441

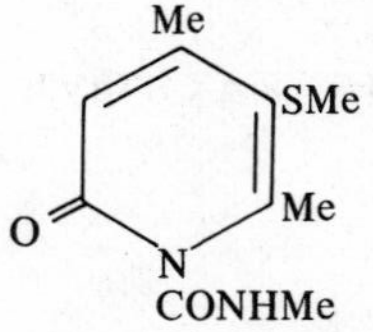

TABLE XV-8. 3-Pyridyl Sulfides (*Continued*)

Substituents	R	M.p. (°C)	Ref.
	—	526
			518, 578, 583

TABLE XV-9. 4-Pyridyl Sulfides

Substituents	R	M.p. (°C)	Ref.
None	Me	46 (picrate, m.p. 170-171°)	61, 267, 446
None	Et	B.p. 123-126/23 mm (n^{18}_D 1.5713; picrate, m.p. 144°)	61, 140, 345
None	HO_2CCH_2	268-269	159, 358, 476, 485
None	$NCCH_2$	178-179 (hydrochloride, m.p. 187-188°)	107
None	$NCCH_2CH_2$	77-78 [picrate, m.p. 152-154° (decomp.)]	106, 187
None	$MeO_2CCH_2CH_2$	B.p. 145-146°/3 mm (n^{20}_D 1.5623; d_{20} 1.2181)	106
None	Pr	B.p. 130-133°/16 mm (n^{25}_D 1.5632, p-toluenesulfonate, m.p. 110-112°; picrate, m.p. 134-136°)	15
None	Bu	B.p. 145-147°/19 mm	14, 345

TABLE XV-9. 4-Pyridyl Sulfides (*Continued*)

Substituents	R	M.p. (°C)	Ref.
None	*t*-Bu	B.p. 64–65°/0.2 mm	412
None	$H_2NCH_2CH_2$	Dihydrochloride, m.p. 236°	105a
None	$MeCONHCH_2CH_2$	95	105a
None	$EtO_2CCH_2CH_2$	164–165	107
None	$CH_2:CHCH_2$	B.p. 86–87°/3 mm (n_D^{18} 1.5872; hydrobromide, m.p. 188–189°; picrate, m.p. 147°)	61, 67, 159
None	$HO_2CCHMeCH_2$	—	187
None	$CH_3O_2CMeCHCH_2$	B.p. 133–134°/2 mm [n_D^{20} 1.5527; d_{20} 1.1800; picrate, m.p. 137–139° (decomp.)]	106
None	C_6H_{13}	—	345
None	C_8H_{17}	B.p. 127–129°/2.0 mm (picrate, m.p. 110–113°)	15
None	$C_{12}H_{25}$	30–31	159

None	C$_{26}$H$_{53}$	52	159
None	Ph	B.p. 160/60 mm (picrate, m.p. 166.5°)	140, 164, 305, 345
None	o-ClC$_6$H$_4$	122-123	260
None	m-ClC$_6$H$_4$	120-121	260
None	p-ClC$_6$H$_4$	105-106 (hydrochloride, m.p. 171-174°)	260, 265
None	o-O$_2$NC$_6$H$_4$	120-122	53
None	o-H$_2$NC$_6$H$_4$	110-111	53
None	p-H$_2$NC$_6$H$_4$	—	53, 239
None	p-MeSO$_2$NHC$_6$H$_4$	199-200	53
None	p-Et$_2$NCH$_2$CH$_2$NHC$_6$H$_4$	B.p. 175-185°/0.01 mm (dihydrochloride, m.p. 193-195°)	53
None	2,3-Cl$_2$C$_6$H$_3$	125-126	260
None	2,4-Cl$_2$C$_6$H$_3$	98-99	260
None	2,5-Cl$_2$C$_6$H$_3$	104-105	260
None	3,5-Cl$_2$C$_6$H$_3$	134-135	260
None	3,4-Cl$_2$C$_6$H$_3$	155-156	260

TABLE XV-9. 4-Pyridyl Sulfides (*Continued*)

Substituents	R	M.p. (°C)	Ref.
None	2,4-O$_2$N(MeO)C$_6$H$_3$	61–63 (B.p. 170–171°/0.1 mm)	53
None	2,4-H$_2$N(MeO)C$_6$H$_3$	96–97	53
None	2,4-ClO$_2$NC$_6$H$_3$	133–135	53
None	2,4-Cl(H$_2$N)C$_6$H$_3$	154–154.5	53
None	2,5,4-Cl$_2$(MeO)C$_6$H$_2$	212–213	260
None	3,4,6-HO(Me$_2$)C$_6$H$_2$	160–161 (hydrochloride, m.p. 215–217°)	205
None	*p*-MeOC$_6$H$_4$	135–136	260
None	2,4-Br(Me)C$_6$H$_3$	119–120	260
None	2,4-Me(Br)C$_6$H$_3$	67–68	260
None	3,4-Cl(Br)C$_6$H$_3$	159–160	260
None	2,5-Br(Cl)C$_6$H$_3$	110–111	260
None	3,4-Cl(BuO)C$_6$H$_3$	100–101	260

None	p-HOC$_6$H$_4$	219–220	260
None	2,4-Cl(HO)C$_6$H$_3$	252–253	260
None	3,4-Cl(HO)C$_6$H$_3$	242–243	260
None	3,5,4-Cl$_2$(HO)C$_6$H$_2$	235–236	260
None	2,5,4-Cl$_2$(HO)C$_6$H$_2$	255–256	260
None	Me$_2$N(piperidine)CH$_2$CH(OH)CH$_2$, Me$_2$	(dihydrate, m.p. 249–251°)	453
None	Cl,Cl-pyrimidinyl (Me)	76–78	455
None	MeO$_2$C ... CO$_2$H, S, N, O, NHCOCH$_2^-$ (cephem)	143–146 (decomp.)	424
None	PhCH$_2$	70 (picrate, m.p. 170°)	61, 67

TABLE XV-9. 4-Pyridyl Sulfides (*Continued*)

Substituents	R	M.p. (°C)	Ref.
None	MeO$_2$C	Hydrochloride, m.p. 194–196°	61
None	EtO$_2$C	Hydrochloride, m.p. 210–212°	61
None	ClCH$_2$CH$_2$O$_2$C	Hydrochloride, m.p. 233–235°	61
None	CH$_2$:CHCH$_2$O$_2$C	Hydrochloride, m.p. 222–224°	61
None	PhCH$_2$O$_2$C	Hydrochloride, m.p. 228–230°	61
None	PhCO	76	61
3-Acetamido	2,4-(O$_2$N)$_2$C$_6$H$_3$	170–171	214
3-Amino	o-BrC$_6$H$_4$	–	315
3-Amino	2,4-(O$_2$N)$_2$C$_6$H$_3$	182–183	214
3-Amino-1-oxide	o-BrC$_6$H$_4$	200–203	120, 410
3-Amino-1-oxide	2,5-Br(Cl)C$_6$H$_3$	229.0–229.5	410
3-Amino-1-oxide	2,5-Br(MeO)C$_6$H$_3$	179.0–180.5	410

2-Amino-3-nitro	Me	220	236
2-Amino-3-cyano-6-benzyl	Me	—	461
2-Benzyl-6-bromo-5-cyano	Me	—	461
2-Benzyl-6-chloro-5-cyano	Me	—	461
2-Benzyl-5-cyano-6-hydroxy	Me	—	461
2-Benzyl-5-cyano-6-iodo	Me	—	461
2-Bromo-3,5-dichloro	Me	—	388
3-Bromo-5-nitro	$o\text{-}H_2NC_6H_4$	—	466
2-t-Butyl	Me	B.p. 105–107/10 mm (picrate, m.p. 113–114°)	267
2-Carbamoyl	Me	—	—
3-Carboxamide	HO_2CCH_2	240–270 (decomp.)	358
3-Carboxamido-1-oxide	Ph	—	120
3-Carboxy	Ph	239–240 (decomp.)	291
3-Carboxy	$HOCH_2CH_2$	190 (decomp.)	291
3-Carboxy	$BrCH_2CH_2$	Hydrobromide	291
3-Carboxy	HO_2CCH_2	225–230 (decomp.)	291

TABLE XV-9. 4-Pyridyl Sulfides (*Continued*)

Substituents	R	M.p. (°C)	Ref.
2-Carboxy-1-oxide	PhCH$_2$	157	275
2-Carboxy-1-oxide	p-O$_2$NC$_6$H$_4$CH$_2$	167	275
2-Carboxy-1-oxide	2,4-(O$_2$N)$_2$C$_6$H$_3$CH$_2$	195–197	275
2-Carboxy-1-oxide	p-O$_2$NC$_6$H$_4$	174–175	275
3-Carboxy-1-oxide	o-BrC$_6$H$_4$	144.5–146	445
3-Carboxy-1-oxide	2,4-Br(Cl)C$_6$H$_3$	62–63	445
3-Carboxy-1-oxide	2,5-Br(MeO)C$_6$H$_3$	158	445
2-Carboxy-3,5-dichloro	C$_8$H$_{17}$	—	335
2-Carboxy-3,5-dichloro	p-ClC$_6$H$_4$	—	335
2-Carboxy-3,5-dichloro	2,4,6-Me$_3$C$_6$H$_2$	—	335
2-Carboxy-3,5-dichloro	p-PhC$_6$H$_4$	—	335
2-Carboxy-3,5,6-trichloro	p-Me$_3$CC$_6$H$_4$	—	335

2-Carboxy-3,5-dichloro	C_6Cl_5	—	335
2-Carboxy-3,5,6-trichloro	Ph	—	335
2-Carboxy-3,5,6-trichloro	Me	119	518, 578
2-Carboxy-3,5,6-trichloro	p-$O_2NC_6H_4$	—	335
2-Carboxy-3,5,6-trichloro	2,4-$(O_2N)_2C_6H_3$	—	335
2-Carbobutoxy-3,5-dichloro	Bu	—	335
2-Carbethoxy-3,5,6-trichloro	Ph	—	335
2-Carbethoxy-3,5,6-trichloro	2,4-$(O_2N)_2C_6H_3$	—	335
2-Carbomethoxy-3,5-dichloro	p-ClC_6H_4	—	335
2-Carbomethoxy-3,5-dichloro	C_8H_{17}	—	335
2-Carboisopropoxy-3,5,6-trichloro	2,4,6-$Me_3C_6H_2$	—	335
2-Chloro-3-nitro	p-PhC_6H_4	—	335
3-(5-Chloro-2-pyridyl)amino	Me	100	236
3-Chloro-3,5-dibromo-6-methyl	Me	214–215 (decomp.)	214
2-Chloro-3,5-dibromo-6-methyl	$NCCH_2CN$	105–108	544
2-Chloro-3,5-dibromo-6-methyl	n-$C_{12}H_{25}$	B.p. 200–205°/0.05 mm	544
2-Chloro-3,5-dibromo-6-methyl	$Me_2N(CH_2)_2$	53–54	544

TABLE XV-9. 4-Pyridyl Sulfides (*Continued*)

Substituents	R	M.p. (°C)	Ref.
2-Chloro-3,5-dibromo-6-methyl	EtO_2CCH_2	68–70	544
2-Chloro-6-phenethyl	Me	—	461
2-Cyano-3,5,6-trichloro	Me	58	518, 578
2-Cyano-3,5,6-trichloro	o-ClC_6H_4	91	518, 578, 583
2-Cyano-3,5,6-trifluoro	Me	B.p. 138–140°/38 mm	518, 578
3-Cyano-5-ethyl-2-hydroxy	Me	—	429, 461
2,6-Dibromo-3,5-dichloro	Me	—	388
2,6-Dicarbohydrazide	Et	211–213	380
2,6-Dicarbohydrazide	Bu	191–193	380
2,6-Dicarbohydrazide	$PhCH_2$	224–228	380
2,6-Dicarbohydrazide	Ph	210–212	380
2,6-Dicarbohydrazide	p-ClC_6H_4	>300	380

2,6-Dicarboxy	Me	102–104	221
2,6-Dicarboxy	Bu	156–158 (decomp.)	221
2,6-Dicarboxy	Ph	(Hydrate) 199–200	221
2,6-Dicarbethoxy	Ph	78–81	221
2,6-Dicarbethoxy	p-ClC$_6$H$_4$	126	221
2,6-Dicarbethoxy	p-BrC$_6$H$_4$	118–120	221
2,6-Dicarbethoxy	Et	47–49	221
2,6-Dicarbethoxy	2-Pr	Oil	221
2,6-Dicarbethoxy	Bu	B.p. 189–199°/1.3 mm	221
2,6-Dicarbethoxy	s-Bu	Oil	221
2,6-Dicarbethoxy	isoC$_5$H$_{11}$	B.p. 205–206°/1.3 mm	221
2,6-Dicarbethoxy	PhCH$_2$	95°	221
2,6-Dicarbomethoxy	Ph	138–140	221
2,6-Dicarbomethoxy	p-Me$_3$CC$_6$H$_4$	117–118	221
2,6-Dicarbomethoxy	o-HO$_2$CC$_6$H$_4$	194–196	221
2,6-Dicarbomethoxy	o-MeO$_2$CC$_6$H$_4$	123–125	221
2,6-Dicarboxamide	Ph	(Hydrate) 275–277	221

TABLE XV-9. 4-Pyridyl Sulfides (*Continued*)

SR

Substituents	R	M.p. (°C)	Ref.
3,5-Dichloro-2,6-dicyano	Me	63–66	518, 578, 583
3,5-Dichloro-2-fluoro	Me	—	388
3,5-Dichloro-2,6-difluoro	Me	—	388
3,5-Dichloro-2-fluoro	Me	—	388
3,5-Dichloro-2-trichloromethyl	Pr	—	388
5,6-Dichloro-3-trichloromethyl	C_6H_{13}	—	163
3,5-Dichloro-2-trichloromethyl	Me	79–81	163, 378
3,5-Dichloro-2-trichloromethyl	C_6Cl_5	—	163
3,5-Dichloro-2-trichloromethyl	Et	—	163
3-5-Dichloro-2-trichloromethyl	C_8H_{17}	—	163
3,5-Dichloro-2-trichloromethyl	p-ClC$_6$H$_4$	—	163
3,5-Dichloro-2-trichloromethyl	p-O$_2$NC$_6$H$_4$	—	163

		M.p./B.p.	References
3,5-Dichloro-2-trichloromethyl	2,4,6-Cl$_2$(Me)C$_6$H$_2$	—	163
3,5-Dichloro-2-trichloromethyl	C$_6$Cl$_5$	—	163
3,5-Dichloro-2-(trifluoromethyl)	Me	—	512, 539, 540, 571, 588
3,5-Dichloro-2-(trifluoromethyl)	p-ClC$_6$H$_4$	—	512, 539, 540, 571, 588
3,5-Dichloro-2-fluoro	Me	—	378
2,6-Dihydroxymethyl	Et	102	156, 309
2,6-Dihydroxymethyl	Ph	142	156, 309
3-(2-Dimethylaminopropyl)amino	o-BrC$_6$H$_4$	B.p. 198–200/0.3 mm	315
3-(3-Dimethylaminopropyl)amino	2,4-Br(Cl)C$_6$H$_3$	B.p. 210–215/0.3 mm	315
3-(2-Dimethylaminopropyl)amino	2,4-Br(Cl)C$_6$H$_3$	B.p. 208–211/0.3 mm	315
3-(3-Dimethylaminopropyl)amino	o-BrC$_6$H$_4$	—	315
2,6-Di-(CH$_2$O$_2$CNHMe)	Me	154	153, 156, 309
2,6-Di-(CH$_2$O$_2$CNHMe)	Et	118	153, 309
2,6-Di-(CH$_2$O$_2$CNHMe)	Ph	127	153, 156, 309
3,5-Dinitro	o-H$_2$NC$_6$H$_4$	—	466
3-(2,4-Dinitrophenyl)amino	Me	209–210	214

TABLE XV-9. 4-Pyridyl Sulfides (Continued)

Substituents	R	M.P. (°C)	Ref.
2-Methyl	Et	B.p. 120°/15 mm (picrate, m.p. 137.5°)	273
3-Methyl	Pr	B.p. 90-91°/0.4 mm	17, 345
2-Methyl	Pr	B.p. 83-84°/0.6 mm	17, 273
2-Methyl	Pr	B.p. 131°/15 mm (picrate, m.p. 124°)	17, 273
2-Methyl	Bu	B.p. 141°/13 mm (picrate, m.p. 124°)	273
2-Methyl	HO_2CCH_2	236-238	273
2-Methyl	$2,4-(O_2N)_2C_6H_3$	163	273
3-Methyl	$C_{26}H_{53}$	40-41	159
3-Methyl-5-nitro	$o-H_2NC_6H_4$	—	466
2-Methyl-1-oxide	Pr	45 (picrate, m.p. 126°)	273
6-Methyl-2,3,5-trichloro	Ph	—	510

312

3-Nitro	o-H$_2$NC$_6$H$_4$	150–151	58, 59, 466
3-Nitro	o-MeCONHC$_6$H$_4$	123–124	58, 59
1-Oxide	Ph	137.5–138	87, 140
1-Oxide	2,5-Br(Me)C$_6$H$_3$	120	87
1-Oxide	3,4-Br(Me)C$_6$H$_3$	168	87
1-Oxide	3,4-Cl(Me)C$_6$H$_3$	151	87
1-Oxide	Et	B.p. 80°/15 mm (picrate, m.p. 140°)	140
1-Oxide	PhCH$_2$	61	207
2-PhCH$_2$NHCO	Me	77–78.5 [B.p. 185–190°/0.3 mm; Cu derivative, m.p. 153–154° (decomp.)]	180
2,3,5,6-Tetrachloro	Me	58.5–59.0	91, 388, 434, 447
2,3,5,6-Tetrachloro	ClCH$_2$	—	529
2,3,5,6-Tetrachloro	Et	—	162, 388
2,3,5,6-Tetrachloro	Bu	—	388
2,3,5,6-Tetrachloro	C$_5$H$_{11}$	—	388

TABLE XV-9. 4-Pyridyl Sulfides (*Continued*)

Substituents	R	M.p. (°C)	Ref.
2,3,5,6-Tetrachloro	C_6H_{13}	35.5	378, 390
2,3,5,6-Tetrachloro	$BrCH_2CHCl$	68.5	378
2,3,5,6-Tetrachloro	$BrCH_2CH_2$	99	378, 388
2,3,5,6-Tetrachloro	Ph	91.5	378
2,3,5,6-Tetrachloro	p-ClC$_6$H$_4$	—	388
2,3,5,6-Tetrachloro	p-ClC$_6$H$_4$CH$_2$	—	388
2,3,5,6-Tetrachloro	s-Bu	40.5	378, 388
2,3,5,6-Tetrachloro	ClCH$_2$	89–91	378, 388
2,3,5,6-Tetrachloro	ClCH$_2$CH$_2$	99.5	378, 388
2,3,5,6-Tetrachloro	ClCH$_2$CH$_2$CH$_2$	61	378
2,3,5,6-Tetrachloro	isoPr	42	378
2,3,5,6-Tetrachloro	Pr	41.0	378

314

Compound	Substituent	m.p.	Ref.
2,3,5,6-Tetrachloro	HO_2CCH_2	—	529
2,3,5,6-Tetrachloro	$ClCOCH_2$	—	529
2,3,5,6-Tetrachloro	$ClCOCHCl$	—	529
2,3,5,6-Tetrachloro	$NCCH_2$	134.5	566
2,3,5,6-Tetrachloro	$NCCH_2CH_2$	110.5	566
2,3,5-Trichloro	$NCCH_2$	75	566
2,3,5,6-Tetrafluoro	Ph	B.p. 129°/9 mm	455
2,3,5,6-Tetrafluoro	O_2N (F F / F F ring structure)	75.5–76.5	394
2,3,5,6-Tetrafluoro	H_2N (F F / F F ring structure)	103–104	394
2,3,5,6-Tetrafluoro	F F / F NH_2 ring structure	99.5–102	394
3,5,6-Tribromo-2-cyano	Me	112	518, 578, 583

TABLE XV-9. 4-Pyridyl Sulfides (*Continued*)

Substituents	R	M.p. (°C)	Ref.
2,3,5-Trichloro	Me	41.2	162, 378
2,3,5-Trichloro	Et	—	91, 388
2,3,5-Trichloro	isoPr	—	388
2,3,5-Trichloro	Bu	—	388
2,3,5-Trichloro	$BrCH_2CH_2$	43.5	378, 388
2,3,5-Trichloro	$ClCH_2CH_2$	37	378, 388
2,3,6-Trichloro	Me	113; 129–130	378, 388, 434
2,3,6-Trichloro	Bu	—	388
2,3,5-Trichloro	$Br(CH_2)_5$	—	388
2,3,5-Trichloro	$C_{12}H_{25}$	—	91, 388
2,3,5-Trichloro	$m\text{-}BrC_6H_4$	—	388
2,3,5-Trichloro	$p\text{-}ClC_6H_4$	118	378, 388

2,3,5-Trichloro	PhCH₂	—	388
2,3,5-Trichloro	p-ClC$_6$H$_4$CH$_2$	—	388
2,3,5-Trichloro-6-trichloromethyl	Me	35.5	378
3,5,6-Trichloro-2-trichloromethyl	Me	—	163, 388
3,5,6-Trichloro-2-trichloromethyl	C$_{18}$H$_{37}$	—	163
3,5,6-Trichloro-2-trichloromethyl	Ph	—	163
3,5,6-Trichloro-2-trichloromethyl	p-MeC$_6$H$_4$	—	163
3,5,6-Trichloro-2-trichloromethyl	p-PhC$_6$H$_4$	—	163

MeNH — N(H) — NH — CN — SMe ring structure   223–224   481

Pyridine (Cl, N, NHCOMe, SCH$_2$COPh)   —   482

Pyridine (Cl, N, NHCOMe, SCH$_2$COPh)   —   482

TABLE XV-9. 4-Pyridyl Sulfides (*Continued*)

Cl, NHCOMe, SCH₂COCH₂CO₂Et	—	482
Cl, NHCOMe, SCH₂COCH₂CO₂Et	—	482

TABLE XV-10. Dipyridyl Sulfides PySPy'

Py	Py'	M.p. (°C)	Ref.
3-Acetamido-4-Py	5-Chloro-2-Py	144–145	214
3-Acetamido-6-ethoxy-2-Py	5-Nitro-2-Py	215–216	214
3-Acetamido-6-methoxy-2-Py	5-Nitro-2-Py	199–200	214
3-Amino-2-Py	3-Amino-2-Py	213–214 (decomp.)	289
3-Amino-4-Py	5-Chloro-2-Py	139–140	214
3-Amino-6-ethoxy-2-Py	5-Nitro-2-Py	167–168	214
3-Amino-6-methoxy-2-Py	5-Nitro-2-Py	172–173	214
2,6-Di-t-butyl-4-Py	2,6-Di-t-butyl-4-Py	125–126	267
6-Hydroxy-3-Py	6-Hydroxy-3-Py	260–263 (di-HgCl₂ complex, m.p. 260–263°)	307
6-Methyl-2-Py-1-oxide	6-Methyl-2-Py-1-oxide	121–122	172
3-Nitro-2-Py	2-Py	—	132
5-Nitro-2-Py	2-Py	—	132
4-Nitro-3-Py-1-oxide	5-Nitro-2-Py	175–177 (decomp.)	49, 132, 262
2-Py-1-oxide	4-Nitro-1-oxido-3-Py	222	332
	1-Oxido-2-Py	174–175 (decomp.)	172
4-Py-1-oxide	1-Oxido-2-Py	228–230 (picrate, m.p. 181–183°)	329
2-Py	2-Py	—	172
4-Py	4-Py	72	61, 67, 203
2,3,5,6-Tetrafluoro-4-Py	2,3,5,6-Tetrafluoro-4-Py	52–53; b.p. 117.5–118° (9 mm)	394, 455
4-Methyl-3-nitro-2-Py	4-Methyl-3-nitro-2-Py	135	443
6-Methyl-3-nitro-2-Py	6-Methyl-3-nitro-2-Py	145	443

TABLE XV-10. Dipyridyl Sulfides PySPy' (Continued)

Py	Py'	M.p. (°C)	Ref.
3-Methyl-5-nitro-2-Py	3-Methyl-5-nitro-2-Py	116	443
4-Methyl-5-nitro-2-Py	4-Methyl-5-nitro-2-Py	129	443
6-Methyl-5-nitro-2-Py	6-Methyl-5-nitro-2-Py	97	443
2,3,5,6-Tetrachloro-4-Py	2,3,5,6-Tetrachloro-4-Py	140–141	460
(pyridine)—S—(pyridine)—CONH—(phenyl: OH, Br)		215–216	437
(pyridine)—S—(pyridine)—CONH—(phenyl: OH, Br)		—	437
EtO$_2$C—(pyridine)—S(CH$_2$)$_2$S—(pyridine: CO$_2$Et, CO$_2$Et)		126–27	221

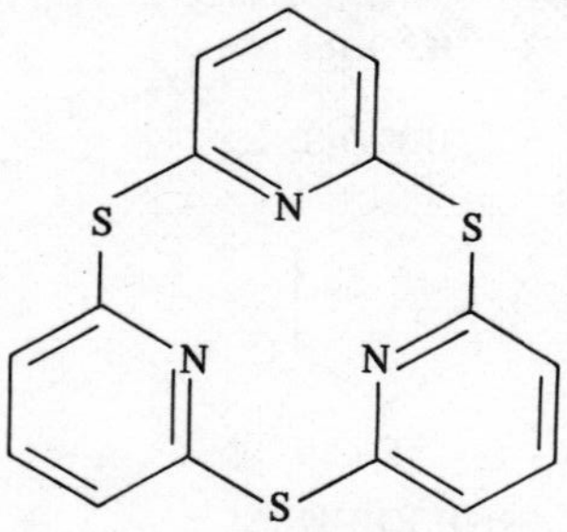 — 549

TABLE XV-11. Quaternary 2-Pyridyl Sulfides

R	R¹	R²	X	M.p. (°C)	Ref.
H	H	H	I	212	323
H	2-Py	Me	Br	211 (decomp.)	323
(2-oxocyclohexyl structure)	Me	H	$2,4,6\text{-}(O_2N)_3C_6H_2O$	158–159	6
OHCCHEt	H	H	$2,4,5\text{-}(O_2N)_3C_6H_2O$	160–161	6
MeCOCHMe	H	H	Cl	161–163	6
(2-methylcyclohexyl structure)	H	H	Cl	192–194	6
EtCOCH₂	H	H	—	—	6
EtCOCH₂	Me	H	Cl	195–196	6

OHCCH$_2$	H	H	—	270 (decomp.)	6
EtCOCH$_2$	Me	H	Cl	195–196	6
OHCCH$_2$	H	H	—	270 (decomp.)	6
MeCOCH$_2$	H	H	2,4,6-(O$_2$N)$_3$C$_6$H$_2$O	146–147	6
MeCO(CO$_2$Et)CH	H	H	Cl	100–102	6
MeCOCHPh	H	H	Br	169–170	6
MeCOCH$_2$	Me	H	Cl	175	6
MeCOCH$_2$	Me	H	2,4,6-(O$_2$N)$_3$C$_6$H$_2$O	147–148	6
PhCOCH$_2$	H	H	Br	185	6
PhCOCHPh	H	H	OH	78	6
PhCOCH$_2$	Me	H	Br	188	6

TABLE XV-12. Quaternary 4-Pyridyl Sulfides

R'	R	R²	X	M.p. (°C)	Ref.
BuO₂CCH₂	C₁₂H₂₅	H	Cl	84	31
C₈H₁₇O₂CCH₂	C₈H₁₇	H	Cl	169	31
C₈H₁₇O₂CCH₂	C₁₂H₂₅	H	Cl	164–166	31
C₁₂H₂₅O₂CCH₂	Et	H	Cl	169–170	31
C₁₂H₂₅	Me	H	Br	105	31
C₁₆H₃₃	p-ClC₆H₄	H	Br	104–105	31
CH₂:CHCH₂	C₁₆H₃₃	H	Br	72–75	31
C₅H₁₁O₂CCH₂	Et	H	Cl	180	31
C₅H₁₁O₂CCH₂	C₁₂H₂₅	H	Cl	184	31
C₅H₁₁O₂CCH₂	C₁₆H₃₃	H	Cl	175–178	31
C₅H₁₁O₂CCH₂	p-ClC₆H₄CH₂	H	Cl	193	31
C₈H₁₇O₂CCH₂	Et	H	Cl	168	31

$C_8H_{17}O_2CCH_2$	$C_{16}H_{33}$	H	Cl	146–148	31
$C_8H_{17}O_2CCH_2$	$p\text{-}ClC_6H_4CH_2$	H	Cl	197	31
$p\text{-}ClC_6H_4CH_2$	$C_{16}H_{33}$	H	Cl	176–178	31
$3,4\text{-}Cl_2C_6H_3CH_2$	$C_{14}H_{29}$	H	Cl	100–101	31
Et_2NCOCH_2	$C_{12}H_{25}$	H	Cl	82	31
EtO_2CCH_2	$C_{16}H_{33}$	H	Cl	150–153	31
EtO_2CCH_2	$p\text{-}ClC_6H_4CH_2$	H	Cl	166	31
EtO_2CCH_2	$C_{12}H_{25}$	H	Cl	170	31
H_2NCOCH_2	$C_{12}H_{25}$	H	Cl	217 (decomp.)	31
$HOCH_2CH_2$	$p\text{-}ClC_6H_4$	H	Br	150–151	31
HO_2CCH_2	$C_{16}H_{33}$	H	Cl	93	31
Me	$C_{12}H_{25}$	H	$p\text{-}MeC_6H_4SO_3$	138–139	31
Me	$C_{12}H_{25}$	H	$MeSO_4$	83	31
Me	$C_{12}H_{25}$	H	$p\text{-}MeC_6H_4SO_3$	125–126	31
Me	$C_{16}H_{33}$	3-Me	$p\text{-}MeC_6H_4SO_3$	52	31
Me	$o\text{-}HOC_6H_4$	H	$p\text{-}MeC_6H_4SO_3$	213	31
Me	$p\text{-}ClC_6H_4$	H	$p\text{-}MeC_6H_4SO_3$	153–155	31

TABLE XV-12. Quaternary 4-Pyridyl Sulfides (Continued)

R'	R	R²	X	M.p. (°C)	Ref.
Me	o-HO$_2$CC$_6$H$_4$	H	p-MeC$_6$H$_4$SO$_3$	173–175	31
Me	o-O$_2$NC$_6$H$_4$	H	p-MeC$_6$H$_4$SO$_3$	114	31
Me	p-O$_2$NC$_6$H$_4$	H	p-MeC$_6$H$_4$SO$_3$	170	31
Me	p-ClC$_6$H$_4$CH$_2$	H	p-MeC$_6$H$_4$SO$_3$	188	31
p-MeC$_6$H$_4$CH$_2$	C$_{12}$H$_{25}$	H	Cl	30–40	31
p-O$_2$NC$_6$H$_4$CH$_2$	C$_{12}$H$_{25}$	H	Cl	120–122	31
PhCH$_2$	C$_{16}$H$_{33}$	H	Cl	113	31
PhO$_2$CCH$_2$	C$_8$H$_{17}$	H	Cl	Wax	31
2-PrO$_2$CCH$_2$	C$_{12}$H$_{25}$	H	Cl	112	31
2-PrO$_2$CCH$_2$CH$_2$	C$_{12}$H$_{25}$	H	Cl	114	31
2,4,5-Cl$_3$C$_6$H$_2$CH$_2$	o-HOC$_6$H$_4$	H	Cl	185–188	31
2,4,5-Cl$_3$C$_6$H$_2$CH$_2$	p-ClC$_6$H$_4$	H	Cl	183–184	31

TABLE XV-13.   Symmetrical Dipyridyl Disulfides PySSPy

Py	M.p. (°C)	Ref.
2-Py	54–55 (molybdenum complex)	117, 203, 319
		513
	COCl$_2$ complex	
	COBr$_2$ complex	
3-Py	Dihydrochloride	203
4-Py	156	117, 203
5-Amino-2-Py	175–176	451
4-Carboxy-2-Py	241–242	79, 80
2-Carboxy-4-Py-1-oxide	178	275
4-Chloro-3-hydroxymethyl-6-Me-5-Py	Dihydrochloride	194
5-Cyano-2-Py	178–180	451
2,6-Dicarbomethoxy-4-Py	259–261	221
2,6-Dicarboethoxy-4-Py	152–153	221
3,5-Dinitro-2-Py	270	262
4,6-Diphenyl-2-Py	205	201
3,4-Bis(hydroxymethyl)-6-Me-5-Py	–	185, 194, 206a, b,
	Dihydrochloride	220
	Diphosphate	
3,4-Bis(hydroxymethyl)-6-Me-5-Py(trisulfide)	Dihydrochloride	194
3-Hydroxymethyl-4,6-Me$_2$-5-Py	Dihydrochloride	194
3-Hydroxymethyl-4-aminomethyl-6-Me-5-Py	Tetrahydrochloride	194, 206b
3-Methyl-5-nitro	184	443
4-Methyl-3-nitro	185	443
4-Methyl-5-nitro	190	443
5-Methyl-3-nitro	250	443
6-Methyl-3-nitro	185	443
6-Methyl-5-nitro	112	443
3-Nitro-2-Py	–	132
5-Nitro-2-Py	175–177 (decomp.)	49, 132, 484
5-Phenyl-3-Py	51–52	48
2-Py-1-oxide	200–201 [SnF$_2$ complex, m.p. 275–276° (decomp.)]	21, 204, 208, 303, 304, 343, 489, 575
4-Py-1-oxide	228–230 (picrate, m.p. 181–183°)	373
2,3,5,6-Tetrafluoro-4-Py	64	455

	186–87	433

	–	526

327

TABLE XV-14. Unsymmetrical Pyridyl Disulfides RSSR'

R	R'	M.p. ($^{\circ}$C)	Ref.
2-Py	3-Nitro-2-Py	138–139	288
2-Py	$H_2NCH_2CH_2$	Hydrochloride, m.p. 144–145°	110
2-Py	Cl_3C	99 (hydrochloride, m.p. 85–119°)	287
2-Py	isoPr	B.p. 67–69°/0.01 mm	438
2-Py	Bu	B.p. 91–92°/0.01 mm	438
2-Py	t-Bu	B.p. 81–82°/0.01 mm	438
2-Py	Cyclohexyl	B.p. 111–113°/0.01 mm	438
2-Py	CPh_3	121–123	438
2-Py	Ph	47–49	438
2-Py	p-MeC_6H_4	38–40	438
2-Py	$2,4$-$(O_2N)_2C_6H_4$	118–121	438
2-Py	$2,4,6$-$Me_3C_6H_2$	(Picrate, m.p. 103–104°)	438
4-Py	Ph	41–43	438
3-Amino-2-Py	3-Me-5-nitro-2-Py	161–161.5	288
3-Amino-2-Py	2-Py	105–108 (picrate, m.p. 159–160.5°)	288
3-Amino-2-Py	5-Nitro-2-Py	176–177 (N-acetyl, m.p. 177–178°)	288
3-Amino-2-Py	3-Nitro-2-Py	167–168 (N-acetyl, m.p. 171–172°)	288
3-Amino-2-Py	3-Me-5-Nitro-2-Py	159–160 (N-acetyl, m.p. 151–152°)	288
3-Amino-2-Py	5-Me-3-Nitro-2-Py	145–146 (N-acetyl, m.p. 197–198°)	288
3-Ethyl-6-Me-2-Py	$2,4,6$-$Me_3C_6H_2$	46–48	439
3-Ethyl-6-Me-2-Py	$2,4,6$-$(i$-$Pr)_3C_6H_2$	(Picrate, m.p. 133–135°)	439
3-Methyl-2-Py	$2,4,6$-$Me_3C_6H_2$	90–92.5	438
5-Nitro-2-Py	$2,4,6$-$Me_3C_6H_2$	86–88	438, 439
2-Py-1-oxide	t-Bu	72–75	438
2-Py-1-oxide	Cl_3C	(Dihydrochloride, m.p. 99°)	248
2-Py-1-oxide	$PhCH_2$	(Picrate, m.p. 129–131°)	438
2-Py-1-oxide	$2,4$-$(O_2N)_2C_6H_4$	172–175	438
2-Py-1-oxide	$2,4,6$-$Me_3C_6H_2$	148–152	438
4-Me-2-Py-1-oxide	2-Py-1-oxide	195–196	21
4-Me-2-Py-1-oxide	6-Me-2-Py-1-oxide	228–229	21

246–247	441

328

TABLE XV-15. 2-Pyridylsulfoxides

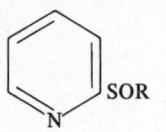

Substituents	R	M.p. (°C)	Ref.
None	2-(N-Morpholino)ethyl	186	84
None	Et$_2$NCH$_2$	–	84
None	2-(N-piperdyl)ethyl	180	84
None	1-(N-Morpholino)-2-propyl	Fumarate, m.p. 130°	84
6-Bromo	2-(N-morpholino)ethyl	200–202	84
6-Chloro	2-(N-morpholino)ethyl	193	84
6-Chloro	Et$_2$NCH$_2$CH$_2$	115	84
6-Chloro	2-(N-piperidyl)ethyl	191–192	84
5-Chloro	2-(N-morpolino)ethyl	216–218	84
5-Chloro	Et$_2$NCH$_2$CH$_2$	Fumarate, m.p. 124–126°	84
3-Chloro	Et$_2$NCH$_2$CH$_2$	204–205	84
3-Cyano	2-(N-morpholino)ethyl	210	84
MeNHCO$_2$ pyridyl SOEt		85.5–88.0	441

329

TABLE XV-16. 3-Pyridylsulfoxides

Substituents	R	M.p. (°C)	Ref.
5-Chloro-2-hydroxy	Et	216–216.5	280
6-Hydroxy	2-Pr	105–107	280
6-Hydroxy	Me	153.5–154.5	280
6-Hydroxy	Et	107.5–110	280

$(n_D^{25}$ 1.5190)   385, 462, 465

47–48.5   385, 462, 465

(Picrate, m.p. 147–148°)   435

Structure		
[pyridyl—SOCH$_2$CH$_2$—]$_2$SO	[Picrate, m.p. 248–250° (decomp.)]	435
[pyridyl—SOCH$_2$CH$_2$—]$_2$NH	[Picrate, m.p. 185–188° (decomp.)]	435
1-methyl-2-oxo-pyridine SOMe, CONHMe	246–247	441
1-methyl-2-oxo-pyridine SOPr-iso, CONHMe	88.5–90.5	441
dichloropyridine CO$_2$H, SOMe	206–209	518

TABLE XV-17.   4-Pyridylsulfoxides

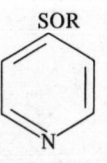

Substituents	R	M.p. (°C)	Ref.
2,5-Dibromo-3,5-dichloro	Me	–	388
2,6-Dichloro-2-fluoro	Me	–	388
3,5-Dichloro-2-fluoro	Me	–	388
3,5-Dichloro-2-trichloromethyl	Me	160–162	388
2,3,5-Trichloro	Me	91	162, 388
2,3,5-Trichloro	m-BrC$_6$H$_4$	–	162, 388
2,3,6-Trichloro	Me	–	388
3,5,6-Trichloro-2-trichloromethyl	Me	–	388
2,3,5,6-Tetrachloro	Me	154–155	434
2,3,5,6-Tetrachloro	Et	113	162, 388
2,3,5,6-Tetrachloro	C$_6$H$_{11}$	62.5	162, 388
2,3,5,6-Tetrachloro	Bu	113	162, 388
2,3,5,6-Tetrachloro	Pr	120.5	162, 388
2,3,5,6-Tetrachloro	ClCH$_2$	113.5	388
2,3,5,6-Tetrachloro	NCCH$_2$	135.5	566
2,3,5,6-Tetrafluoro	2,3,5,6-Tetrafluoro	131–132	455
2,5,6-Trichloro-3-cyano	Me	179	518, 578, 583
3,5,6-Trichloro-2-cyano	Me	124	518, 578, 583

TABLE XV-18. 2-Pyridyl Sulfones

Substituents	R	M.p. (°C)	Ref.
None	Et	(n_D^{25} 1.5332)	462, 465
None	2-(N-morpholino)ethyl	Hydrochloride, m.p. 222–225°	84
None	Et, NCH$_2$CH$_2$	Hydrochloride, m.p. 208–210°	105a
None	(CH$_3$CO)$_2$NCH$_2$CH$_2$	120	546
None	Ph	89–90	546
None	p-ClC$_6$H$_4$	120–121.5	371, 372
None	2,4-(O$_2$N)$_2$C$_6$H$_3$	188–194	438
None	2,4,5-Cl$_3$C$_6$H$_2$	160–161	371, 372
1-Oxide	Ph	164–165	546
3-Acetamido-5-chloro	2,4-(O$_2$N)$_2$C$_6$H$_3$	200–202	216
3-Acetamido-6-ethoxy	5-Nitro-2-Py	—	216
5-Amino	2,4-Cl$_2$C$_6$H$_3$	166–167 (N-acetyl, m.p. 195–196°)	218, 1
5-Amino	3,4-Cl$_2$C$_6$H$_3$	242–243 (N-acetyl, m.p. 153–154°)	1
5-Amino	3,5-Cl$_2$C$_6$H$_3$	104–105 (N-acetyl, m.p. 138–139°)	1
5-Amino	2,4-Cl(HO)C$_6$H$_3$	216–217	1
5-Amino	3,4-Cl(HO)C$_6$H$_3$	232–233 (acetate, m.p. 107–108°)	1
5-Amino	2,5,4-Cl$_2$(HO)C$_6$H$_2$	230–231 (acetate, m.p. 157–158°)	1
5-Amino	3,5,4-Cl$_2$(HO)C$_6$H$_2$	232–233 (acetate, m.p. 227–228°)	1
6-Bromo	2-(N-morpholino)ethyl	Hydrochloride, m.p. 207–210°	84

TABLE XV-18. 2-Pyridyl Sulfones (*Continued*)

and

Substituents	R	M.p. (°C)	Ref.
4-Carbamoyl-3,5,6-trichloro	Me	211.5	518
6-Carbomethoxy-3,4,5-tribromo	Me	131.0–132.5	503
6-Carbomethoxy-3,4,5-tribromo	p-MeC$_6$H$_4$	127–129	503
6-Carbomethoxy-3,4,5-tribromo	p-O$_2$NC$_6$H$_4$	173–175	503
6-Carbomethoxy-3,4,5-trichloro	p-ClC$_6$H$_4$	107–109	503
6-Carbomethoxy-3,4,5-trichloro	p-MeC$_6$H$_4$	105.0–106.5	503
6-Carbomethoxy-3,4,5-trichloro	p-O$_2$NC$_6$H$_4$	135–137	503
4-Carboxy-3,5,6-trichloro	Me	196	518, 583
3-Chloro	2-(N-Morpholino)ethyl	Hydrochloride, m.p. 219°	84
5-Chloro	2-(N-Morpholino)ethyl	Hydrochloride, m.p. 237–240°	84
5-Chloro	Et$_2$NCH$_2$CH$_2$	Hydrochloride, m.p. 138–140°	84
5-Chloro	p-ClC$_6$H$_4$	147–148	371, 372
5-Chloro	2,4,5-Cl$_3$C$_6$H$_2$	169–171	371, 372
6-Chloro	Et$_2$NCH$_2$CH$_2$	Hydrochloride, m.p. 151–153°	84
6-Chloro	2-(N-Morpholino)ethyl	Hydrochloride, m.p. 211°	84
3-Chloro-5-amino	2,4-Cl$_2$C$_6$H$_3$	185–186	1
3-Chloro-5-amino	3,4-Cl$_2$C$_6$H$_3$	212–213 (N-acetyl, m.p. 183–184°)	1
3-Chloro-5-amino	3,5-Cl$_2$C$_6$H$_3$	184–185	1
3-Chloro-5-amino	p-HOC$_6$H$_4$	214–215	1
3-Chloro-5-amino	2,4-Cl(HO)C$_6$H$_3$	276–277	1
3-Chloro-5-amino	3,4-Cl(HO)C$_6$H$_3$	211–213 (N-acetyl, m.p. 263–264°)	1
3-Chloro-5-amino	3,5,4-Cl$_2$,(HO)C$_6$H$_2$	238–239 (N-acetyl, m.p. 213–214°)	1
3-Chloro-5-nitro	2,4-Cl$_2$C$_6$H$_3$	149–150	1
3-Chloro-5-nitro	3,4-Cl$_2$C$_6$H$_3$	158–160	1

3-Chloro-5-nitro	3,5-Cl$_2$C$_6$H$_3$	157–158	1
3-Chloro-5-nitro	p-HOC$_6$H$_4$	207–208 (N-acetyl, m.p. 131–135°)	1
3-Chloro-5-nitro	2,4-Cl(HO)C$_6$H$_3$	227–228 (N-acetyl, m.p. 139–140°)	1
3-Chloro-5-nitro	3,4-Cl(HO)C$_6$H$_3$	178–179	1
3-Chloro-5-nitro	3,5,4-Cl$_2$(HO)C$_6$H$_2$	210–211 (N-acetyl, m.p. 149–150°)	1
5-Chloro-3-nitro	2-MeCONHC$_6$H$_4$	—	218
5-Chloro-1-oxide	p-ClC$_6$H$_4$	190–192	371, 372
5-Chloro-1-oxide	2,4,5-Cl$_3$C$_6$H$_2$	233.5–237	371, 372
6-Cyano-3,4,5-trichloro	Me	157.5	518, 583
4-Cyano-3,5,6-trichloro	Me	237	518, 583
4-Cyano-3,5,6-trichloro	Pr	137	518, 583
3,5-Dichloro-4-nitro	Me	—	545
3,5-Dichloro-4-nitro		158	545
2,3-Dichloro-4-nitroso		147	568
3-(2,4-Dinitrophenylamino)-5-chloro		211–213	216
4-C$_7$H$_{15}$		68–71	2
3-Methyl	Ph	188–194	438
4-Methyl	Ph	118–120	2
4-Methyl-1-oxide	Ph	180–182	2
5-Nitro	2,4-Cl$_2$C$_6$H$_3$	158–159	1
5-Nitro	3,4-Cl$_2$C$_6$H$_3$	175–176	1
5-Nitro	3,5-Cl$_2$C$_6$H$_3$	153–154	1
5-Nitro	p-HOC$_6$H$_4$	177–178 (acetate, m.p. 126–127°)	1
5-Nitro	2,4-Cl(HO)C$_6$H$_3$	210–211	1
5-Nitro	3,4-Cl(HO)C$_6$H$_3$	179–180 (acetate, m.p. 144–145°)	1
5-Nitro	2,5,4-Cl$_2$(HO)C$_6$H$_2$	213–214 (acetate, m.p. 178–180°)	1
5-Nitro	3,5,4-Cl$_2$(HO)C$_6$H$_2$	195–196	1
5-Nitro	p-MeC$_6$H$_4$	159.5–160.5	74
5-Nitro	NCCH$_2$CH$_2$	141–142	346
1-Oxide	Et	104–105	138
1-Oxide	p-ClC$_6$H$_4$	179.5–180	371, 372
1-Oxide	2,4,5-Cl$_3$C$_6$H$_2$	209.5–210	371, 372

TABLE XV-18. 2-Pyridyl Sulfones (Continued)

Substituents	R	M.p. (°C)	Ref.
1-Oxide	Ph	139–140	475
2,3,5-Tribromo-4-nitro	Me	173	545
3,5,6-Trichloro	Me	—	586
EtSO$_2$ ⟶ OPS(OEt)$_2$ (structure)		(n_D^{25} 1.5207)	385
MeNHCO$_2$ ⟶ SO$_2$Et (structure)		79–80	441
CONHMe / SO$_2$Pr (structure)		—	441
NC / F / F / SO$_2$Pr (structure)		183–186	518

TABLE XV-19. 3-Pyridyl Sulfones

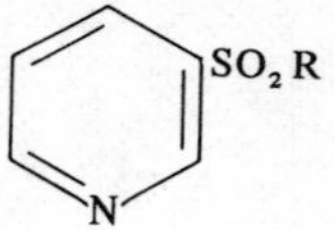

SO_2R and

R'

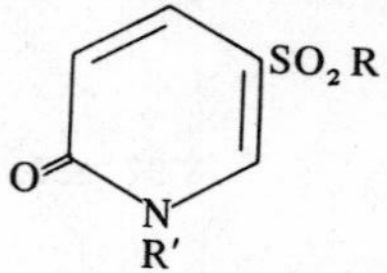

Substituents	R	M.p. (°C)	Ref.
None	Ph	106–107	546
None	$2,4,5\text{-}Cl_3C_6H_2CH{:}CH$	119–121	338
4-Cyano-2,5,6-trichloro	Me	145–150	518–583
4-Cyano-2,5,6-trichloro	Pr	110–112	583
5,6-Dichloro-2,4-dicyano	Me	208–214	518
6-Hydroxy	Me	247–248.5	280
6-Hydroxy	Et	122–123.5	280
6-Hydroxy	isoPr	165–167	280
1-Oxide	Ph	160–161	546
4-Nitro-1-oxide	4-Nitro-3-Py-1-oxide	237	332
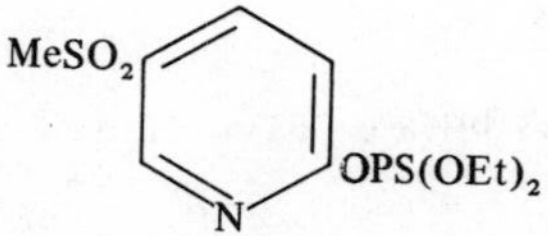		56.5–58	385

$MeSO_2$ ⬡ $OPS(OEt)_2$ (N)

TABLE XV-19. 3-Pyridyl Sulfones (Continued)

Substituents	R	M.p. (°C)	Ref.
EtSO₂ — pyridine OPS(OEt)₂		(n_D^{25} 1.5332)	385, 462, 465
isoPrSO₂ — pyridine OPS(OEt)₂		(n_D^{25} 1.5221)	385, 462, 465
[pyridine — SO₂CH₂CH₂ —]₂ CH₂		144–144.5 (Picrate, m.p. 171–172°; dihydrochloride, m.p. 200–202°)	435
pyridone — SO₂Pr-iso		127–128.5	441

$R = $
SO₂R (pyridine) and R′N (pyridone) SO₂R

TABLE XV-20. 4-Pyridyl Sulfones

Substituents	R	M.p. (°C)	Ref.
None	Et	Oil (n_D^{18} 1.5340)	61
None	Ph	129-129.5	140
None	Ph	125-126	546
None	o-ClC$_6$H$_4$	204-205	260
None	m-ClC$_6$H$_4$	203-204	260
None	p-ClC$_6$H$_4$	176-177	260
None	2,3-Cl$_2$C$_6$H$_3$	223-224	260
None	2,4-Cl$_2$C$_6$H$_3$	185-186	260
None	2,5-Cl$_2$C$_6$H$_3$	214-215	260
None	3,5-Cl$_2$C$_6$H$_3$	254-253	260
None	3,4-Cl$_2$C$_6$H$_3$	207-208	260
None	2,5,4-Cl$_2$(HO)C$_6$H$_2$	—	260
None	2,5,4-Cl$_2$(MeO)C$_6$H$_2$	228-229	260
None	p-MeOC$_6$H$_4$	168-169	260
None	2,4-BrMeC$_6$H$_3$	129-130	260
None	2,4-MeBrC$_6$H$_3$	186-187	260
None	3,4-ClBrC$_6$H$_3$	207-208	260
None	2,5-BrClC$_6$H$_3$	—	260
None	3,4-Cl(BuO)C$_6$H$_3$	161-162	260
None	p-HOC$_6$H$_4$	260-261	260
None	2,4-Cl(HO)C$_6$H$_3$	—	260
None	3,4-Cl(HO)C$_6$H$_3$	255-256	260
None	3,5,4-Cl$_2$(HO)C$_6$H$_2$	254-255	260
None	MeCONHCH$_2$CH$_2$	124	105a

TABLE XV-20. 4-Pyridyl Sulfones (Continued)

Substituents	R	M.p. (°C)	Ref.
1-Oxide	Ph	139.5-140.5	546
2-Bromo-3,5-dichloro	BrCH₂CH₂	121	91, 388
2-Bromo-3,5-dichloro	Me	92	91, 162, 388
2,6-Dicarboethoxy	Et	—	162
2,6-Dicarboethoxy	isoPr	—	221
2,6-Dicarboethoxy	Bu	—	221
2,6-Dicarboethoxy	s-Bu	—	221
2,6-Dicarboethoxy	isoC₅H₁₁	—	221
2,6-Dicarboethoxy	Ph	155-157	221
2,6-Dicarboethoxy	PhCH₂	139-140	221
2,6-Dicarboethoxy	p-ClC₆H₄	112.5-114.5	221
2,6-Dicarboethoxy	p-BrC₆H₄	113.5-114.5	221
2,6-Dicarbomethoxy	Ph	205-206	221
2,6-Dicarbohydrazide	Et	215-217 (picrate, m.p. 161°)	380
2,6-Dicarbohydrazide	Bu	192-192	380
2,6-Dicarbohydrazide	Ph	256	380
3,5-Dichloro-2,6-difluoro	Me	118-123	162
3,5-Dichloro-2-fluoro	Me	88-90	91, 162, 388
3,5-Dichloro-2-trichloromethyl	Me	160-162	91
3,5-Dichloro-2-trichloromethyl	Pr	107.5	91, 162, 388
3,5-Dichloro-2-trifluoromethyl	Me	—	539, 571, 588
3,5-Dichloro-2-trifluoromethyl	p-ClC₆H₄	102.5-105.5	539, 571, 588

Compound	Substituent	bp/mp	References
2,6-Difluoro-3,4-dichloro	Me	118-123	91
2-Methyl	Pr	31-32	273
2-Me-1-oxide	Pr	87	273
1-Oxide	Ph	139-140	140
1-Oxide	2,5-Br(Me)C$_6$H$_3$	175	87
1-Oxide	3,4-Cl(Me)C$_6$H$_3$	199	87
1-Oxide	3,4-Cl(Me)C$_6$H$_3$	205	87
2,3,5,6-Tetrabromo	Me	165	388
2,3,5,6-Tetrachloro	Me	138-140; 146-148	91, 162, 388, 447, 434
2,3,5,6-Tetrachloro	Et	129-131	91, 162, 388, 447
2,3,5,6-Tetrachloro	Pr	163-172	91, 162, 388, 447
2,3,5,6-Tetrachloro	Bu	113	91
2,3,5,6-Tetrachloro	C$_6$H$_{13}$	62.5	91
2,3,5,6-Tetrachloro	ClCH$_2$CH$_2$	182	91, 162, 388
2,3,5,6-Tetrachloro	p-ClC$_6$H$_4$	228-234	91, 162, 388
2,3,5,6-Tetrachloro	p-O$_2$NC$_6$H$_4$	—	529
2,3,5,6-Tetrachloro	p-ClC$_6$H$_4$CH$_2$	144-146	91, 162, 388
2,3,5,6-Tetrachloro	C$_5$H$_{11}$	92	91, 162, 388
2,3,5,6-Tetrachloro	Et(Me)CH	119	91, 162, 388
2,3,5,6-Tetrachloro	BrCH$_2$CH$_2$	199	91, 388
2,3,5,6-Tetrachloro	CH:CH$_2$	—	577
2,3,5,6-Tetrachloro	ClCH$_2$	113.5	91, 162
2,3,5,6-Tetrachloro	NCCH$_2$	121	91, 162
2,3,5,6-Tetrachloro	BrCH$_2$CH$_2$	199	566
2,3,5,6-Tetrachloro	NCCH$_2$CH$_2$	151	566
2,3,5-Trichloro	Me	105.5-106	162
2,3,5-Trichloro	C$_{12}$H$_{25}$	46	91, 162, 434
2,3,5-Trichloro	Et	87	91, 162, 388
2,3,5-Trichloro	Bu	78.5	91, 162, 388

TABLE XV-20. 4-Pyridyl Sulfones (Continued)

Substituents	R	M.p. (°C)	Ref.
2,3,5-Trichloro	BrCH₂CH₂	121	162, 388
2,3,5-Trichloro	2-Pr	84	91, 162, 388
2,3,6-Trichloro	Me	121; 129.5– 130.5	91, 162, 388 434
2,3,6-Trichloro	Bu	–	388
3,5,6-Trichloro	o-ClC₆H₄	189–190	518, 578
2,3,5-Trichloro	p-ClC₆H₄CH₂	–	162
3,5,6-Trichloro-2-cyano	Me	138.5	518, 578, 583
3,5,6-Trichloro-2-cyano	Pr	166	518, 578, 583
3,5,6-Trichloro-2-cyano	C₁₀H₂₁	67.5	518, 578, 583
3,5,6-Trifluoro-2-cyano	Me	75–77	518, 578
3,5,6-Trichloro-2-cyano	Me	211.5	583
3,5,6-Trichloro-2-carbamoyl	Me	–	583
3,5,6-Trichloro-2-carboxy	Me	–	583
3,5,6-Trichloro-2-cyano	2-ClC₆H₁₀	189–190	583
	(position unknown; either 2- or 4-)	227	518, 578, 583
2,3,5-Trichloro	p-ClC₆H₄	136.5–137.5	91, 162, 388
2,3,5-Trichloro	PhCH₂	111	91, 162, 388
3,5,6-Trichloro-2-trichloromethyl	Me	127.5	91, 162, 388

CMe₃ structure text: CMe₃, N, CMe₃, CMe₃, SO₂

CMe_3 ... N ... CMe_3 ... SO_2 (bracketed)₂

2,3,5,6-Tetrafluoro

EtO_2C, N, EtO_2C, $\text{SO}_2\text{CH}_2\text{CH}_2\text{O}_2\text{S}$, CO_2Et, N, CO_2Et

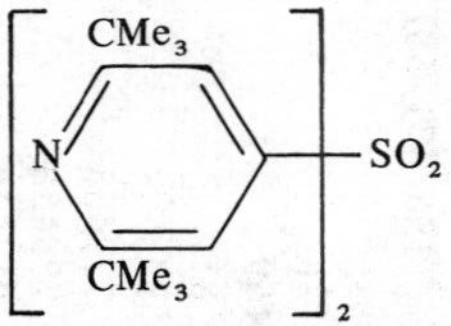		250–251	267
Ph		149	455
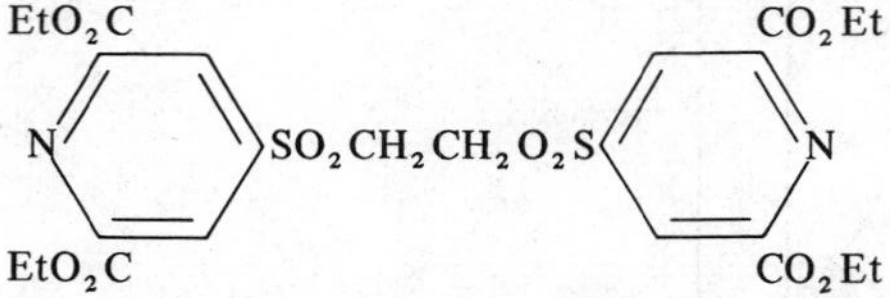 —			221

TABLE XV-21. 2-Pyridinesulfonic Acids

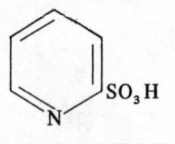

Substituents	M.p. (°C)	Ref.
None	–	102
3-Carboxy	275–278	81
4-Carboxy	296–297	79
6-Carboxy	260–262	79
3-Hydroxy	300–302	68, 69
4-Methyl	220–223	79
6-Methyl	295–298	79
3-Nitramino	184	68, 69
5-Nitramino	217–220	68, 69

TABLE XV-22. 3-Pyridinesulfonic Acids

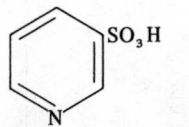

Substituents	M.p. (°C)	Ref.
None	–	18, 52, 92, 94, 96, 102, 118, 119, 123, 124, 168, 189, 190, 196, 471, 587
6-Amino	–	126
6-Amino-5-methyl	–	469
2-Carboxy	334–337	81
4-Carboxy	315	77
5-Carboxy	333–335	77, 458
6-Carboxy	285	77, 458
2-Chloro	276–280	81, 520
4-Chloro-2,6-di-*t*-butyl	Na salt	267
6-Cl₂CHCONH	>270	367
2,6-Di-*t*-butyl	325–327 (decomp.) *S*-benzylisothiuronium salt, m.p. 213–215°	102, 266, 267
2,6-Dimethyl	–	102
4-Ethoxy-2,6-di-*t*-butyl	*S*-Benzylisothiuronium salt, m.p. 215–216°	267

344

TABLE XV-22.   3-Pyridinesulfonic Acids *(Continued)*

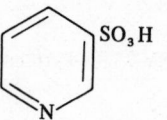

Substituents	M.p. (°C)	Ref.
4(1*H*)-one	256.5–258	267
6-Linoleamide	260	126
4-Methyl	355	77
5-Methyl	312	77
6-Methyl	340	77, 119
5-Nitro-2(1*H*)-one	Na salt	197, 198, 199, 210, 238, 242, 270, 282, 283, 296, 100
4-Nitro-1-oxide	Na salt	332
2-[*m*-(Trifluoromethyl)phenyl] amino	–	520
2-(*m*-Chlorophenyl)amino	252–253°; Na salt, H$_2$O	469
4-(*m*-Chlorophenyl)amino	Na salt, m.p. > 300°	490, 577
6-[*m*-(Trifluoromethyl)phenyl] amino	257–258° (decomp.)	469, 577
6-Hydroxy-5-methyl	Na salt, m.p. > 300°	469

Ph ⎵ SO$_3$H (O=, N–Me)	–	535
SO$_3$H (O=, N–H, OH)	–	521
1-SO$_3$–	–	196
2-Thiono	235	81

TABLE XV-23. 4-Pyridinesulfonic Acids

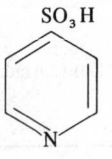

SO$_3$H

Substituents	M.p. (°C)	Ref.
None	36–38	102, 103, 292, 328, 353, 358
2-Carboxy	272–273	78
2-Carboxy-1-oxide	260; Na salt, m.p. 317°	78, 275
2,6-Di-*t*-butyl	344–346 (decomp.)	267
2,6-Dimethyl-1-oxide	>270	327
2-Ethyl	Na salt	121
2-Methyl	263–265	78, 328
3-Methyl	306	78, 328
2-Methyl-1-oxide	270–272	78, 327
3-Methyl-1-oxide	>270	327
1-Oxide	326–328	3
2,3,5,6-Tetrachloro	>360° (*S*-benzylthiuronium salt, m.p. 208–209°)	460, 529
2,3,5,6-Tetrafluoro	35 (Hygroscopic) (Na salt, m.p. 300–301°) (*S*-benzylthiuronium salt, m.p. 178°)	455

TABLE XV-24. Pyridinesulfinic Acids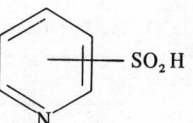

Compound	M.p. (°C)	Ref.
	Unstable (Na salt, m.p. 292–296°)	438
	Unstable (Na salt, m.p. 250–270°)	438
	90 (decomp.)	460
	–	457

TABLE XV-25. Pyridinesulfobetaines

Compound	M.p. (°C)	Ref.
SO_3^- pyridinium, N-$CH_2CH_2:CH_2$	–	283
Ph, SO_3^- pyridinium, N-Me	–	535
SO_3^- Me, Me pyridinium, N-Me	270	397
SO_3^- Ph, Ph pyridinium, N-Me	>360	397
SO_3^- Me, Me pyridinium, N-Et	248	397
SO_3^- Ph, Ph pyridinium, N-Et	310	397

348

TABLE XV-25. Pyridinesulfobetaines (*Continued*)

Compound	M.p. (°C)	Ref.
	260	97
	257	97
	–	117, 118
	Na salt	125, 127, 128
	70–90 (mono-hydrate)	125, 127, 128, 392
	182 (decomp.); 194	420, 423

349

TABLE XV-26. Bipyridinesulfonic Acids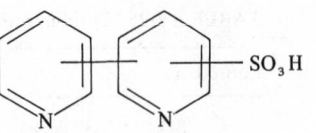

Compound	M.p. °C	Ref.
2,2′-Bipyridine-5-sulfonic acid	–	252
3,3′-Bipyridine-5-sulfonic acid	–	251
3,4′-Bipyridine-3′-sulfonic acid	>350	253, 270
2,2′-Bipyridine-5,5′-disulfonic acid	–	250

TABLE XV-27. Pyridinesulfonyl chlorides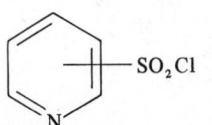

Substituents	Position of SO₂Cl	M.p. (°C)	Ref.
None	2	–	35, 376, 384
5-Chloro	2	–	109
5-MeCONH	2	–	35, 249
None	3	–	100, 473, 587
2-Chloro	3	–	100
6-Chloro	3	–	100
6-Chloro-5-methyl	3	56–57	469
2-Chloro-5-nitro	3	(Hydrochloride, m.p. 190–220°)	118, 191, 258
1-Oxide	4	–	4, 376
2,3,5,6-Tetrachloro	4	53–54 (sulfonyl fluoride, m.p. 57.0–58.5°)	460, 522, 570
2-[m-(Trifluoromethyl-phenyl)amino	3	–	520

350

TABLE XV-28. 2-Pyridinesulfonamides

Substituents	R	R'	M.p. (°C)	Ref.
None	H	H	144-145	35, 376, 384
None	H	Me₃C (1,3,4-thiadiazol-2-yl)	167 (decomp.)	35, 384
None	H	MeCOCH₂CO	143	89
None	H	PhCH₂CO	218 (decomp.)	89
None	H	MeCO	220-221 (decomp.)	89
None	H	(HO, CO₂H azophenyl)	—	440
5-Acetamido	H	H	242-243	249

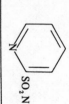

TABLE XV-28. 2-Pyridinesulfonamides (*Continued*)

Substituents	R	R'	M.p. (°C)	Ref.
5-Acetamido	H	Me₃C (thiadiazole)	210 (decomp.)	35, 384
5-Bromo	H	H	172–173	109
5-Chloro	H	H	130–131	109
4-Methyl	H	H	163.5–164.5	376
6-Hydroxymethyl	H	Me	B.p. 213–220°/0.15 mm; *p*-nitrobenzoate, m.p. 159–160°	558, 559, 581
6-Hydroxymethyl	Me	Me	—	581
6-Hydroxymethyl	Et	Et	—	581
6-Phenylcarbamoylmethyl	H	Me	142.5–143.0	558
4-Me-1-oxide	H	H	222–223	376
4-Me-1-oxide	H	MeCOCH₂CO	153–154	376

6-Me-1-oxide	H	Me	—	559, 581
1-Oxide	H	H	228	376
1-Oxide	H	MeCOCH$_2$CO	152–153 (decomp.)	90, 376
1-Oxide	H	PhCH$_2$CO	190 (decomp.)	90, 237
1-Oxide	H	ClCH$_2$CO	—	90
1-Oxide	H	NCCH$_2$CO	—	90

TABLE XV-29. 3-Pyridinesulfonamides

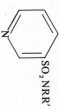

Substituents	R	R'	M.P. (°C)	Ref.
None	isoPr	isoPr	121–123	473
None	isoBu	isoBu	75–77°; b.p. 125–127°/1 mm	473
None	$(CH_2)_4$		78.5–79.5	473
None	$(CH_2)_5$		94–96	473
None	$O(CH_2)_4$		119–121	473
None	cyclohexyl	cyclohexyl	121	473
None	phenyl	phenyl	138	473
None	$3\text{-PySO}_2\text{NHCH}_2\text{CH}_2\text{NHSO}_2\text{-3-Py}$		234–236	473
None	$3\text{-PySO}_2\text{N}$ [piperazine] $\text{NSO}_2\text{-3-Py}$		296–298	473
6-Amino	H	H	—	230
6-Acetamido	H	H	202–204	230
6-Acetamido	H	H	202–204	26
6-Chloro	H	H	—	305
6-Chloro	H	H	—	26

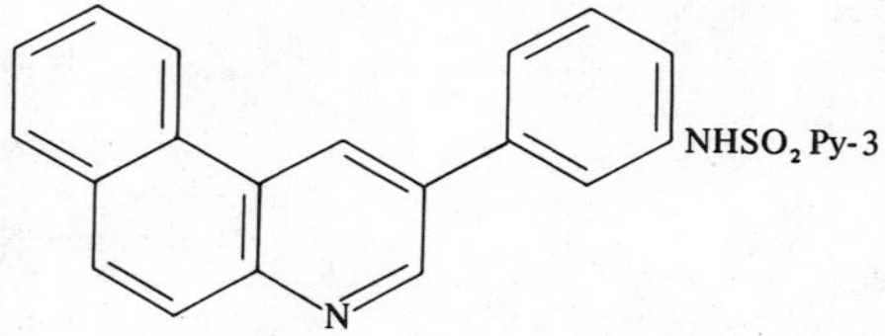

EtO

NHSO$_2$Py-3

NHSO$_2$Py-3

NHSO$_2$Py-3

NHSO$_2$Py-3

NHSO$_2$Py-3

164–175   191

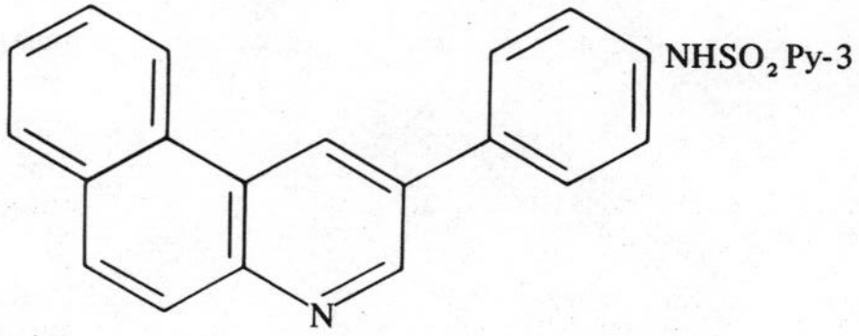

241 (decomp.)   191

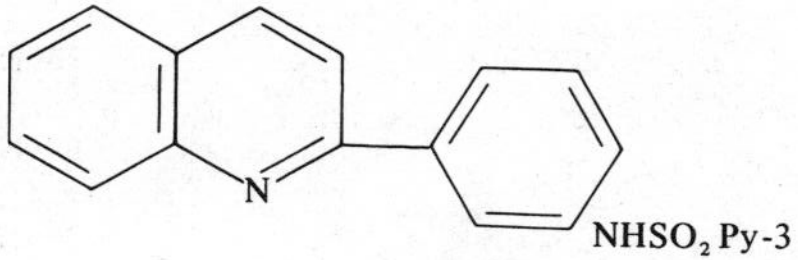

190   191

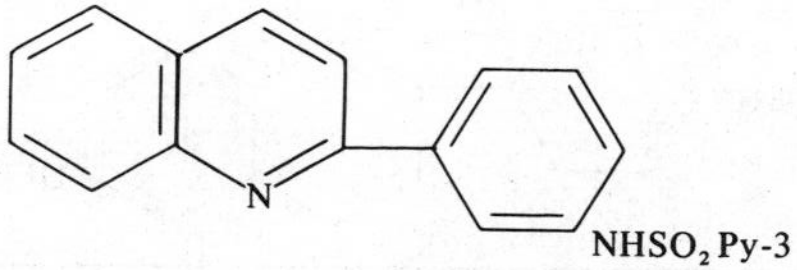

224–225   191

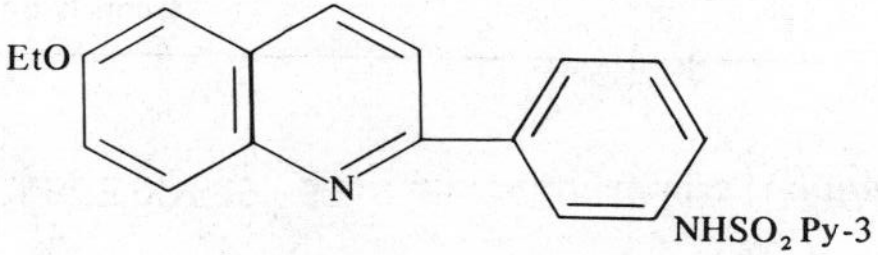

209–210   191

TABLE XV-29. 3-Pyridinesulfonamides (*Continued*)

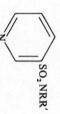

Substituents	R	R'	M.p. (°C)	Ref.
EtO (NHSO₂Py-3)			215–215.5	191
None	H	H	—	10, 139, 587
None	Me	H	114.5–116	258, 290
None	Et	H	76–77	258, 290
None	Pr	H	65–66; 152–153	258, 290
None	isoPr	H	62–63	258
None	Bu	H	82–83.5	258
None	Me₃C	H	89–90	258
None	Me	Me	96–98	290
None	Et	Et	51–52	290
None	Pr	Pr	61–63	473
None	Bu	Bu	8–9, b.p. 159–161°/1 mm.	473

357

2-Chloro-5-nitro		H	—	100
2-Chloro-5-nitro		H	—	100
2-Chloro-5-nitro		H	—	100
6-Chloro-5-methyl		H	H	192–193	469
4-Methyl		H	H	135	290
4-Methyl		Me	Me	—	290
4-Methyl		Me	H	—	290
2-(p-H₂NSO₂C₆H₄)		H	H	200	305
6-		H	H	162	19

TABLE XV-29. 3-Pyridinesulfonamides (*Continued*)

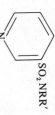

Substituents	R	R'	M.p. (°C)	Ref.
6-HO₂CCH₂S	H	H	162–164	26, 230
6-HS	H	H	240 (decomp.)	26, 230
6-H₂NCSNH	H	H	183–184 (decomp.)	26, 230
6-p-HO₂CC₆H₄SO₂NH	H	H	289–290	26, 230
6-Stearamido	H	H	115	126
6- (CF₃ phenyl NH)	H	H	176–178°	469, 502, 577
6- (CF₃ phenyl NH)-5-Me	H	H	192–193°	469, 502, 577
2- (Me, Me phenyl NH)	H	H	186–187°; N¹ salt, m.p. 232–233°	469, 577

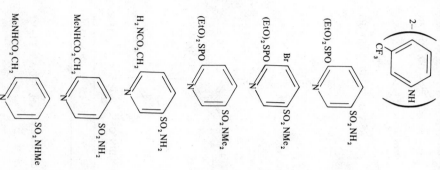

	H	Me
	—	520

2-		mp	Ref.
(EtO)$_2$SPO — SO$_2$NH$_2$		—	418
(EtO)$_2$SPO — Br — SO$_2$NMe$_2$		(n_D^{25} 1.5442)	418
(EtO)$_2$SPO — SO$_2$NMe$_2$		(n_D^{25} 1.5177)	418
H$_2$NCO$_2$CH$_2$ — SO$_2$NH$_2$		183-184	442, 478
MeNHCO$_2$CH$_2$ — SO$_2$NH$_2$		168-169	442
MeNHCO$_2$CH$_2$ — SO$_2$NHMe		128-129	442

TABLE XV-29. 3-Pyridinesulfonamides (*Continued*)

Substituents	R	R'	M.p. (°C)	Ref.
MeNHCO$_2$CH$_2$ (pyridine) SO$_2$NHEt			99–100	442
MeNHCO$_2$CH$_2$ (pyridine) SO$_2$NMe$_2$			145–146	442
MeNHCO$_2$CH$_2$ (pyridine) SO$_2$NEt$_2$			128–129	442
Me$_2$CHNHCO$_2$CH$_2$ (pyridine) SO$_2$NMe$_2$			111–112	442
CH$_2$:CHCH$_2$NHCO$_2$CH$_2$ (pyridine) SO$_2$NMe$_2$			108–109	442
Me$_2$NCO$_2$CH$_2$ (pyridine) SO$_2$NMe$_2$			109–110	442

Structure	mp (°C)	Ref.
PhCH₂NHCO₂CH₂—(pyridine)—SO₂NMe₂	123–124	442
H₂NCO₂CH₂—(pyridine)—SO₂NHCO₂Et	145–146	442
MeNHCO₂CH₂—(pyridine)—SO₂NHCO₂Et	135–136	442
MeNHCO₂CH₂—(pyridine)—SO₂NHCONHMe	96–97	442
Me₂NCO₂CH₂—(furan)—CH₂NHCO₂CH₂—(pyridine)—SO₂NMe₂	88–89	442
Me₂NCO₂CH₂—(pyridine)—SO₂NH₂	143–144	442
Me–CH₂–NCO₂CH₂ / CH₂—(pyridine)—SO₂NMe₂	93–94	442

TABLE XV-29. 3-Pyridinesulfonamides (Continued)

Substituents	R	R'	M.p. (°C)	Ref.
Me₂NCO₂CH₂ ... SO₂NHCO₂Et			136–137	442
Me₂NCO₂CH₂ ... SO₂NHCONHMe			140–141	442
NCO₂CH₂ ... SO₂NH₂ (cyclohexyl)			129–130	442
NCO₂CH₂ ... SO₂NH₂ (morpholine)			161–162	442
NCO₂CH₂ ... SO₂NMe₂ (cyclohexyl)			121–122	442

132–133	442
159–160	442
142–143	442

TABLE XV-30. 4-Pyridinesulfonamides and 4-Pyridinesulfonylhydrazides

SO_2NRR'

Substituents	R	R'	M.p. (°C)	Ref.
None	H	H	—	255
1-Oxide	H	H	228 (decomp.) (diacetyl deriv., m.p. 192°)	4, 255, 376
1-Oxide	H	isoPr	210–212	3
1-Oxide	H	Bu	175	3
1-Oxide	H	$PhCH_2$	192	4
1-Oxide	H	$PhCH_2CH_2$	235	4
1-Oxide	H		164	4
1-Oxide	H	$-(CH_2)_5-$	143; 190	3,4
1-Oxide		$-(CH_2)_2-O-(CH_2)_2-$	164	4
1-Oxide	H	Ph	178	4
1-Oxide	H	$o\text{-}MeC_6H_4$	181	4
1-Oxide	Et	$p\text{-}MeC_6H_4$	144	4
1-Oxide	H	Ph	176	4
1-Oxide	H	$o\text{-}ClC_6H_4$	228	4
1-Oxide	H	$m\text{-}ClC_6H_4$	203	4
1-Oxide	H	$p\text{-}ClC_6H_4$	207	4
1-Oxide	H	$m\text{-}HOC_6H_4$	164; 191	3,4
1-Oxide	H	$p\text{-}EtO_2CC_6H_4$	185	4
1-Oxide	H	$m\text{-}EtO_2CC_6H_4$	238–240 (decomp.)	4
1-Oxide	H	$4,3\text{-}MeO_2C(HO)C_6H_3$	148 (decomp.)	4
1-Oxide	H	H_2N	200–205	3
1-Oxide	Et	Et	172	3

			mp	Ref.
1-Oxide	H	PhCH₂	124	3
1-Oxide	H	PhCHMe	195	3
1-Oxide	H	o-MeOC₆H₄	125	3
1-Oxide	H	p-EtOC₆H₄	126	3
1-Oxide	H	3,4-HO(EtO₂C)C₆H₃	145	3
1-Oxide	H	3,4-HO(H₂NSO₂)C₆H₃	177	3
1-Oxide	H	Me₃COCH₂CO	228	3
1-Oxide	H	PhCH₂CO	183–184	3
1-Oxide	H	4-C₅H₄NCONH	198–199 (decomp.)	90, 376
1-Oxide	H	N:CEt₂	205 (decomp.)	90, 237
1-Oxide	H	N:CEtPr	183	4
1-Oxide	H	N:CMe(CH₂CH₂CH:CH₂)	175	4
1-Oxide	H	N:C(CH₂)₅	177	4
1-Oxide	H	N:CMePh	165	4
1-Oxide	H	N:CMe(C₆H₄Me-p)	199	4
1-Oxide	H	N:CHCH:CHPh	212	4
1-Oxide	H	N:CHPh	193	4
1-Oxide	H	N:CHC₆H₄.NMe₂-p	191	4
1-Oxide	H	N:CH[C₆H₃(OMe)₂-3,4]	202	4
1-Oxide	H	N:CH[C₆H₃(O₂CH₂)-3,4]	206	4
1-Oxide	H	N:CH[C₆H₃OMe(OH)-3,4]	—	4
1-Oxide	H	N:CMe(CH₂CO₂Et)	220	4
1-Oxide	H	—(CH₂)₆—	163	4

Structure (pyridine 1-oxide, O⁻ N⁺, bearing SO₂N-linked dimethylpyrazole, Me, Me): mp 165, Ref. 4

TABLE XV-30. 4-Pyridinesulfonamides and 4-Pyridinesulfonylhydrazides (*Continued*)

SO$_2$NRR'

Substituents	R	R'	M.p. (°C)	Ref.
2,3,5,6-Tetrachloro	H	H	196–197	460
2,3,5,6-Tetrachloro	H	Me	185	460
2,3,5,6-Tetrachloro	Me	Me	203–205	460
2,3,5,6-Tetrachloro	H	Ph	129–130	460

TABLE XV-31. Pyridinethiosulfinates, Pyridinethiosulfonates

$-SO_2SR$ and Pyridylthiosulfonates

Compound	M.p. (°C)	Ref.
2,4,6-Trimethylphenyl-(3-methyl-2-pyridine)thiosulfinate	113–115	439
2,4,6-Trimethylphenyl-(3-ethyl-6-methylpyridine)thiosulfinate	98	439
Allyl 3-pyridinethiosulfonate	B.p. 86–88°/0.001 mm; d_{20}^{20} 1.2986; n_D^{20} 1.5838	337
Butyl 3-pyridinethiosulfonate	B.p. 98–99/0.001 mm d_{20}^{20} 1.2231; n_D^{20} 1.5517 (oxalate, m.p. 97–99°)	337
isoButyl 3-pyridinethiosulfonate	B.p. 95–96°/0.001 mm; d_{20}^{20} 1.2263; n_D^{20} 1.5543	337
Ethyl 3-pyridinethiosulfonate	B.p. 90–91°/0.001 mm; d_{20}^{20} 1.3125; n_D^{20} 1.5751 (oxalate, m.p. 112–114°)	337
Ethyl 4-pyridinesulfonate-1-oxide	242	4
Methyl 3-pyridinethiosulfonate	53–54	337
Potassium 3-pyridinethiosulfonate	93.5–94	337
Propyl 3-pyridinethiosulfonate	B.p. 94–96°/0.001 mm; d_{20}^{20} 1.2644; n_D^{20} 1.5627 (oxalate, m.p. 107–109°)	337
isoPropyl 3-pyridinethiosulfonate	B.p. 92–94°/0.001 mm; d_{20}^{20} 1.2688; n_D^{20} 1.5652	337
2,4,6-Trimethylphenyl-2-pyridinethiosulfonate	110–111	438
2,4,6-Trimethylphenyl-(3-methyl-2-pyridine)thiosulfonate	122–126	438
2,4,6-Trimethylphenyl-(5-nitro-2-pyridine)thiosulfonate	163–165	438
2,4,6-Trimethylphenyl-2-pyridinethiosulfonate-1-oxide	157–159	438
2-Pyridyl isopropylthiosulfonate	32.5–35	438
2-Pyridyl t-butanethiosulfonate	89–91.5	438
2-Pyridyl cyclohexylthiosulfonate	52–54.5	438

TABLE XV-32. 3-Pyridinesulfonylazide

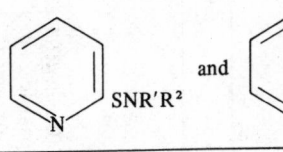

Substituents	M.p. (°C)	Ref.
None	–	146

SNR'R²

TABLE XV-33. 2- And 4-Pyridinesulfenamides and

SNR'R²

Substituent	NR'R²	M.p. (°C)	Ref.
	2-Pyridinesulfenamides		
None	NH₂	79–80	148, 286
None	N=CHMe	B.p. 106–108°/3 mm	286
None	N=CHEt	B.p. 115–116°/3 mm	286
None	N=CHPr	B.p. 128–131°/4 mm	286
None	N(CH₂)₄O	58–59	234
None	N=CHPy-2	90–93	286
None	N=C(Me)Py-2	–	286
4-Methyl	N=CHPy-2	–	286
4-Methyl	N=C(Me)Py-2	–	286
4-Methyl	NH₂	79–80	148, 286
1-Oxide	NH₂	146–148	148, 286
1-Oxide	N=CHPy-2	–	286
1-Oxide	N=C(Me)Py-2	–	286

4-Pyridinesulfenamides

(structure with SN-piperidine on tetrafluoropyridine)	52–53	455
(structure SNHPh on tetrachloropyridine)	59–60	460

368

TABLE XV-34. 4-Pyridinesulfenyl Chlorides

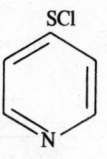

Substituents	M.p., °C	Ref.
6-Cyano-2,3,5-trichloro	–	522
2,3,5,6-Tetrachloro	59–60	460, 522, 524
2,3,5-Trichloro	Oil	522

TABLE XV-35. Pyridinedisulfonic Acids

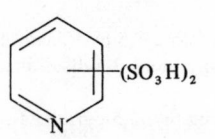

Compound	M.p. (°C)	Ref.
2,3-Pyridinedisulfonic acid	–	81
3,5-Pyridinedisulfonic acid	235–237	95
2-Hydroxypyridine-3,5-disulfonic acid	210–220	100

TABLE XV-36. Pyridinedisulfonyl Chlorides

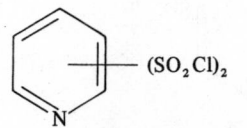

Compound	M.p. (°C)	Ref.
3,5-Pyridinedisulfonyl chloride	129	95
2-Amino-3,5-pyridinedisulfonyl chloride	137.5–139 (decomp.) 204–207 (decomp.)	66, 67, 22, 100, 325, 356
2-Amino-6-methyl-3,5-pyridinedisulfonyl chloride	132–133	66, 67, 22, 356
2-Amino-6-pyridone-3,5-disulfonyl chloride	–	66, 67
2-Amino-6-chloro-3,5-pyridinedisulfonyl chloride	–	66
4-Amino-3,5-pyridinedisulfonyl chloride	123–125	66, 67
2-Butyrylamino-6-methyl-3,5-pyridinedisulfonyl chloride	–	22
2-Chloro-3,5-pyridinedisulfonyl chloride	69–70	100
2,6-Diamino-3,5-pyridinedisulfonyl chloride	213–214.5 (decomp.)	66, 67

TABLE XV-37. Pyridinedisulfonamides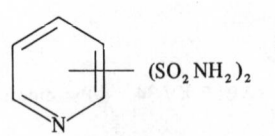

Compound	M.p. (°C)	Ref.
2,5-Pyridinedisulfonamide	213–214 (decomp.)	26, 230
2-Amino-3,5-pyridinedisulfonamide	232–234 (decomp.)	22, 54, 66, 67, 308, 325, 256, 363, 364
4-Amino-3,5-pyridinedisulfonamide	284.5–285.5 (decomp.)	66, 67
2-Amino-6-chloro-3,5-pyridinedisulfonamide	275 (decomp.)	66, 67
2-Amino-6-pyridone-3,5-disulfonamide	282–283.5 (decomp.)	66, 67
2-Amino-6-methyl-3,5-pyridinedisulfonamide	249–251 (decomp.)	22, 54, 66, 67, 356, 363
2-Amino-N,N'-dimethyl-6-methyl-3,5-pyridinedisulfonamide	–	66
2,6-Diamino-3,5-pyridinedisulfonamide	246.5–248 (decomp.)	66, 67
2-Methylamino-3,5-pyridinedisulfonamide	–	22
H_2NO_2S —⟨pyridine, Me⟩— SO_2NH_2, $NHCOCH_2SCH_2Ph$	237–239	66
H_2NO_2S —⟨pyridine, Me⟩— SO_2NHMe, NH_2	–	66

TABLE XV-38. Pyridylalkanethiols

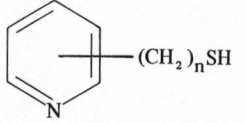

$(CH_2)_nSH$

Compound	M.p. (°C)	Ref.
2-PyCH$_2$SH	B.p. 87–89°/10 mm (Sn$_2$I$_6$ complex; PtCl$_2$ complex; PdCl$_2$ complex; thiocarbamate, m.p. 141–143°)	60, 83, 132, 195, 212, 245, 326
3-PyCH$_2$SH	(N, N-Dimethyldithiocarbamate, oil; Me p-toluenesulfonate, m.p. 161–162.5°)	177
4-PyCH$_2$SH	(N, N-Dimethyldithiocarbamate, m.p. 67.5–69°; Me p-toluenesulfonate, m.p. 161–162.5°)	177
6-Methyl-2-PyCH$_2$SH	[Thiolacetate, b.p. 112–114°/5 mm; picrolonate, m.p. 164–165° (decomp.)]	172
6-Methyl-2-PyCH$_2$SH-1-oxide	Hydrochloride, m.p. 133–134°	172
2-PyCH$_2$SH-1-oxide	Hydrochloride, m.p. 114–115°	172
2-PyCH$_2$CH$_2$SH	(Dithiocarbamate; K$_3$[MoBr$_6$] complex; N, N-diethyldithiocarbamate hydrochloride, m.p. 155°; Ni (ClO$_4$)$_2$ complex; Pt (ClO$_4$)$_2$ complex; Pd (ClO$_4$)$_2$ complex	60, 132, 211, 272, 340, 360, 379
6-Methyl-2-PyCH$_2$CH$_2$SH	–	416, 417
3-Py(CH$_2$)$_3$SH	(N, N-dimethyldithiocarbamate, oil, perchlorate, m.p. 76–76.5°)	177
4-Py(CH$_2$)$_2$SH	–	132
4-Py(CH$_2$)$_3$SH	(N, N-Dimethyldithiocarbamate, m.p. 62–64°; N, N-diethyldithiocarbamate, b.p. 177°/0.2 mm; perchlorate, m.p. 101–102°; Me p-toluenesulfonate (hygroscopic), m.p. 109.5–111°	177
2,6-Py(CH$_2$SH)$_2$	B.p. 94–96°/0.35 mm; di-(methylcarbamate), m.p. 120.5–102°; di-(ethylcarbamate) m.p. 106–107°; di-(cyclopropylcarbamate), m.p. 146–148°) di-(isopropylcarbamate),– di-(phenylcarbamate), – di-(piperidylcarbamate), –	259, 516, 555, 582
6-HOCH$_2$PyCH$_2$SH-2	–	523, 538, 551
![structure] CH$_2$OH HO—[ring]—CH$_2$SH Me—N	–	530, 574

TABLE XV-39. Pyridylalkylsulfides, Pyridylalkyldisulfides, and Pyridylalkenylsulfides

Compound	M.p. (°C)	Ref.
2-PyCH$_2$SMe	B.p. 97°/20 mm (picrate, m.p. 99° d 1.08)	169, 459
2-PyCH$_2$SPh	B.p. 123–127°/0.4 mm (n_D^{28} 1.6210, hydrochloride, m.p. 137–138°)	16
4-PyCH$_2$SPh	B.p. 126–128°/1 mm (n_D^{24} 1.6194, hydrochloride, m.p. 142–144°; picrate, m.p. 131–132°)	16
2-PyCH$_2$SPr	B.p. 90–93°/4 mm (n_D^{31} 1.5373)	16
4-PyCH$_2$SC$_6$H$_4$Cl-p	B.p. 136–141°/0.3 mm (hydro-chloride, m.p. 212–213°)	16
4-PyCH$_2$SC$_6$H$_4$CMe$_3$-p	B.p. 152–157°/0.06 mm (hydro-chloride, m.p. 215–217°)	16
4-PyCH$_2$SPr	B.p. 100–105°/3 mm (p-toluene-sulfonate, m.p. 115–117°; picrate, m.p. 129–131°)	16
4-PyCH$_2$SCMe$_3$	B.p. 70–71°/0.25 mm	412
3-PyCH$_2$SC$_6$Cl$_5$	134–136	269
6-Methyl-2-PyCH$_2$SCMe$_3$	B.p. 71–72°/0.4 mm	412

Compound	Properties	Ref.
6-Methyl-2-PyCH$_2$SEt-1-oxide	B.p. 143–146°/3 mm (picrolonate, m.p. 120.5–121°)	172
2-PyCH$_2$SEt-1-oxide	B.p. 134–137°/6 mm (picrolonate, m.p. 137°)	172
(2-PyCH$_2$)$_2$S	B.p. 148°/0.1 mm	45, 132, 272
(6-Me-3-PyCHMe)$_2$S	98–99	299
2-Chloro-3-PyCH$_2$SC$_6$H$_4$NH$_2$-o	B.p. 150–160°/0.02 mm	122
2-PyCH$_2$SCH$_2$CH$_2$Py-2	Hydrochloride, m.p. 149–151° n_D^{25} 1.6113, b.p. 150°/0.15 mm)	45
2-PyCH$_2$CH$_2$SPy-2	—	45
4-PyCH$_2$CH$_2$SPh	B.p. 190°/0.4 mm	155
2-PyCH$_2$CH$_2$SCH$_2$CH$_2$CHCO$_2^-$ $\overset{+}{N}H_3$	HgCl$_2$ complex	548, 554
2-PyCH$_2$CH$_2$SPh	B.p. 118–121°/0.2 mm (hydrochloride, m.p. 111–113°)	155
4-PyCH$_2$CH$_2$SCMe$_3$	B.p. 83–86°/0.4 mm (n_D^{25} 1.5273, hydrochloride, m.p. 126–128°)	155
6-Methyl-2-PyCH$_2$SCH$_2$CH$_2$O$_2$CMe	B.p. 143–144°/5 mm (picrate, m.p. 105–107°)	172
4-PyCH$_2$CH$_2$SPr	B.p. 95–100°/0.2 mm (picrate, m.p. 105–107°)	155
4-PyCH$_2$CH$_2$SPh	B.p. 140°/0.2 mm (n_D^{26} 1.5966, hydrochloride, m.p. 139–142°)	155

TABLE XV-39. Pyridylalkylsulfides, Pyridylalkyldisulfides, and Pyridylalkenylsulfides (*Continued*)

Compound	M.p. (°C)	Ref.
4-PyCH$_2$CH$_2$S	40–43 (monohydrate, m.p. 67–70°; hydrochloride, m.p. 141–143°)	155
2-PyCH$_2$CH$_2$S	89–91	155
4-PyCH$_2$CH$_2$SCH$_2$CH$_2$OH	B.p. 120°/0.5 mm (n_D^{25} 1.5685, hydrochloride, m.p. 127–129°)	155
4-PyCH$_2$CH$_2$SCH$_2$CH$_2$O$_2$CNHPh	86.5–89.5 (hydrochloride, m.p. 118–120°)	155
2-PyCH$_2$CH$_2$SCH$_2$CH$_2$OH	—	295
2-PyCH$_2$CH$_2$SCH$_2$CH$_2$Cl	—	295

5-Ethyl-2-Py(CH$_2$)$_2$S [thiophene ring]	—	271
2-PyCHPhCH$_2$SBu	B.p. 176–179°/4 (n_D^{20} 1.5741, picrate, m.p. 139°)	188
(2-PyCH$_2$CH$_2$)$_2$S	B.p. 168°/0.03 mm	45, 340
[pyridine] CHSPh / Ph	—	508
[pyridine] CHS-[Ph]-Cl / Ph	—	508
HO, Me [pyridine] CH$_2$S(CH$_2$)$_2$Cl	140–142 (decomp.) (hydrochloride, m.p. 180–182°)	134

TABLE XV-39. Pyridylalkylsulfides, Pyridylalkyl Disulfides, and Pyridylalkenylsulfides (*Continued*)

Compound	M.p. (°C)	Ref.
![structure] Me, HO pyridine ring, CH₂S(CH₂)₂Cl	Hydrochloride, m.p. 150.5–152.5°	134
![structure] Me, HO pyridine ring, CH₂SCHMeCH₂OH	143–146	134
![structure] Me, HO pyridine ring, CH₂SCHMeCH₂Cl	Hydrochloride, m.p. 128.5–131.5°	134
![structure] Me, HO pyridine ring, CH₂SC₆H₄Cl-*p*	164.5–167.5 (hydrochloride, m.p. 199.5–202°)	134

CH₂S(CH₂)₂OH $CH_2S(CH_2)_2OH$ 101–103.5 134

$CH_2S(CH_2)_2Cl$ Hydrochloride, m.p. 179–182° 134

$CH_2SC_6H_4Cl\text{-}p$ 91–94 134

170 (decomp.) 137

TABLE XV-39. Pyridylalkylsulfides, Pyridylalkyl Disulfides, and Pyridylalkenylsulfides (Continued)

Compound	M.p. (°C)	Ref.
	161–162 (decomp.)	137
	—	428
	Tetrahydrochloride, m.p. 290–291° (decomp.)	300
	Tetrahydrochloride, m.p. 252° (decomp.)	300

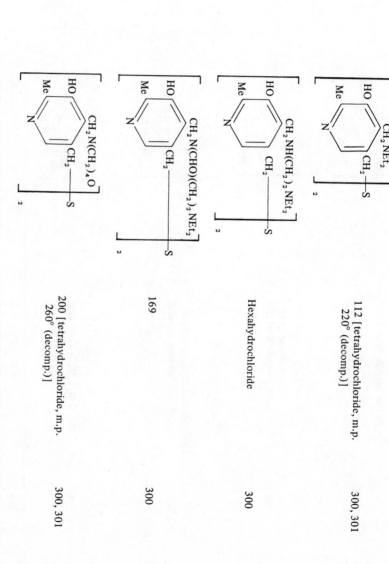

112 [tetrahydrochloride, m.p. 220° (decomp.)]   300, 301

Hexahydrochloride   300

169   300

200 [tetrahydrochloride, m.p. 260° (decomp.)]   300, 301

TABLE XV-39. Pyridylalkylsulfides, Pyridylalkyl Disulfides, and Pyridylalkenylsulfides (*Continued*)

Compound	M.p. (°C)	Ref.
	O-Diacetate, 2HCl, m.p. 196°	301, 302, 509 531–535
	Dihydrochloride, m.p. 205° (hexamethylenetetramine salt)	301, 302
	150 [tetrahydrochloride, m.p. 140° (decomp.)]	301
	Tetrahydrochloride, m.p. 220°	301

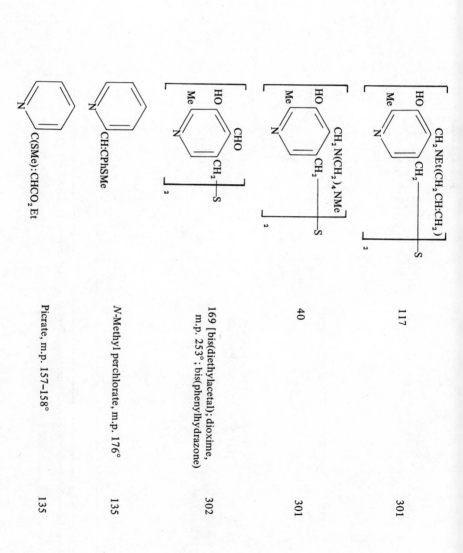

117   301

40   301

169 [bis(diethylacetal); dioxime, m.p. 253°; bis(phenylhydrazone)]   302

N-Methyl perchlorate, m.p. 176°   135

Picrate, m.p. 157–158°   135

TABLE XV-39. Pyridylalkylsulfides, Pyridylalkyldisulfides, and Pyridylalkenylsulfides (*Continued*)

Compound	M.p. (°C)	Ref.
(structure: EtO$_2$C, 2-Py, S, Py-2, CO$_2$Et — pyranone)	118	137
(pyridine; Me, HO, CH$_2$OH, CH$_2$SMe)	135° [hydrochloride, m.p. 155° (decomp); methiodide, m.p. 165°]	415, 454
(pyridine; Me, HO, CH$_2$OH, CH$_2$SPr-iso)	Hydrochloride, m.p. 135–136°	454
(pyridine; Me, HO, CH$_2$OH, CH$_2$SCH$_2$CH:CH$_2$)	111 (hydrochloride, m.p. 136°)	454

Me, HO, CH₂OH pyridine, CH₂SCH₂C:CH	140 (hydrochloride, m.p. 135°)	454
Me, HO, CH₂OH pyridine, CH₂SBu	Hydrochloride, m.p. 124°	454
Me, HO, CH₂OH pyridine, CH₂SCH₂OH	Hydrochloride, m.p. 157°	454
Me, HO, CH₂OH pyridine, CH₂SCH₂CH₂SMe	—	454
Me, HO, CH₂OH pyridine, CH₂SCH₂CH₂CH₂NH₂	Dihydrochloride, m.p. 212°	454

TABLE XV-39. Pyridylalkylsulfides, Pyridylalkyldisulfides, and Pyridylalkenylsulfides (*Continued*)

Compound	M.p. (°C)	Ref.
(pyridine ring) HO, Me, CH_2OH, $CH_2SCH_2CH_2NMe_2$	Dihydrochloride, m.p. 200° (decomp.)	454
(pyridine ring) HO, Me, CH_2OH, $CH_2SCH_2CH(NH_2)CO_2H$	230° (decomp.)	454
(pyridine ring) HO, Me, CH_2OH, $CH_2SCH_2CO_2Et$	96 (hydrochloride, m.p. 175°)	454
(pyridine ring) HO, Me, CH_2OH, $CH_2SCH_2CH_2CO_2Me$	Hydrochloride, m.p. 224°	454

Structure		
HO, Me pyridine, CH$_2$OH / CH$_2$SCOCH$_3$	Hydrochloride, m.p. 160°	454
HO, Me pyridine, CH$_2$OH / CH$_2$SCOC$_{11}$H$_{23}$	118–119 (hydrochloride, m.p. 125°)	454
HO, Me pyridine, CH$_2$OH / CH$_2$SCONMe$_2$	188	454
Pyridine, CH$_2$CH$_2$SCH$_2$CHCO$_2$H / NH$_2$	—	494
CHS= pyridine (N–H), O$_2$N-phenyl	158	82

TABLE XV-39. Pyridylalkylsulfides, Pyridylalkyldisulfides, and Pyridylalkenylsulfides (*Continued*)

Compound	M.p. (°C)	Ref.
[4-PyCH$_2$]$_2$S	—	132
[4-PyCH$_2$ CH$_2$]$_2$ S	—	132
2-PyCH$_2$ SPy-2	—	132
3-PyCH$_2$ SPy-2	—	132
4-PyCH$_2$ SPy-2	—	132
2-PyCH$_2$ CH$_2$ S	—	132

TABLE XV-40   Pyridylalkyl Sulfoxides, Pyridylalkyl Sulfones, and Pyridylalkenyl Sulfones

Compound	M.p. (°C)	Ref.
$(2\text{-PyCH}_2)_2\text{SO}$	78	45
$(2\text{-PyCH}_2)_2\text{SO}_2$	96–97	45
$2\text{-PyCH}_2\text{SO}_2\text{CH}_2\text{CH}_2\text{Py-2}$	112–113	45
$[6\text{-Me-3-PyCHMe}]_2\text{SO}_2$	160.5–162	299
$[1\text{-Oxide-2-PyCH}_2]_2\text{SO}_2$	247	45
$2\text{-PyCH}_2\text{CH}_2\text{SO}_2\text{Py-2}$	61–62	45
$(2\text{-PyCH}_2\text{CH}_2)_2\text{SO}_2$	67	45
6-Methyl-3-PyCHMeSO$_2$CHMe-		
3'-Py'-6'-Me-1-oxide	97	299
$2\text{-PyCH:CHSO}_2\text{Et}$	51–53 (b.p. 137–138°/0.05 mm)	374
$2\text{-PyCH:CHSO}_2\text{Bu}$	–	374
$3\text{-PyCH:CHSO}_2\text{Et}$	68.5–70	374
$3\text{-PyCH:CHSO}_2\text{Ph}$	85–86	9
$4\text{-PyCH:CHSO}_2\text{Ph}$	190–191	9
$3\text{-PyCH:CHSO}_2\text{C}_6\text{H}_4\text{Me-}p$	84–85	9
$4\text{-PyCH:CHSO}_2\text{C}_6\text{H}_4\text{Me-}p$	215–216	9
$4\text{-PyCH:CHSO}_2\text{Et}$	–	374
CH(OH)CH$_2$SOMe	111.5–112.5 } two 74–75° } isomers	561, 571
CH(OMe)CH$_2$SOMe	65–66 } two 164–166 } isomers	561, 571
CH(O$_2$CPh)CH$_2$SOMe	160–161	561, 571
CH(OH)CH$_2$SO$_2$Me	154–155	561, 571

TABLE XV-41. Pyridylalkyl Disulfides

Compound	M.p. (°C)	Ref.
(2-PyCH₂S)₂	58.5–59.5	245, 438
(3-PyCH₂S)₂	Dihydrochloride; picrate, m.p. 190–192°	194, 438
(1-Oxide-2-PyCH₂S)₂	Hydrochloride, m.p. 162–163° (decomp.)	172
(4-PyCH₂S)₂	77–78	137, 438

| | 132 | 231, 475 |

| | 218–220, 223–225 (dihydrochloride, m.p. 184°; dihydrobromide, m.p. 198–199°; tetra-O-acetate, m.p. 134°) | 158, 192, 229, 231, 264, 368, 369, 421, 431, 470, 474, 477–479, 500, 515, 522, 531, 532–534, 553, 562, 563 |

Structure		
$\left[\begin{array}{c} \text{HO} \\ \text{Me} \end{array}\ \text{CH}_2\text{NH}_2,\ \text{CH}_2\text{S}-\right]_2$ (pyridine)	—	158
$\begin{array}{c} \text{HO} \\ \text{Me} \end{array}$ pyridine, CH_2NH_2, $\text{CH}_2\text{SSCH}_2\text{Ph}$	140–142 [hydrochloride, m.p. 208–210° (decomp.)]	456
$\begin{array}{c} \text{HO} \\ \text{Me} \end{array}$ pyridine, CH_2NH_2, $\text{CH}_2\text{SSCMe}_3$	126–128	456
$\begin{array}{c} \text{HO} \\ \text{Me} \end{array}$ pyridine, CH_2OH, CH_2SSEt	128 (hydrochloride, m.p. 121–122°)	232
$\begin{array}{c} \text{HO} \\ \text{Me} \end{array}$ pyridine, CH_2OH, CH_2SSPr	145 (hydrochloride, m.p. 117–118°)	232, 383

TABLE XV-41. Pyridylalkyl Disulfides (*Continued*)

Compound	M.p. (°C)	Ref.
CH_2OH / Me / $CH_2SSCH_2CH:CH_2$	118	232
CH_2OH / HO / Me / CH_2SSBu	111–113 (diace- tate, m.p. 37–38°)	232, 383
CH_2OH / HO / Me / CH_2SSBu-iso	(Hydrochloride, m.p. 97–100°)	232
CH_2OH / HO / $CH_2SSC_5H_{11}$-iso	114–115	232

Structure		
Me, HO, CH_2OH, $CH_2SSC_{12}H_{25}$ (pyridine)	114–115	232, 383
Me, HO, CH_2OH, $CH_2SSC_{16}H_{33}$ (pyridine)	110	232
Me, HO, CH_2OH, $CH_2SSC_6H_{13}$ (pyridine)	90–92	456
Me, HO, CH_2OH, CH_2SSCH_2Ph (pyridine)	140	232, 383, 456
Me, HO, CH_2OH, $CH_2SSCH_2C_6H_4Br\text{-}p$ (pyridine)	156–157	232

TABLE XV-41. Pyridylalkyl Disulfides (*Continued*)

Compound	M.p. (°C)	Ref.
Me, HO, CH$_2$OH, CH$_2$SS(CH$_2$)$_4$O$_2$CCH$_3$	71–72	232
Me, HO, CH$_2$OH, CH$_2$SSCH$_2$CH$_2$O$_2$CEt	71	232
Me, HO, CH$_2$OH, CH$_2$SSCH$_2$CH$_2$C$_6$H$_4$Cl-*p*	145–146	232
Me, HO, CH$_2$OH, CH$_2$SSPh	108	232, 383

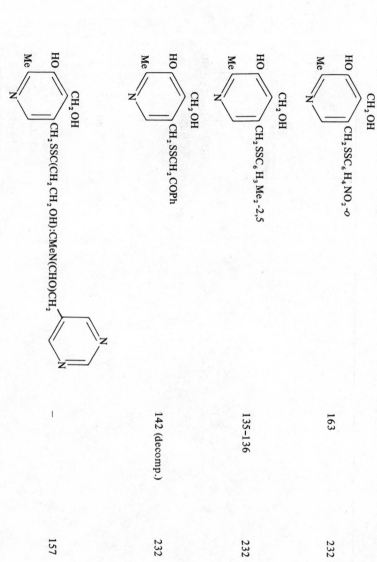

HO— pyridine (Me, CH₂OH, CH₂SSC₆H₄NO₂-o)	163	232
HO— pyridine (Me, CH₂OH, CH₂SSC₆H₃Me₂-2,5)	135–136	232
HO— pyridine (Me, CH₂OH, CH₂SSCH₂COPh)	142 (decomp.)	232
HO— pyridine (Me, CH₂OH, CH₂SSC(CH₂CH₂OH):CMeN(CHO)CH₂ - pyrimidine)	—	157

TABLE XV-41. Pyridylalkyl Disulfides (Continued)

Compound	M.p. (°C)	Ref.
Me, HO, pyridine, CH_2SSBu, CH_2OH	112-114	419
Me, HO, pyridine, $CH_2SSC_{14}H_{29}$, CH_2OH	105-107	419
Me, HO, pyridine, CH_2SSCH_2Ph, CH_2OH	123-126	419
$[$ Me, HO, pyridine, CH_2S—, CH_2OH $]_2$	—	470

TABLE XV-42. Pyridylalkylsulfonic Acids

Compound	M.p. (°C)	Ref.
3-PyCH$_2$SO$_3$H	>300	496, 543
2-PyCH$_2$CH$_2$SO$_3$H	–	56, 115
4-PyCH$_2$CH$_2$SO$_3$H	–	115
2-PyCH(NO$_2$)SO$_3$Na	–	427
2-PyCHPhCH$_2$SO$_3$H	–	188
	–	313

TABLE XV-43. Pyridylalkyl Xanthates

Compound	M.p. (°C)	Ref.
	–	386
	–	472

TABLE XV-44. Pyridylthiocyanates and Pyridylalkylthiocyanates

Compound	M.p. (°C)	Ref.
2-Pyridylthiocyanate	B.p. 122°/12 mm (n_D^{20} 1.5826)	117, 306
3-Pyridylthiocyanate	28–31	117
4-Pyridylthiocyanate	54–55	117
2-Acetamido-5-pyridylthiocyanate	202–204	117
2-Amino-5-pyridylthiocyanate	110–112 [hydrothio-cyanate, m.p. 175° (decomp.)]	117
3-Bromo-4-pyridylthiocyanate-1-oxide	159–160	329, 373
5-Bromo-2-pyridylthiocyanate	78–79	117
2-Chloro-3-pyridylthiocyanate	90.5–91.5	117
5-Chloro-2-pyridylthiocyanate	74–76	117
2,6-Diamino-3-pyridylthiocyanate	140–142	117
2,6-Diamino-3,5-pyridyldithiocyanate	210 (decomp.)	117
2,6-Dicarboethoxy-4-pyridylthiocyanate	207–209	221
2,6-Dicarboxy-4-pyridylthiocyanate	162–164	221
4,6-Dimethyl-2-pyridylthiocyanate	72–76	306
3-Nitro-4-pyridylthiocyanate-1-oxide	169–170 (decomp.)	306, 329, 373
4-Nitro-3-pyridylthiocyanate-1-oxide	134	332
2-Pyridylmethylthiocyanate	B.p. 109–110°/1 mm	83, 244
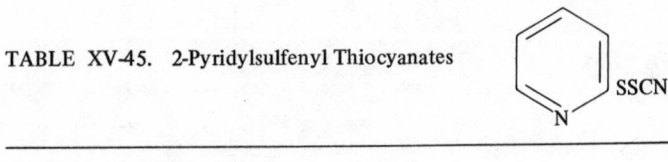	184 (decomp.)	313
HO / CH$_2$OH / Me / N / CH$_2$SCN	141–143	431

TABLE XV-45. 2-Pyridylsulfenyl Thiocyanates

Substituent	M.p. (°C)	Ref.
None	125–130 (decomp.)	117

TABLE XV-46. Mercaptopyridoxines

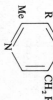

R	R¹	R²	M.p. (°C)	Ref.
H	HS	HO	—	182
HS	HO	HO	Hydrochloride	182, 183, 184 185, 194, 276
HO	HS	HO	180–181 (isothiuronium chloride, m.p. 122–124°)	182, 185, 206b, 422, 456
HO	HS	PhCO₂	Hydrochloride, m.p. 117–119° (decomp.)	313
HS	H	HO	—	182, 185
HS	HS	HO	Hydrochloride	185, 194
HS	HS	NH₂	Dihydrobromide, m.p. 198.5	298
HO	HS	PhCO₂	Hydrochloride, m.p. 182–184° (decomp.)	298
HO	MeCOS	MeCO₂	116.5	298
HO	HS	HO	Hydrochloride, m.p. 179–180°	298
HO	MeO	HS	Hydrochloride, m.p. 169–170° (decomp.)	298
HO	HS	MeO	Hydrochloride, m.p. 144–146° (decomp.)	298
HO	—CH₂—S—CH₂—		232 [Hydrochloride, m.p. 242–245° (decomp.)]	298
HO	HO	PrS	116	330
HO	HO	EtOCS₂	170–171	368

TABLE XV-46. Mercaptopyridoxines (*Continued*)

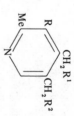

R	R¹	R²	M.p. (°C)	Ref.
HO	HO	HS	181 (hydrochloride, m.p. 132–133°; sulfamate, m.p. 148°)	368–369
H_2N	HS	NH_2	(Dihydrobromide, m.p. 242–246°)]	298
$MeCO_2$	$MeCO_2$	NaO_3S_2	238 (decomp.)	231, 232
MeCOS	MeCOS	HO	Hydrobromide	182, 185, 194
MeCOS	HO	HO	Hydrobromide	194
$MeCO_2$	MeCOS	MeCONH	178.5	298
$MeCO_2$	MeCOS	$PhCO_2$	142–143	298
$MeCO_2$	MeCOS	$MeCO_2$	64.5	298
$MeCO_2$	MeCOS	HO	153	298
$MeCO_2$	MeCOS	MeO	88.5	298
$MeCO_2$	$-CH_2-S-CH_2-$		100	330
$MeCO_2$	HO	PrS	116	298
MeCONH	MeCOS	MeCONH	206–208	298
HO	$-SCH_2S-$		—	182
	$-SCH_2CH_2S-$	HO	Hydrochloride	194
HO	HO	$MeSO_2$	219°	396

CH$_2$SCOMe   145–146 (hydrochloride, m.p. 180°)   231, 422

CH$_2$SCOPh   193–194 (decomp.)   422

TABLE XV-47. Pyridylisothiuronium Salts PySC(:NH)NH$_2$ · 2HX

R in R-Sc(:NH)NH$_2$	HX	M.p. °C	Ref.
2,6-Dicarbomethoxy-4-Py	HCl	178–180 (decomp.)	221
3-Nitro-2-Py	HCl	187–189 (decomp.)	37
2-Py-1-oxide	HCl	166–168	138, 172

TABLE XV-48. Pyridylmethylisothiuronium Salts PyCH$_2$SC(:NH)NH$_2$·2HX

Compound	HX	M.p. (°C)	Ref.
	HBr	184–185° (decomp.)	172
	HBr	191–192	172
	HBr	203–205	172, 562
	HCl	165–167 (decomp.)	264
	HCl	163	231

TABLE XV-48. Pyridylmethylisothiuronium Salts $PyCH_2SC(:NH)NH_2 \cdot 2HX$ (*Continued*)

Compound	HX	M.p. (°C)	Ref.
	HCl 2HCl	175–176 (decomp.) 205–206 (decomp.)	231, 422
	HCl	147	231
	HBr	170–172	264
	HBr	224–226	264

TABLE XV-49. Bis(4-Pyridylthio)alkanes

Substituents	n	M.p. (°C)	Ref.
None	1	148–149	335
None	3	79–80	335
None	$(CH_2)_2CHMe(CH_2)_2$	B.p. 225–230°/0.15 mm (picrate, m.p. 133°)	335
None	10	88	335
None	12	76–78	335
None	$CH_2C:CCH_2$	128–130	335

TABLE XV–50. Quaternary Bis(4-Pyridylthio)alkanes (160)

R	n	X	M.p. ($^{\circ}$C)
Me	1	p-MeC$_6$H$_4$SO$_3$	163–166
Me	2	p-MeC$_6$H$_4$SO$_3$	209–211
Me	2	I	220–222
Me	3	p-MeC$_6$H$_4$SO$_3$	138
Me	4	p-MeC$_6$H$_4$SO$_3$	180–182
Me	5	p-MeC$_6$H$_4$SO$_3$	168–170
Me	5	p-MeC$_6$H$_4$SO$_3$	141–143
Me	6	p-MeC$_6$H$_4$SO$_3$	20–25
Me	9	p-MeC$_6$H$_4$SO$_3$	150
Me	10	p-MeC$_6$H$_4$SO$_3$	188–189
Me	10	I	165–170
Me	10	Br	166–170
Me	12	p-MeC$_6$H$_4$SO$_3$	210–213
Me	(CH$_2$)$_2$CHMe(CH$_2$)$_2$	p-MeC$_6$H$_4$SO$_3$	80–82
Me	CH$_2$C:CCH$_2$	p-MeC$_6$H$_4$SO$_3$	208–210
Me	(CH$_2$)$_4$O(CH$_2$)$_4$	Cl	104–106
EtS(CH$_2$)$_2$	2	Cl	203–205
BuO$_2$CCH$_2$	6	Cl	189
HO(CH$_2$)$_2$	19	Cl	148
CH$_2$:CHCH$_2$	10	Br	112–115
Bu	10	Br	108–110
BuO$_2$CCH$_2$	10	Cl	162–165
MeS(CH$_2$)$_2$	10	Cl	70
MeO(CH$_2$)$_2$	10	Br	128–132
Me$_2$CH(CH$_2$)$_2$	10	Br	140–142
CH$_2$:CHCH$_2$	CH$_2$CEt$_2$CH$_2$	Br	40
n-C$_{16}$H$_{33}$	4	Br	138
PhCH$_2$	10	Cl	183–185
ClC$_6$H$_4$CH$_2$	10	Cl	149–150
MeC$_6$H$_4$CH$_2$	10	Cl	144–150
p-MeOC$_6$H$_4$CH$_2$	10	Cl	120–125
p-O$_2$NC$_6$H$_4$CH$_2$	2	Cl	215–216
2,4,5-Cl$_3$C$_6$H$_2$CH$_2$	1	Cl	163–166
2,4,5-Cl$_3$C$_6$H$_2$CH$_2$	3	Cl	160–162
2,4,5-Cl$_3$C$_6$H$_2$CH$_2$	4	Cl	156–159
2,4,5-Cl$_3$C$_6$H$_2$CH$_2$	5	Cl	127–129
2,4,5-Cl$_3$C$_6$H$_2$CH$_2$	6	Cl	127–128
2,4,5-Cl$_3$C$_6$H$_2$CH$_2$	10	Cl	107–110
2,4,5-Cl$_3$C$_6$H$_2$O(CH$_2$)$_2$	2	Cl	148
p-ClC$_6$H$_4$S(CH$_2$)$_2$	2	Cl	197–199

402

Compound	M.p. (°C)	Ref.
S_2COEt (tetrachloropyridyl)	95°	519
$S_2COCHMe_2$ (tetrachloropyridyl)	93.5–95.5	519
$S_2COCHMeEt$ (tetrachloropyridyl)	64.5–65.5	519
$S_2COCH_2CHMe_2$ (tetrachloropyridyl)	n_D^{20} 1.6041	519
$S_2COC_5H_{11}$-n (tetrachloropyridyl)	n_D^{20} 1.6059	519
$S_2COCHEt_2$ (tetrachloropyridyl)	69–74	519
$S_2COC_{12}H_{25}$-n (tetrachloropyridyl)	n_D^{20} 1.5650	519

Compound	M.p. (°C)	Ref.
$S_2COCHMe_2$ (3,4,6-trichloropyridin-2-yl)	74–76	519
$S_2COCH_2CHMe_2$ (3,4,6-trichloropyridin-2-yl)	n_D^{20} 1.6044	519
$S_2COCHMeEt$ (3,4,6-trichloropyridin-2-yl)	n_D^{20} 1.5911	519
$S_2COC_5H_{11}$ (3,4,6-trichloropyridin-2-yl)	n_D^{20} 1.5574	519
$S_2COC_6H_{11}$ (3,4,6-trichloropyridin-2-yl)	n_D^{20} 1.5627	519
$CH_2SCONEt_2$ (pyridin-2-yl)	112–122°/0.4 mm n_D^{20} 1.5506	536
$CH_2SCON(Pr\text{-}iso)_2$ (pyridin-2-yl)	n_D^{20} 1.5358	536
$CH_2SCON(Pr\text{-}iso)_2$ (pyridin-3-yl)	n_D^{20} 1.5374	536

Compound	M.p. (°C)	Ref.
CH$_2$SCON(Pr-iso)$_2$ (pyridine)	n_D^{20} 1.5363	536
(pyridine) CH$_2$SCON$\left(-\triangleleft\right)_2$	115.5–121/0.07 mm n_D^{20} 1.5403	536
(pyridine) CH$_2$SCON$\left(-\triangleleft\right)_2$	118.5–127.0/0.07 mm n_D^{20} 1.5401	536
CH$_2$SCON$\left(-\triangleleft\right)_2$ (pyridine)	119–133/0.06 mm m.p. 62.0–64.5°	536
(pyridine) CH$_2$SCONPr$_2$	146.0–156.5/0.07 mm n_D^{20} 1.5918	536
(pyridine) CH$_2$SCONPr$_2$	157.0–160.5/0.07 mm n_D^{20} 1.5931	536
CH$_2$SCONPr$_2$ (pyridine)	162.0–163.5/0.06 mm n_D^{20} 1.5934	
(pyridine) CH$_2$SCON(CH$_2$CH:CH$_2$)$_2$	119–133/0.03 mm n_D^{20} 1.5601	536

405

TABLE XV-51. Pyridylxanthates and Pyridylalkylthio- and dithiocarbamates (*Continued*)

Compound	M.p. (°C)	Ref.
(pyridine) CH$_2$SCON(Bu-iso)$_2$	n_D^{20} 1.5240	536
(pyridine) CH$_2$SCON(Bu-iso)$_2$	n_D^{20} 1.5259	536
CH$_2$SCON(Bu-iso)$_2$ (pyridine)	n_D^{20} 1.5246	536
(pyridine) CH$_2$SCONBu$_2$	n_D^{20} 1.5260	536
(pyridine) CH$_2$SCONBu$_2$	n_D^{20} 1.5278	536
CH$_2$SCONBu$_2$ (pyridine)	n_D^{20} 1.5270	536
(pyridine) CH$_2$SCON(C$_5$H$_{11}$)$_2$	n_D^{20} 1.5195	536
(pyridine) CH$_2$SCON(Et)(Bu)	n_D^{20} 1.5354	536

TABLE XV-51. Pyridylxanthates and Pyridylalkylthio- and dithiocarbamates (*Continued*)

Compound	M.p. (°C)	Ref.
(pyridyl) CH_2SCON(Et)(Bu)	123–133.5/0.03 mm n_D^{20} 1.5376	536
CH_2SCON(Et)(Bu) (pyridyl)	n_D^{20} 1.5362	536
(pyridyl) CH_2SCON(Me)(C_6H_{11})	136–148/0.05 mm n_D^{20} 1.5628	536
(pyridyl) CH_2SCON(Me)(C_6H_{11})	145–150/0.05 mm n_D^{20} 1.5638	536
CH_2SCON(Me)(C_6H_{11}) (pyridyl)	n_D^{20} 1.5597	536
(pyridyl) CH_2SCON(Et)(C_6H_{11})	n_D^{20} 1.5534	536
(pyridyl) CH_2SCON(Et)(C_6H_{11})	n_D^{20} 1.5540	536

407

TABLE XV-51. Pyridylxanthates and Pyridylalkylthio- and dithiocarbamates (*Continued*)

Compound	M.p. (°C)	Ref.
CH$_2$SCON(Et)(C$_6$H$_{11}$) pyridine	n_D^{20} 1.5561	536
CH$_2$SCON(piperidine) pyridine	137–146/0.04 mm n_D^{20} 1.5762	536
CH$_2$SCON(Me, Et piperidine) pyridine	n_D^{20} 1.5530	536
CH$_2$SCON(Me, Et piperidine) pyridine	n_D^{20} 1.5540	536
CH$_2$SCON(Me, Et piperidine) pyridine	n_D^{20} 1.5535	536
CH$_2$SCON(azepane) pyridine	n_D^{20} 1.5612	536
CH$_2$SCON(azepane) pyridine	162–172/0.04 mm n_D^{20} 1.5731	536

TABLE XV-51. Pyridylxanthates and Pyridylalkylthio- and dithiocarbamates (*Continued*)

Compound	M.p. (°C)	Ref.
[pyridine]$CH_2S_2N(Bu\text{-}iso)_2$	n_D^{20} 1.5745	536
[pyridine]$CH_2S_2N(Bu\text{-}iso)_2$	n_D^{20} 1.5766	536
[pyridine]$CH_2CH_2SCONHPh$	114–115	411
[pyridine]$CH_2CH_2SCONHC_6H_4Cl\text{-}m$	130–131	411
[pyridine]$CH_2CH_2S_2CNHCH_2Ph$	84–85	411
[pyridine]$CH_2CH_2S_2CN(Me)Ph$	79–81	411
[pyridine]$CH_2CH_2S_2CNHPh$	137–138	411
[pyridine]$CH_2CH_2S_2CNHC_6H_4Me\text{-}o$	121.5–122.5	411
[pyridine]$CH_2CH_2S_2CNHC_6H_4Me\text{-}m$	137.5–138.5	411

TABLE XV-51. Pyridylxanthates and Pyridylalkylthio- and dithiocarbamates (*Continued*)

Compound	M.p. (°C)	Ref.
$CH_2 CH_2 S_2 CNHC_6 H_4 Me\text{-}p$	153–154	411
$CH_2 CH_2 S_2 CNHC_6 H_4 OMe\text{-}o$	114.5–115.5	411
$CH_2 CH_2 S_2 CNHC_6 H_4 OMe\text{-}p$	128–129	411
$CH_2 CH_2 S_2 CNHC_6 H_4 OEt\text{-}p$	103–104	411
$CH_2 CH_2 S_2 CNHC_6 H_4 OH\text{-}o$	134–135	411
$CH_2 CH_2 S_2 CNHC_6 H_4 OH\text{-}p$	144–145	411
$CH_2 CH_2 S_2 CNHC_{10} H_7 \text{-}2$	129–131	411
$CH_2 CH_2 S_2 CNHPy\text{-}3$	141–142	411
$CH_2 CH_2 S_2 CNHNHPh$	148–149	411

TABLE XV-51. Pyridylxanthates and Pyridylalkylthio- and dithiocarbamates

Compound	M.p. (°C)	Ref.
$CH_2CH_2S_2CNHNHC_6H_4NO_2$-$m$ (3-pyridyl)	–	411
$CH_2CH_2S_2CNHNHCOPy$-3 (3-pyridyl)	126–127	411
$CH_2CH_2S_2CNPh_2$ (3-pyridyl)	148–149	411
$CH_2CH_2S_2CNHCOPh$ (3-pyridyl)	124–124.5	411
$MeNHCS_2CH_2$ ... CH_2S_2CNHMe (pyridyl)	–	538
$HOCH_2$ ... CH_2S_2NHMe (pyridyl)	–	523, 538, 551, 567
$MeNHCO_2CH_2$ ... CH_2S_2NHMe (pyridyl)	–	514, 523, 538, 551, 567
$MeNHCO_2CH_2$ ... $CH_2S_2NHCHMe_2$ (pyridyl)	–	514
$MeNHCO_2CH_2$ ... CH_2S_2N (piperidino) (pyridyl)	–	514

411

TABLE XV-51. Pyridylxanthates and Pyridylalkylthio- and dithiocarbamates

Compound	M.p. (°C)	Ref.
EtNHCO$_2$CH$_2$ —⟨pyridine⟩— CH$_2$S$_2$NHMe	–	514, 523, 538, 551, 567
PhNHCO$_2$CH$_2$ —⟨pyridine⟩— CH$_2$S$_2$NHMe	–	514, 523, 538, 551, 567
MeNHCO$_2$CH$_2$ —⟨pyridine⟩— CH$_2$S$_2$NHEt	–	514, 523, 538, 551, 567
EtNHCO$_2$CH$_2$ —⟨pyridine⟩— CH$_2$S$_2$NHEt	–	523, 538, 551, 567
PhNHCO$_2$CH$_2$ —⟨pyridine⟩— CH$_2$S$_2$NHPh	–	523, 538, 551, 567
HO$_2$C —⟨pyridine⟩— CH$_2$S$_2$CNH$_2$	–	524, 538, 551, 567

412

TABLE XV-52. Sugar Derivatives of Sulfur Compounds of Pyridine

Substituents	Position of S in Py	M.p. (°C)	Ref.
		A. (β-D-Glucopyranosyl)thiopyridines	
None	2	93–101 (monohydrate) $[\alpha]_D^{23}$ -56.4° (c, 2.5, HCONMe$_2$)	349, 350, 352, 353, 355
None	3	177.5–178.5 $[\alpha]_D^{23}$ -55.0° (c, 2.5, HCONMe$_2$)	349, 350, 352, 353, 355
None	4	177.5–178.5 $[\alpha]_D^{22}$ -74.7° (c, 2.5, HCONMe$_2$)	349, 350, 352, 353
1-Oxide	2	323–324 $[\alpha]_D^{24}$ -130.6° (c, 2.5, H$_2$O)	350, 352, 353, 355
1-Oxide	4	206–208 $[\alpha]_D^{19}$ -73.8° (c, 2.5, HCONMe$_2$)	349, 350, 352, 353, 355
		B. (2,3,4,6-Tetra-O-acetyl-α-D-glucopyranosyl)thiopyridines	
None	2	127–128 $[\alpha]_D^{21}$ -3.8° (c, 5.0, CHCl$_3$)	353
None	3	113.5–114.5 $[\alpha]_D^{22}$ -18.1° (c, 5.0, CHCl$_3$)	353
None	4	140–141 $[\alpha]_D^{21}$ -23.7° (c, 5.0, CHCl$_3$)	353
None	2,4	242–245 $[\alpha]_D^{18}$ -66.1° (c, 2.5, CHCl$_3$)	353
1-Oxide	4	179–181 $[\alpha]_D^{19}$ -43.5° (c, 5.0, CHCl$_3$)	353
1-Oxide	2	175 $[\alpha]_D^{19}$ -62.9° (c, 4.2, CHCl$_3$)	353
		C. (2,3,4,6-Tetra-O-acetyl-β-D-glucopyranosyl)thiopyridines	
None	2	127–128 $[\alpha]_D^{19}$ -3.8° (c, 5.0, CHCl$_3$)	255
None	4	140–141 $[\alpha]_D^{21}$ -23.7° (c, 5.1, CHCl$_3$)	355
3,5-Dinitro	2	193–197 $[\alpha]_D^{17}$ 62.9° (c, 2.5, CHCl$_3$)	354
3-Nitro	2	132 $[\alpha]_D^{17}$ 70.9° (c, 2.5, CHCl$_3$)	354
5-Nitro	2	187–187.5 (sublimes 170°) $[\alpha]_D$ 17.4° (c, 2.5, CHCl$_3$)	354
1-Oxide	2	175 $[\alpha]_D^{19}$ 62.9° (c, 4.2, CHCl$_3$)	355
1-Oxide	4	180–181 $[\alpha]_D^{19}$ -43.5° (c, 5.1, CHCl$_3$)	355

TABLE XV-53. 2- And 4-Pyridylmercaptoketones

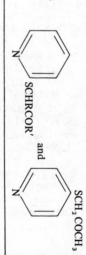

SCHRCOR' and SCH$_2$COCH$_3$

Substituents	R	R'	M.p. (°C)	Ref.
			2-Pyridylmercaptoketones	
None	H	Me	B.p. 89–93°/0.4 mm (methiodide, m.p. 116–117°)	37
None	H	Ph	B.p. 161–169°/0.4 mm (methiodide, m.p. 166–168°)	37
None	H	p-BrC$_6$H$_4$	113–115°	37
None	Me	Me	B.p. 125–133°/0.2 mm (picrate, m.p. 140–141°)	37
None	Me	Ph	B.p. 144–148°/0.4 mm (picrate, m.p. 148–149°)	37
None	Ph	Ph	79.5–81	37
None	—(CH$_2$)$_4$—		43–44	37
None	—(CH$_2$)$_3$—		B.p. 116–122°/0.6 mm (picrate, m.p. 140–143°)	37

			B.p. 103-109°/0.2 mm (methiodide, m.p. 121-123°)	
None	COMe	Me	B.p. 103-109°/0.2 mm (methiodide, m.p. 121-123°)	37
3-Chloro	H	Ph	105.5-106.5	37
5-Chloro	H	Me	57-57.5	37
5-Chloro	H	p-BrC$_6$H$_4$	107-107.5	37
3-Nitro	H	Me	70-71	37
3-Nitro	H	p-BrC$_6$H$_4$	138.5-140	37
5-Nitro	H	p-BrC$_6$H$_4$	88-89	37
5-Nitro	—(CH$_2$)$_4$—		113-114	37
5-Nitro	H	p-BrC$_6$H$_4$	147-149	37
5-Nitro	COMe	Me	131.5-132.5	37

4-Pyridylmercaptoketones

2-Chloro-3,5-dibromo-6-methyl	—	—	94-96	544
2,3,6-Trichloro	—	—	136-138	434

TABLE XV-54. 2-Pyridylmercaptoacetaldehyde Acetals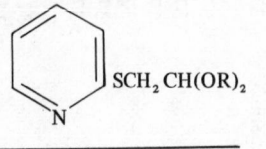

Substituents	R	M.p. (°C)	Ref.
None	Me	B.p. 100–103°/0.2 mm (methiodide, m.p. 118–120°)	37
None	Et	B.p. 101°/0.5 mm (methiodide, m.p. 89–91°)	37
3-Chloro	Et	50.5–51.5; b.p. 115–120°/0.5 mm	37

TABLE XV-55. Pyridylalkylthioketones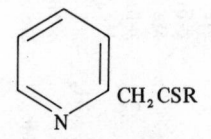

R	M.p. (°C)	Ref.
1-C$_{10}$H$_7$	153–154 (hexachloroplatinate, m.p. 230–232°; Cd chelate, m.p. 201–203°)	339
Ph	133–134 (hexachloroplatinate, m.p. 244–246°; Ni chelate, m.p. 216–218°)	339
isoPropyl	B.p. 115–120°/0.6 mm (Ni chelate, m.p. 170°)	339

TABLE XV-56. Pyridinesulfonylureas

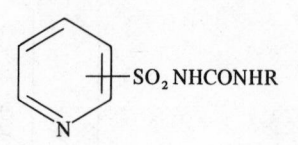

Substituents	Position of S in Py	R	M.p. (°C)	Ref.
None	2	Bu	164	35, 384
None	3	Bu	104–105	35, 384

TABLE XV-57. Pyridylthiophosphates

Compound	M.p. (°C)	Ref.
SPS(OEt)₂	B.p. 97–98°/0.01 mm	213
O₂N — SPS(OEt)₂	78	213

417

TABLE XV-58. Pyridylalkanethiophosphates

$(CH_2)_n(SCH_2)_m SP(S)OR^1(OR^2)$

Substituent	Position of side chain in Py	n	m	R	R¹	M.p. (°C)	Ref.
None	2	1	0	Me	Me	Oil	104
None	2	1	0	Et	Et	B.p. 105–106°/0.01 mm (n_D^{30} 1.5302, d_{20} 1.0571; hydrochloride, m.p. 54°; picrate, m.p. 120–121°)	99, 104, 375
None	2	1	0	Pr	Pr	(n_D^{30} 1.5455)	99
None	2	1	0	isoPr	isoPr	Oil	104
None	3	1	0	Et	Et	[Hydrochloride, m.p. 100° (decomp.)]	104
None	4	1	0	Et	Et	[Hydrochloride, m.p. 100° (decomp.)]	104
None	4	1	0	Et	Et	[Hydrochloride, m.p. 127° (decomp.)]	104

None	2	2	1	Et	Et	(n_D^{30} 1.5568)	99
None	3	1	1	Et	Et	(n_D^{30} 1.5658)	99
None	2	1	1	Et	Et	(n_D^{30} 1.5628)	99
None	2	1	1	Pr	Pr	(n_D^{30} 1.5601)	99
None	2	2	1	Pr	Pr	(n_D^{30} 1.5499)	99
None	4	2	1	Et	Et	(n_D^{30} 1.5575)	99, 295
None	2	—CH$_2$CHNO$_2$	0	Et	Et	—	357
None	3	—CH$_2$CHNO$_2$	0	Et	Et	—	357
None	2	2	0	Et	Et	—	7
None	2	2	0	Me	Me	—	7
None	2	2	0	isoPr	isoPr	—	7
None	2	2	0	MeO(CH$_2$)$_2$	MeO(CH$_2$)$_2$	Oil	7
6-Methyl	2	1	1	Et	Et	(n_D^{30} 1.5593)	99
3-Ethyl	2	2	0	Et	Et	—	7
3-Ethyl	2	2	0	Cl(CH$_2$)$_2$	Cl(CH$_2$)$_2$	(n_D^{30} 1.5542)	99

TABLE XV-59. Tetrahydropyridines

Substituents	R	M.p. (°C)	Ref.

A. 1,2,3,6-Tetrahydro-2-thionopyridines

None	PhCO	97–98	5
None	MeCO	78–80	5
4(or 5-)-Methyl	PhCO	99–101	5
4(or 5-)-Methyl	MeCO	82–83	5
4,5-Dimethyl	MeCO	93.5–95.0	5
4,5-Dimethyl	PhCO	146	5

B. 1-Substituted-1,2,3,4-tetrahydropyridines

2-MeO	$PhSO_2$	59–60	113
2-$PhSO_2NH$	$PhSO_2$	162–163	113

C. 1-Substituted-1,2,3,6-tetrahydropyridines

None	$p\text{-}MeC_6H_4SO_2$	102–104	142
None	$p\text{-}ClC_6H_4SO_2$	67–68	142

	–	365
	116–117	413

420

TABLE XV-60. Heavy Metal Derivatives of Sulfur Compounds of Pyridine

Compound	M.p. (°C)	Ref.
SHgPh	–	149
S BiPh	218	505

TABLE XV-61. Selenium Compounds of Pyridine

Compound	M.p. (°C)	Ref.
	132–137	193, 224
	72.5–73	224
	68–78	224
	B.p. 43–44°/0.25 mm (n_D^{25} 1.6190)	224
	121–124	224
	64	569
	215	569
	82	569

TABLE XV-61. Selenium Compounds of Pyridine (*Continued*)

Compound	M.P. (°C)	Ref.
pyridine with CONH$_2$ and SeMe	218	569
pyridine with COMe and SeMe	61	569
pyridine with CH$_2$OH and SeMe	58	569
pyridine with CHO and SeMe	49	569
pyridine with CO$_2$Et and SeCl	150	569
pyridine with CO$_2$Et and SeCH$_2$CO$_2$H	98	569
pyridine with CO$_2$H and SeCH$_2$CO$_2$H	195	569
pyridine with CO$_2$Et and SeCH$_2$CO$_2$Et	70	569

TABLE XV-61. Selenium Compounds of Pyridine (*Continued*)

Compound	M.p. (°C)	Ref.
Pyridine, CH:CHCO$_2$Et, SeMe	64	569
Pyridine, CH:CHCO$_2$H, SeMe	230	569
Pyridine, CH:CHCOMe, SeMe	62	569
[Pyridine-Se]$_2$	47.5–48	224
[Pyridine, CO$_2$Et, Se]$_2$	234	569
[Pyridine, CO$_2$H, Se]$_2$	226	569
H$_3$C-Pyridine-CH:CHSeH (*cis*)	62–63	173, 174

TABLE XV-61. Selenium Compounds of Pyridine (*Continued*)

Compound	M.p. (°C)	Ref.
SePh on pyridine, MeO_2C and CO_2Me at 2,6-positions	142–143	221
SePh on pyridine, HO_2C and CO_2H at 2,6-positions	203–205 (decomp.)	221
SePh on pyridine (2-SePh pyridine)	B.p. 143–145°/1.5 mm	221
O_2N, NO_2 pyridine with SeZ (Z = 1-(2,3,4,6-tetra-*O*-acetyl-β-D-gluco-pyranosyl)	127–129 $[\alpha]_D^{20}$ −88.4° (c, 2.5, $CHCl_3$)	348, 351, 352
O_2N pyridine with SeZ	176–178 $[\alpha]_D^{20}$ −3.6° (c, 5.0, $CHCl_3$)	348, 351, 352
O_2N, NO_2 pyridine with SeZ	211.5–213.5 $[\alpha]_D^{20}$ 22.7° (c, 2.5, $CHCl_3$)	348, 351, 352
MeCONH pyridine with SeZ	134–136 $[\alpha]_D^{20}$ −28.2° (c, 5.0, $CHCl_3$)	348
MeCONH pyridine with Se-β-D-glucosyl	$[\alpha]_D^{20}$ −71.8 (c, 2.5, H_2O)	348

TABLE XV-61. Selenium Compounds of Pyridine (*Continued*)

Compound	M.p. (°C)	Ref.
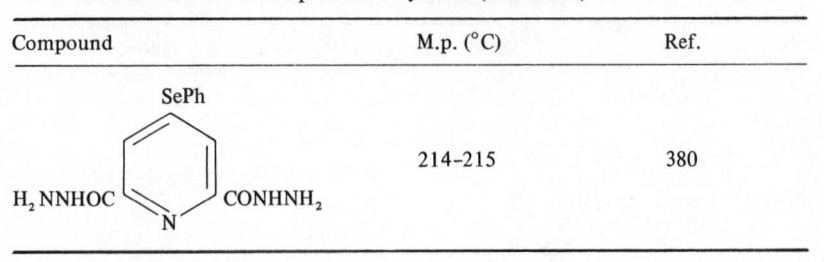	214–215	380

References

1. S. P. Acharya and K. S. Nargund, *J. Sci. Ind. Res.*, **21B**, 483(1962); *Chem. Abstr.*, **58**, 5623c(1963).
2. D. E. Ames and B. T. Warren, *J. Chem. Soc., Suppl. No. 1*, 5518(1964).
3. J. Angulo and A. M. Muñicio, *An. Real. Soc. Españ. Fís. Quím., Series B.*, **55**, 527(1959); *Chem. Abstr.*, **54**, 3292d(1960).
4. J. Angulo and A. M. Muñicio, *An. Real. Soc. Españ. Fís. Quím.*, **55**, 7409d(1961).
5. B. A. Arbuzov and N. N. Zobova, *Dokl. Akad. Nauk SSSR*, **167**, 815(1966); *Chem. Abstr.*, **65**, 3826e(1966).
6. F. S. Babichev and V. N. Bubnovskaya, *Ukr. Khim. Zh.*, **30**, 848(1964); *Chem. Abstr.*, **62**, 1766b(1965).
7. J. W. Baker and K. L. Godfrey, U.S. Patent 2,889,330; *Chem. Abstr.*, **54**, 585a(1960).
8. D. A. T. Baldry, *Bull. Entomol. Res.*, **55**, 49(1964); *Chem. Abstr.*, **63**, 6260f (1965).
9. V. Baliah and M. Seshapathirao, *J. Org. Chem.*, **24**, 867(1959).
10. S. J. Ball, E. W. Warren, and E. W. Parnell, *Nature*, **208**, 397(1965).
11. R. W. Balsiger, J. A. Montgomery, and T. P. Johnston, *J. Heterocycl. Chem.*, **2**, 97(1965).
12. G. B. Barlin and W. V. Brown, *J. Chem. Soc., C*, 2473(1967).
13. G. L. Barnes and R. S. Zerkel, *Plant Disease Rep.*, **45**, 426(1961); *Chem. Abstr.*, **55**, 22690g(1961).
14. L. Bauer and T. E. Dikerhofe, *J. Org. Chem.*, **29**, 2183(1964).
15. L. Bauer and L. A. Gardella, *J. Org. Chem.*, **28**, 1320(1963).
16. L. Bauer and L. A. Gardella, *J. Org. Chem.*, **28**, 1323(1963).
17. L. Bauer and A. L. Hirsch, *J. Org. Chem.*, **31**, 1210(1966).
18. R. M. Bechtle and G. H. Scherr, *Antibiotics Chemotherapy*, **9**, 715(1959); *Chem. Abstr.*, **54**, 12260e(1960).
19. R. Behnisch, F. Hoffmeister, H. Horstmann, E. Schraufstaetter, and W. Worth, *Med. Chem., Abhandl. Med.-Chem. Forschungsstaetten Farbwerke Hoechst A.-G.*, **7**, 296(1963); *Chem. Abstr.*, **61**, 5628c(1964).
20. L. J. Bellamy and P. E. Rogasch, *Proc. Roy. Soc., Series A.*, **257**, 98 (1960); *Chem. Abstr.*, **54**, 23831a(1960).
21. J. Bernstein and K. A. Losee, German Patent 1,224,744; *Chem. Abstr.*, **66**, 10853z(1967).
22. J. Bernstein and W. A. Lott, U.S. Patent 3,063,995; *Chem. Abstr.*, **58**, 9106e(1963).

23. J. Bernstein, W. A. Lott, and K. A. Losee, German Patent 1,140,578; *Chem. Abstr.*, **58**, 13920h(1963).
24. H. Berthod and A. Pullman, *C. R. Acad. Sci., Paris, Series, C*, **262**, 76 (1966); *Chem. Abstr.*, **64**, 18685b(1966).
25. H. C. Beyerman and W. Maassen van den Brink, *Proc. Chem. Soc.*, 266(1963).
26. J. B. Bicking, U.S. Patent 3,276,958; *Chem. Abstr.*, **65**, 20105g (1966).
27. J. F. Biellmann and H. Callot, *Tetrahedron Lett.*, 3991(1966).
28. F. Bjorksten, *Biochim. Biophys. Acta*, **127**, 265(1966).
29. E. J. Blanz, Jr., and F. A. French, *J. Med. Chem.*, **6**, 185(1963).
30. H. M. Blatter and H. Lukaszewski, *Tetrahedron Lett.*, 1087(1964).
31. C. H. Boehringer Sohn, Belgian Patent 621,644; *Chem. Abstr.*, **60**, 1712d(1964).
32. Boeringer Ingelhein, G. m. b. H., British Patent 1,018,805; *Chem. Abstr.*, **64**, 11251a(1966).
33. E. V. Bogacheva and V. Delo, USSR Patent 110,108; *Chem. Abstr.*, **64**, 2396b(1966).
34. S. V. Bogachova, *Fiziol. Zh., Akad. Nauk Ukr. RSR*, **11**, 686(1965); *Chem. Abstr.*, **64**, 13280e(1966).
35. I. B. G. Boggiano, V. Petrow, O. Stephenson, and A. M. Wild, *J. Pharm. Pharmacol.*, **13**, 567(1961).
36. F. Bonati and R. Sala, *Boll. Soc. Ital. Biol. Sper.*, **35**, 1749(1959); *Chem. Abstr.*, **56**, 13237h(1962).
37. C. K. Bradsher and D. F. Lohr, Jr., *J. Heterocycl. Chem.*, **3**, 27(1966).
38. C. K. Bradsher and J. W. McDonald, *J. Org. Chem.*, **27**, 4475(1962).
39. C. K. Bradsher and J. W. McDonald, *J. Org. Chem.*, **27**, 4478(1962).
40. C. K. Bradsher and J. W. McDonald, *J. Org. Chem.*, **27**, 4482(1962).
41. C. K. Bradsher, L. D. Quinn, R. E. LeBleu, and J. W. McDonald, *J. Org. Chem.*, **26**, 4944(1961).
42. B. A. Bridges, *Intern. J. Radiat. Biol.*, **3**, 49(1961); *Chem. Abstr.*, **58**, 8214a(1963).
43. B. B. Brown, U.S. Patent 3,207,760; *Chem. Abstr.*, **64**, 708e(1966).
44. J. S. A. Brunskill, *J. Chem. Soc., C*, 960(1968).
45. G. Buchmann and H. Franz, *Z. Chem.*, **6**, 107(1966).
46. D. H. Busch, D. C. Jicha, M. C. Thompson, J. W. Wrathall, and E. Blinn, *J. Am. Chem. Soc.*, **86**, 3642(1964).
47. J. R. Campbell, *J. Org. Chem.*, **29**, 1830(1964).
48. J. R. Campbell and R. E. Hatton, U.S. Patent 3,119,877; *Chem. Abstr.*, **60**, 10603c(1964).
49. R. N. Castle, K. Kaji, G. A. Gerhardt, W. D. Guiter, C. Weber, M. P. Malm, R. R. Shoup, and W. D. Rhoads, *J. Heterocycl. Chem.*, **3**, 79(1966).
50. J. F. Cavalla, *J. Chem. Soc.*, 4664(1962).
51. W. A. Chandler, *Plant Disease Rep.*, **49**, 419(1965); *Chem. Abstr.*, **63**, 6259f(1965).
52. L. Chevillard and M. C. Laury, *C. R. Acad. Sci., Paris, Ser. C*, **250**, 3746 (1960); *Chem. Abstr.*, **54**, 21484a(1960).
53. (a) Ciba Ltd, British Patent 879,194; *Chem. Abstr.*, **56**, 15490h(1962); (b) J. Druey and K. Meier, Swiss Patent 372,052; *Chem. Abstr.*, **60**, 13229g(1964).
54. Ciba Ltd, British Patent 881,506; *Chem. Abstr.*, **56**, 11605c(1962).
55. Ciba Ltd, British Patent 884,847; *Chem. Abstr.*, **58**, 3401e(1963).
56. F. E. Cislak and W. H. Rieger, U.S. Patent 3,219,667; *Chem. Abstr.*, **64**, 8150f(1966).
57. J. Clark, R. K. Grantham, and J. Lydiate, *J. Chem. Soc., C*, 1122(1968).
58. F. H. Clarke, U.S. Patent 3,118,884; *Chem. Abstr.*, **60**, 10696c(1964).

59. F. H. Clarke, G. B. Silverman, C. M. Watnick, and N. Sperber, *J. Org. Chem.*, **26**, 1126(1961).

60. P. Clechet and J. C. Merlin, *Bull. Soc. Chim. Fr.*, 2644(1964).

61. A. M. Comrie, *J. Chem. Soc.*, 688(1963).

62. (a) L. H. Conover, A. R. English, and C. E. Larrabee; U.S. Patent 2,921,073; *Chem. Abstr.*, **54**, 8860h(1960); (b) Chas. Pfizer and Co., Inc., British Patent 847,701; *Chem. Abstr.*, **55**, 11446e(1961).

63. L. H. Conover, U.S. Patent 3,027,379; *Chem. Abstr.*, **56**, 15619g(1962).

64. D. Cook, *Can. J. Chem.*, **42**, 2292(1964).

65. J. E. Cotton, *Diss. Abstr.*, **24**, 2673(1964); *Chem. Abstr.*, **60**, 15001b(1964).

66. E. J. Cragoe, U.S. Patent 3,155,653; *Chem. Abstr.*, **62**, 4043a(1965).

67. E. J. Cragoe, Jr., J. A. Nicholson, and J. M. Sprague, *J. Med. Chem.*, **4**, 369(1961).

68. W. Czuba, *Bull. Acad. Polon. Sci., Ser. Sci. Chim.*, **8**, 281(1960); *Chem. Abstr.*, **60**, 2883g(1964).

69. W. Czuba, *Rocz. Chem.*, **35**, 1347(1961); *Chem. Abstr.*, **57**, 8543g(1962).

70. J. A. W. Dalziel and M. Thompson, *Analyst* (London), **89**, 707(1964).

71. J. A. W. Dalziel and M. Thompson, *Analyst* (London), **91**, 90(1966).

72. J. J. D'Amico and A. G. Weiss, U.S. Patent 3,151,024; *Chem. Abstr.*, **62**, 9157h(1965).

73. J. J. D'Amico, U.S. Patent 3,155,671; *Chem. Abstr.*, **62**, 4015a(1965).

74. A. Deavin and D. W. Rees, *J. Chem. Soc.*, 4970(1961).

75. J. Dekker and P. A. Ark, *Physiol. Plant.*, **12**, 888(1959); *Chem. Abstr.*, **60**, 3275e(1964).

76. C. S. Delahunt, R. B. Stebbins, J. Anderson, and J. Bailey, *Toxicol. Appl. Pharmacol.*, **4**, 286(1962).

77. J. Delarge, *Farmaco, Ed. Sci.*, **20**, 629(1965); *Chem. Abstr.*, **64**, 3467e(1966).

78. J. Delarge, *Farmaco, Ed. Sci.*, **22**, 99(1967); *Chem. Abstr.*, **66**, 94891d(1967).

79. J. Delarge, *Farmaco, Ed. Sci.*, **22**, 1069(1967); *Chem. Abstr.*, **69**, 2830u(1968).

80. J. Delarge, *J. Pharm. Belg.*, **22**, 257(1967); *Chem. Abstr.*, **68**, 104913v(1968).

81. J. Delarge and C. L. Lapiere, *Bull. Soc. Chim. Belges*, **75**, 32(1966).

82. V. Denes, M. Farcasan, and G. Ciurdaru, *Chem. Ber.*, **96**, 174(1963).

83. G. DeStevens, M. Sklar, H. Lukaszewski, and L. Witkin, *J. Med. Chem.*, **5**, 919(1962).

84. Deutsche Gold-und Silber-Scheideanstalt vorm. Roessler, Netherlands Patent 6,407,791; *Chem. Abstr.*, **63**, 1797h(1965).

85. Z. Deyl and J. Rosmus, *J. Chromatogr.*, **8**, 537(1962); *Chem. Abstr.*, **58**, 9400c(1963).

86. T. E. Dickerhofe, *Diss. Abstr. B*, **28**, 1870(1967).

87. V. K. N. Dixit, S. D. Jolad, and S. Rajagopal, *Monatsh.*, **94**, 414(1963); *Chem. Abstr.*, **59**, 7485c(1963).

88. B. Dmuchovsky, F. B. Zienty, and W. A. Vredenburgh, *J. Org. Chem.*, **31**, 865(1966).

89. R. Dohmori, *Chem. Pharm. Bull.* (Tokyo), **12**, 591, 601 (1964); *Chem. Abstr.*, **61**, 4248h, 4718 (1964).

90. R. Dohmori, *Chem. Pharm. Bull.* (Tokyo), **12**, 595 (1964); *Chem. Abstr.*, **61**, 4305 (1964).

91. Dow Chemical CO., Netherlands Patent 6,515,950; *Chem. Abstr.*, **65**, 15338e(1966).

92. Dreizen, C. N. Spirakis, and R. E. Sloan, *Int. Z. Vitaminforsch.*, **33**, 321(1963); *Chem. Abstr.*, **60**, 16349d(1964).

93. R. S. Driscoll, *U.S. Dept. Agr. Forest Serv. Pacific Northwest Forest Range Exptl. Sta. Res.* Note PNW-5, 1963, 8 pp.; *Chem. Abstr.*, **61**, 15287b(1964).

94. B. F. Duesel and S. Emmanuele, U.S. Patent 3,218,330; *Chem. Abstr.*, **64**, 3503c(1966).
95. B. F. Duesel and L. S. Scarano, U.S. Patent 3,267,109; *Chem. Abstr.*, **65**, 16947b(1966).
96. E. L. Dulaney and E. O. Stapley, *Appl. Microbiol.*, **7**, 276(1959); *Chem. Abstr.*, **54**, 6871b(1960).
97. M. A.-F. Elkaschef, M. H. Nosseir, and A. Abdel-Kader, *J. Chem. Soc.*, 4647(1963).
98. I. E. El-Kholy, F. K. Rafla, and G. Soliman, *J. Chem. Soc.*, 4490(1961).
99. P. F. Epstein and M. E. Brooke, U.S. Patent 3,304,226; *Chem. Abstr.*, **67**, 32598a(1967).
100. Etablissements Kuhlmann, Netherlands Patent 6,510,350; *Chem. Abstr.*, **65**, 2393c(1966).
101. R. F. Evans and H. C. Brown, *J. Org. Chem.*, **27**, 1665(1962).
102. R. F. Evans and H. C. Brown, *J. Org. Chem.*, **27**, 3127(1962).
103. R. F. Evans, H. C. Brown, and H. C. van der Plas, *Org. Synth.*, **43**, 97(1963).
104. Farbenfabriken Bayer A.-G, British Patent 1,018,317; *Chem. Abstr.*, **64**, 17554a(1966).
105. I. Kh. Fel'dman and A. V. Voropaeva, *Tr. Leningr. Khim.-Farmatsevt. Inst.*, 17(1962); *Chem. Abstr.*, **61**, 638g(1964).
105a. I. Kh. Fel'dman and A. V. Voropaeva, *Tr. Leningr. Khim.-Farmatsevt. Inst.* 20(1962); *Chem. Abstr.*, **61**, 639b(1964).
106. I. Kh. Fel'dman and A. V. Voropaeva, *Zh. Obshch. Khim.*, **32**, 1738(1962); *Chem. Abstr.*, **58**, 4505g(1963).
107. I. Kh. Fel'dman and A. V. Voropaeva, *Zh. Obshch. Khim.*, **33**, 269(1963); *Chem. Abstr.*, **58**, 13910c(1963).
108. I. Kh. Fel'dman and A. V. Voropaeva, *Zh. Obshch. Khim.*, **34**, 1547(1964); *Chem. Abstr.*, **61**, 5605a(1964).
109. Ferrosan Aktieselskabet, Danish Patent 107,422; *Chem. Abstr.*, **69**, 10376r(1968).
110. L. Field, H. Haerle, T. C. Owen, and A. Ferretti, *J. Org. Chem.*, **29**, 1632(1964).
111. K. Flemming, *Naturwiss.*, **48**, 555(1961); *Chem. Abstr.*, **56**, 5079i(1962).
112. W. O. Foye, M. C. M. Solis, J. W. Schermerhorn, and E. L. Prien, *J. Pharm. Sci.*, **54**, 1365(1965); *Chem. Abstr.*, **63**, 13614b(1965).
113. J. E. Franz, M. W. Dietrich, A. Henshall, and C. Osuch, *J. Org. Chem.*, **31**, 2847(1966).
114. J. E. Franz and C. Osuch, *J. Org. Chem.*, **29**, 2592(1964).
115. M. Freifelder and H. B. Wright, *J. Med. Chem.*, **7**, 664(1964).
116. W. J. French and A. W. Helton, *Phytopathology*, **52**, 810(1962); *Chem. Abstr.*, **58**, 887h(1963).
117. F. Friedrich and R. Pohloudek-Fabini, *Arch Pharm.* (Weinheim), **298**, 162(1965); *Chem. Abstr.*, **63**, 6971f(1965).
118. H. Fuerst, M. Heltzig, and W. Goebel, *J. Prakt. Chem.*, **36**, 160(1967).
119. H. Fuerst and W. Kretzschmann, German (East) Patent 23,315; *Chem. Abstr.*, **58**, 9088d(1963).
120. (a) Fujisawa Pharmaceutical Company, Ltd., Japan Patent 9227/68; (b) Fujisawa Pharmaceutical Company, Ltd., Japan Patent 9228/68.
121. Y. Funatsukuri, Japan Patent 18,041/64; *Chem. Abstr.*, **62**, 5257d(1965).
122. F. Gadient, E. Jucker, A. Lindenmann, and M. Taeschler, *Helv. Chim. Acta*, **45**, 1860(1962).
123. A. W. Galston, R. Kaur, N. Maheshwari, and S. C. Maheshwari, *Amer. J. Bot.*, **50**, 487(1963); *Chem. Abstr.*, **59**, 10465d(1963).
124. W. Gamble and L. D. Wright, *Proc. Soc. Exp. Biol. Med.*, **107**, 160(1961).
125. J. R. Geigy A.-G., British Patent 1,077,036; *Chem. Abstr.*, **68**, 8716u(1968).

126. J. R. Geigy A.-G., Netherlands Patent 6,509,930; *Chem. Abstr.*, **64**, 19571b(1966).
127. J. R. Geigy A.-G., French Patent 1,477,997; *Chem. Abstr.*, **68**, 39483p(1968).
128. J. R. Geigy A.-G., French Patent 1,477,998; *Chem. Abstr.*, **68**, 29606s(1968).
129. K. Gerlach and F. Kröhnke, *Chem. Ber.*, **95**, 1124(1962).
130. K. Gerwald, *J. Prakt. Chem.*, **31**, 205(1966).
131. R. Graf, D. Guenther, H. Jensen, and K. Matterstock, German Patent 1,144,718; *Chem. Abstr.*, **59**, 6368c(1963).
132. D. R. Grassetti, M. E. Brokke, and J. F. Murray, Jr., *J. Med. Chem.*, **8**, 753(1965).
133. J. L. Greene, Jr., and J. A. Montgomery, *J. Med. Chem.*, **7**, 17(1964).
134. J. L. Greene, Jr., A. M. Williams, and J. A. Montgomery, *J. Med. Chem.*, **7**, 20(1964).
135. J. Grosselck, L. Beress, H. Schenk, and G. Schmidt, *Angew. Chem. Intern. Ed. Engl.*, **4**, 1080(1965).
136. T. J. Haley, A. M. Flesher, and L. Mavis, *Arch. Int. Pharmacodyn. Ther.*, **138**, 133(1962).
137. R. Haller, *Arch. Pharm.* (Weinheim), **298**, 306(1965); *Chem. Abstr.*, **63**, 5613h(1965).
138. M. Hamana and M. Yamazaki, *Yakugaku Zasshi*, **81**, 612(1961); *Chem. Abstr.*, **55**, 24742f(1961).
139. Harshaw Chemical Company, German Patent 1,150,855; *Chem. Abstr.*, **59**, 10993b(1963).
140. E. Hayashi, H. Yamanaka, and C. Iijima, *Yakugaku Zasshi*, **80**, 1145(1966); *Chem. Abstr.*, **55**, 546d(1961).
141. A. L. Hayden and M. Maienthal, *J. Ass. Offic. Agr. Chem.*, **48**, 596(1965); *Chem. Abstr.*, **63**, 6793g(1965).
142. D. F. Hayman, V. Petrow, and O. Stephenson, *J. Pharm. Pharmacol.*, **14**, 522(1962).
143. A. W. Helton and W. J. French, *Phytopathology*, **52**, 1050(1962); *Chem. Abstr.*, **58**, 2794a(1963).
144. A. W. Helton and A. E. Harvey, *Phytopathology*, **53**, 895(1963); *Chem. Abstr.*, **60**, 3279h(1964).
145. A. W. Helton and K. G. Rohrbach, *Phytopathology*, **56**, 939(1966); *Chem. Abstr.*, **65**, 17626h(1966).
146. Hercules Powder Company, British Patent 979,658; *Chem. Abstr.*, **63**, 3136f(1965).
147. A. G. Hovey, U.S. Patent 2,909,459; *Chem. Abstr.*, **54**, 1796g(1960).
148. T. J. Hurley and M. A. Robinson, *J. Med. Chem.*, **8**, 888(1965).
149. T. Iida and E. Tomitori, Japanese Patent 21,362/64; *Chem. Abstr.*, **62**, 10460b(1965).
150. Imperial Chemical Industries Ltd., Netherlands Application 6,516,409; *Chem. Abstr.*, **65**, 18564e(1966).
151. Imperial Chemical Industries Ltd., Netherlands Application 6,611,714; *Chem. Abstr.*, **68**, 21842v(1968).
152. Imperial Chemical Industries Ltd., Netherlands Application 6,611,766; *Chem. Abstr.*, **68**, 59438f(1968).
153. M. Inoue, French Patent 1,396,624; *Chem. Abstr.*, **63**, 5610a(1965).
154. T. Irikura, K. Shirai, and S. Sato, *Yakugaku Zasshi*, **84**, 793 (1964); *Chem. Abstr.*, **62**, 2775g(1965).
155. Irwin, Neisler and Company, British Patent 842,995; *Chem. Abstr.*, **55**, 7437i(1961).
156. M. Ishikawa, H. Ishikawa, and M. Inoue, Japanese Patent 8620/67; *Chem. Abstr.*, **68**, 2815x(1968).
157. M. Iwanami, I. Osawa, and M. Murakami, *Bitamin*, **36**, 119(1967); *Chem. Abstr.*, **68**, 12817p(1968).

158. M. Iwanami, I. Osawa, and M. Murakami, *Bitamin*, 36, 122(1967); *Chem. Abstr.*, 68, 12817p(1968).
159. D. Jerchel, German Patent 1,225,178; *Chem. Abstr.*, 65, 18565d(1966).
160. D. Jerchel and K. Thomas, U.S. Patent 3,121,088; *Chem. Abstr.*, 60, 13229c(1964).
161. F. Johnson, U.S. Patent 3,247,214.
162. H. Johnston, U.S. Patent 3,296,272; *Chem. Abstr.*, 66, 104903a(1967).
163. H. Johnston and M. S. Tomita, U.S. Patent 3,244,722; *Chem. Abstr.*, 65, 8884b(1966).
164. J. E. Jones, U.S. Patent 2,977,229; *Chem. Abstr.*, 55, 14136d(1961).
165. J. P. Jones, *Plant Disease Reptr.*, 45, 376(1961); *Chem. Abstr.*, 55, 21446f(1961).
166. W. Jucker and A. Ebnoether, U.S. Patent 3,086,972; *Chem. Abstr.*, 59, 10057c(1963).
167. L. F. Judge and D. J. Kooyman, German Patent 1,183,203; *Chem. Abstr.*, 63, 2003a(1965).
168. Ch. Sh. Kadyrov and E. V. Bicheroy, USSR Patent 172,157; *Chem. Abstr.*, 63, 17071e(1965).
169. K. Kahmann, H. Sigel, and H. Erlenmeyer, *Helv. Chim. Acta*, 47, 1754(1964).
170. K. S. Karsten and W. S. Taylor, British Patent 957,458; *Chem. Abstr.*, 61, 5911h(1964).
171. K. S. Karsten, W. S. Taylor, and J. J. Purran, U.S. Patent 3,236,733.
172. S. Kasuga and T. Taguchi, *Chem. Pharm. Bull.* (Tokyo), 13, 233(1965); *Chem. Abstr.*, 63, 6961c(1965).
173. E. G. Kataev and V. N. Petrov, *Zh. Obshch. Khim.*, 32, 3699(1962); *Chem. Abstr.*, 59, 481f(1963).
174. L. M. Kataeva, I. V. Anonimova, L. K. Yuldasheva, and E. G. Kataev, *Zh. Obshch. Khim.*, 32, 3965(1962); *Chem. Abstr.*, 59, 407b(1963).
175. H. Kato, K. Tanaka, and M. Ohta, *Bull. Chem. Soc. Japan*, 39, 1248(1966); *Chem. Abstr.*, 65, 13681a(1966).
176. H. T. Kemp, Jr., C. W. Cooper, and R. M. Kell, *J. Paint Technol.*, 38, 363(1966); *Chem. Abstr.*, 65, 10800h(1966).
177. K. C. Kennard and D. M. Burness, U.S. Patent 2,937,090; *Chem. Abstr.*, 54, 22119a(1960).
178. (a) E. Kinkel, H. Seiz, and H. Jerenz, German Patent 1,140,813; *Chem. Abstr.*, 58, 7543b(1963). (b) Koepff and Soehne, G. M. B. H., British Patent 911,886; *Chem. Abstr.*, 58, 4086f(1963).
179. A. H. M. Kirby and E. L. Frick, *Ann. Rept. East Malling Res. Sta., Kent, 1962*, 117(1963); *Chem. Abstr.*, 60, 3431c(1964).
180. R. W. Kluiber and W. De W. Horrocks, Jr., *Inorg. Chem.*, 5, 1816(1966).
181. W. Knoth and R. C. Knoth-Born, *Arzneim.-Forsch.*, 11, 981(1961).
182. R. Koch, D. Klemm, and I. Seiter, *Arzneim.-Forsch.*, 10, 683(1960).
183. R. Koch and K. Markau, *Z. Naturforsch.*, 166, 586(1961); *Chem. Abstr.*, 56, 8193d(1962).
184. R. Koch, I. Onderka, and I. Seiter, *Arzneim.-Forsch.*, 12, 265(1962).
185. R. Koch and U. Schmidt, *Strahlentherapie*, 113, 89(1960); *Chem. Abstr.*, 55, 774g(1961).
186. D. J. Kooyman, U.S. Patent 3,235,455.
187. V. S. Korobkov, A. V. Voropaeva, and I. Kh. Fel'dman, *Zh. Obshch. Khim.*, 31, 3136(1961); *Chem. Abstr.*, 56, 15059b(1962).
188. A. N. Kost, P. B. Terent'ev, and M. A. Chernova, *Vestn. Mosk. Univ., Ser. II, Khim.*, 19, 59(1964); *Chem. Abstr.*, 61, 3064e(1964).

432   Sulfur and Selenium Compounds of Pyridine

189. L. B. Kotnis, M. V. Narurkar, and M. B. Sahasrabudhe, *Brit. J. Cancer*, **16**, 541(1962); *Chem. Abstr.*, **58**, 9496d(1963).
190. L. M. Kozloff, U.S. Patent 2,945,735; *Chem. Abstr.*, **54**, 21627b(1960).
191. O. K. Koz'minykh, *Uchenye Zapiski, Perm. Gosudarst. Univ. A. M. Gor'kogo*, **15**(4), 111(1958); *Chem. Abstr.*, **55**, 27313c(1961).
192. M. Kozu, Japanese Patent 2706/67; *Chem. Abstr.*, **67**, 73531s(1967).
193. M. H. Krackov, C. M. Lee, and H. G. Mautner, *J. Amer. Chem., Soc.*, **87**, 892(1965).
194. H. G. Kraft, L. Fiebig, and R. Hotvy, *Arzneim.-Forsch.*, **11**, 922(1961).
195. M. M. Kreevoy, B. E. Eichinger, F. E. Stary, E. A. Katz, and J. H. Sallstedt, *J. Org. Chem.*, **29**, 1641(1964).
196. W. Kretzschmann and H. Fuerst, *Chem. Tech.* (Berlin), **15**, 559(1963); *Chem. Abstr.*, **59**, 15291f(1963).
197. D. Kritchevsky and S. A. Tepper, *Arch. Int. Pharmacodyn. Ther.*, **147**, 564(1964).
198. D. Kritchevsky and S. A. Tepper, *J. Nutr.*, **82**, 157(1964); *Chem. Abstr.*, **60**, 12560b(1964).
199. D. Kritchevsky and S. A. Tepper, *Proc. Soc. Exptl. Biol. Med.*, **115**, 841(1964).
200. A. F. Krivis, E. S. Gazda, G. R. Supp, and M. A. Robinson, *Anal. Chem.*, **35**, 966(1963).
201. F. Kroehnke and K. Gerlach, *Chem. Ber.*, **95**, 1108(1962).
202. F. Kroehnke, K. Gerlach, and K.-E. Schnalke, *Chem. Ber.*, **95**, 1118(1962).
203. S. Kubota and T. Akita, *Yakugaku Zasshi*, **81**, 511(1961); *Chem. Abstr.*, **55**, 19926b(1961).
204. M. M. Kulik and E. C. Tims, *Plant Disease Rep.*, **44**, 379(1960); *Chem. Abstr.*, **54**, 25484d(1960).
205. G. Kunesch and F. Wessely, *Monatsh. Chem.*, **96**, 1547(1965).
206. (a) T. Kuroda, *Bitamin.*, **30**, 431(1964); *Chem. Abstr.*, **62**, 10402c(1965); (b) T. Kuroda and M. Masaki, *Bitamin.*, **30**, 436(1964); *Chem. Abstr.*, **62**, 10402c(1965).
207. H. Kwart and R. W. Body, *J. Org. Chem.*, **30**, 1188(1965).
208. A. F. Langlykke, J. Bernstein, and A. L. La Via, U.S. Patent 3,346,578.
209. E. Laville, *Fruits*, **16**, 442(1961); *Chem. Abstr.*, **56**, 10625d(1962).
210. Z. H. Levinson and E. D. Bergmann, *J. Insect Physiol.*, **3**, 293(1959); *Chem. Abstr.*, **54**, 13465b(1960).
211. L. F. Lindoy, S. E. Livingstone and T. N. Lockyer, *Aust. J. Chem.*, **18**, 1549(1965); *Chem. Abstr.*, **64**, 4564g(1966).
212. B. Loev and J. F. Olin, U.S. Patent 3,069,472; *Chem. Abstr.*, **59**, 7495g(1963).
213. W. Lorenz, Belgian Patent 635,443; *Chem. Abstr.*, **61**, 16096d(1964).
214. Y. Maki, *Yakugaku Zasshi*, **77**, 862(1957); *Chem. Abstr.*, **52**, 1174b(1958).
215. Y. Maki, K. Kawasaki, and K. Sato, *Gifu Yakka Daigaku Kiyo*, **13**, 34(1963); *Chem. Abstr.*, **60**, 13220e(1964).
216. Y. Maki, Y. Okada, Y. Yoshida, and K. Obata, *Gifu Yakka Daigaku Kiyo*, **12**, 54(1962); *Chem. Abstr.*, **59**, 11479b(1963).
217. Y. Maki, M. Sato, and K. Yumane, *Yakugaku Zasshi*, **85**, 429(1965); *Chem. Abstr.*, **63**, 5477f(1965).
218. Y. Maki, K. Yamane, and T. Masugi, *Gifu Yakka Daigaku Kiyo*, (15) 31(1965); *Chem. Abstr.*, **68**, 21796h(1968).
219. Y. Maki, K. Yamane, and M. Sato, *Yakugaku Zasshi*, **86**, 50(1966); *Chem. Abstr.*, **64**, 11165a(1966).
220. O. Manousek and P. Zuman, *Collect. Czech. Chem. Commun.*, **29**, 1432(1964).
221. D. G. Markees, *J. Org. Chem.*, **28**, 2530(1963).
222. E. Maruszewska-Wieczorkowska and J. Michalski, *Rocz. Chem.*, **38**, 625(1964); *Chem. Abstr.*, **61**, 10702h(1964).

223. A. Matta and A. Vittone, *Notiz. Malattie Piante,* **52,** 91(1960); *Chem. Abstr.,* **55,** 14795a(1961).
224. H. G. Mautner, S. H. Chu, and C. M. Lee, *J. Org. Chem.,* **27,** 3671(1962).
225. C. E. Maxwell and P. N. Gordon, U.S. Patent 3,056,798; *Chem. Abstr.,* **58,** 5645c(1963).
226. R. E. McClure and A. Ross, *J. Org. Chem.,* **27,** 304(1962).
227. R. E. McClure and D. A. Shermer, U.S. Patent 3,159,640; *Chem. Abstr.,* **62,** 7732e(1965).
228. E. Merck A.-G., British Patent 852,398; *Chem. Abstr.,* **55,** 10477g(1961).
229. E. Merck A.-G., French Patent M 948; *Chem. Abstr.,* **58,** 9032f(1963).
230. Merck and Company, Inc., British Patent 944,506; *Chem. Abstr.,* **60,** 15843e(1964).
231. E. Merck A.-G., Belgian Patent 659,401; *Chem. Abstr.,* **64,** 3500c(1966).
232. E. Merck A.-G., Netherlands Patent 6,412,891; *Chem. Abstr.,* **64,** 8154c(1966).
233. N. V. Monich and A. F. Vompe, USSR Patent 154,280; *Chem. Abstr.,* **60,** 5466g(1964).
234. Monsanto Chemicals Ltd., French Patent 1,368,623; *Chem. Abstr.,* **62,** 2903e(1965).
235. Monsanto Co., Netherlands Patent 286,724; *Chem. Abstr.,* **63,** 8264e(1964).
236. J. Montgomery and K. Hewson, *J. Med. Chem.,* **9,** 354(1966).
237. T. Naito, R. Dohmori, and T. Kotake, *Chem. Pharm. Bull.* (Tokyo), **12,** 588(1964).
238. M. Nakamura and B. L. Pitsch, *Rev. Biol. Trop. Univ. Costa Rica,* **10,** 35(1962); *Chem. Abstr.,* **58,** 9433b(1963).
239. R. Neher and F. W. Kahnt, *Experientia,* **21,** 310(1965).
240. D. Neely, E. B. Himelick, and G. L. DeBarr, *Phytopathology,* **56,** 203(1966); *Chem. Abstr.,* **65,** 11265(1966).
241. V. Nigrovic, *Arzneim.-Forsch.,* **13,** 787(1963).
242. K. Nordstrom, *J. Inst. Brew.,* **70,** 209(1964); *Chem. Abstr.,* **61,** 8658e(1964).
243. L. Novacek, K. Palat, M. Celadnik, and E. Matuskova, *Cesk. Farm.,* **11,** 76(1962); *Chem. Abstr.,* **57,** 15067d(1962).
244. L. Nutting, R. M. Silverstein, and C. M. Himel, U.S. Patent 2,905,701; *Chem. Abstr.,* **54,** 12163e(1960).
245. L. Nutting, R. M. Silverstein, and C. M. Himel, U.S. Patent 2,951,848; *Chem. Abstr.,* **55,** 4542b(1961).
246. C. O. Obenland, U.S. Patent 3,310,568; *Chem. Abstr.,* **67,** 82111g(1967).
247. T. Okamoto and M. Itoh, *Chem. Pharm. Bull.* (Tokyo), **11,** 785(1963).
248. Olin Mathieson Chemical Corp., British Patent 897,900; *Chem. Abstr.,* **57,** 15077b(1962).
249. Omikron-Ctagliardi Société di Fatto, British Patent 833,049; *Chem. Abstr.,* **55,** 1665a(1961).
250. O. S. Otroschenko, Yu. V. Kurbatov, and A. S. Sadykov, *Nauchn. Tr., Tashkentsk. Gos. Univ.,* **263,** 27(1964); *Chem. Abstr.,* **63,** 4248c(1965).
251. O. S. Otroshcenko, Y. V. Kurbatov, and A. S. Sadykov, *Nauchn. Tr. Tashkentsk. Gos. Univ.,* **286,** 88(1966); *Chem. Abstr.,* **67,** 99969r(1968).
252. O. S. Otroshchenko, Yu. V. Kurbatov, A. S. Sadykov and F. Pirnazarova, *Nauchn. Tr. Tashkentsk. Gos. Univ.,* **263,** 33(1964); *Chem. Abstr.,* **63,** 4248e(1965).
253. O. S. Otroshchenko, A. A. Ziyaev, and A. S. Sadykov, *Nauchn. Tr. Tashkentsk. Gos. Univ.,* **286,** 85(1966); *Chem. Abstr.,* **67,** 82057u(1967).
254. G. F. Ottmann and I. M. Voynick, U.S. Patent 3,347,863; *Chem. Abstr.,* **68,** 87164g(1968).
255. A. S. Palacios, *Arch. Inst. Farmacol. Exptl.* (Madrid), **16,** 1(1964); *Chem. Abstr.,* **62,** 4482d(1965).

256. K. Palat, M. Celadnik, L. Novacek, M. Polster, R. Urbancik, and E. Matuskova, *Acta Fac. Pharm. Brun. Bratislav.*, **4**, 65(1962); *Chem. Abstr.*, **57**, 4769a(1962).

257. L. J. Pandya and B. D. Tilak, *J. Sci. Ind. Res.*, **18B**, 371(1959); *Chem. Abstr.*, **54**, 17391b(1960).

258. E. W. Parnell, British Patent 1,055,266; *Chem. Abstr.*, **68**, 39488u(1968).

259. R. A. Partyka, U.S. Patent 3,290,319; *Chem. Abstr.*, **66**, 115609k(1967).

260. V. D. Patil, S. P. Acharya, and K. S. Nargund, *J. Sci. Ind. Res.*, **21B**, 451(1962); *Chem. Abstr.*, **58**, 2433c(1963).

261. D. Perlman, *Antimicrobial Agents Chemotherapy*, 114(1964); *Chem. Abstr.*, **63**, 3359f(1965).

262. V. G. Pesin and I. G. Vitenberg, *Zh. Obshch. Khim.*, **26**, 1268(1966); *Chem. Abstr.*, **65**, 16956g(1966).

263. L. J. Peterson, R. Baker, and R. E. Skiver, *Plant Disease Rep.*, **43**, 1204(1959); *Chem. Abstr.*, **54**, 15807a(1960).

264. L. A. Petrova and N. N. Bel'tsova, *Zh. Obshch. Khim.*, **32**, 274(1962); *Chem. Abstr.*, **57**, 16543g (1962).

265. N. V. Philips'Gloeilampenfabrieken, Belgian Patent 618,679; *Chem. Abstr.*, **59**, 9999b(1963).

266. H. C. van der Plas and T. H. Crawford, *J. Org. Chem.*, **26**, 2611(1961).

267. H. C. Van der Plas and H. J. den Hertog, *Rec. Trav. Chim. Pays-Bas.*, **81**, 841(1962).

268. T. B. Platt, J. Gentile, and M. J. George, *Ann. N. Y. Acad. Sci.*, **130**, 664(1965).

269. R. Ponci, A. Baruffini, F. Gialdi, and P. Borgna, *Farmaco, Ed. Sci.*, **19**, 489(1964); *Chem. Abstr.*, **61**, 8217g(1964).

270. B. Prescott and J. H. Stone, *Farmaco, Ed. Sci.*, **21**, 471(1966); *Chem. Abstr.*, **65**, 11182d(1966).

271. E. Profft, *Wiss. Z. Tech. Hochsch. Chem. Leuna-Merseburg*, **2**, 237 (1959-1960); *Chem. Abstr.*, **55**, 889b(1961).

272. E. Profft and G. Busse, *Z. Chem.*, **1**, 19(1961); *Chem. Abstr.*, **55**, 25939h(1961).

273. E. Profft and W. Rolle, *Wiss. Z. Tech. Hochsch. Chem. Leuna-Merseburg*, **2**, 187(1959-1960); *Chem. Abstr.*, **55**, 1609h(1961).

274. E. Profft and R. Schmuck, *Arch. Pharm.* (Weinheim), **296**, 209(1963); *Chem. Abstr.*, **59**, 6354f(1963).

275. E. Profft and W. Steinke, *J. Prakt. Chem.*, **13**, 58(1961).

276. G. Quadbeck, H. R. Landmann, W. Sachsse, and I. Schmidt, *Med. Exptl.*, **7**, 144(1962); *Chem. Abstr.*, **59**, 7274a(1963).

277. F. C. Quebral and L. N. Gibe, *Philippine Agriculturist*, **43**, 271(1959); *Chem. Abstr.*, **55**, 22691g(1961).

278. H. L. Rawlings and J. J. D'Amico, U.S. Patent 3,107,994; *Chem. Abstr.*, **60**, 6828h(1964).

279. H. L. Rawlings and J. J. D'Amico, U.S. Patent 3,150,145; *Chem. Abstr.*, **61**, 13288a(1964).

280. W. Reifschneider and J. S. Kelyman, U.S. Patent 3,335,149; *Chem. Abstr.*, **68**, 87177p(1968).

281. J. Renault and J. C. Cartron, *Chim. Ther.*, **1966**, 337.

282. V. M. Reznidov and V. I. Bliznyukov, *Nekotorye Vopr. Emission. Molekulyarh. Spektroskopii, Kras noyarsk, Sb. 1960*, 193; *Chem. Abstr.*, **58**, 9768c(1963).

283. C. Richter and G. Feth, U.S. Patent 3,336,324; *Chem. Abstr.*, **68**, 87169n(1968).

284. M. A. Robinson, U.S. Patent 3,146,237; *Chem. Abstr.*, **61**, 10661c(1964).

285. M. A. Robinson, *J. Inorg. Nucl. Chem.*, **26**, 1277(1964).

286. M. A. Robinson and T. J. Hurley, *Inorg. Chem.*, **4**, 1716(1965).

287. J. Rockett and B. B. Brown, U.S. Patent 2,922,790; *Chem. Abstr.*, **54**, 8857g(1960).
288. O. R. Rodig, R. E. Collier, and R. K. Schlatzer, *J. Org. Chem.*, **29**, 2652(1964).
289. O. R. Rodig, R. E. Collier, and R. K. Schlatzer, *J. Med. Chem.*, **9**, 116(1966).
290. E. F. Rogers and R. L. Clark, U.S. Patent 3,202,576; *Chem. Abstr.*, **63**, 14641e(1965).
291. W. C. J. Ross, *J. Chem. Soc., C*, 1816(1966).
292. L. Rossi and E. Pasargiklian, *Studi Sassaresi, Sezione I*, **39**, 93(1961); *Chem. Abstr.*, **57**, 7868b(1962).
293. H. Saikachi and T. Hisano, *Chem. Pharm. Bull.* (Tokyo), **7**, 716(1959).
294. Sandoz Ltd., Belgian Patent 611,216; *Chem. Abstr.*, **59**, 1461g(1963).
295. G. A. Saul, U.S. Patent 2,961,445; *Chem. Abstr.*, **55**, 8441c(1961).
296. G. H. Scherr and R. M. Bechtle, *Antibiotics Ann., 1958-1959*, 855(1959); *Chem. Abstr.*, **54**, 13452a(1960).
297. R. K. Schlatzer, Jr., *Diss. Abstr. B*, **27**, 2658(1967).
298. U. Schmidt and G. Giesselmann, *Ann. Chem.*, **657**, 162(1962).
299. A. M. Schnitzer, U.S. Patent 3,121,725; *Chem. Abstr.*, **61**, 4323h(1964).
300. G. Schorre, German Patent 1,238,473; *Chem. Abstr.*, **68**, 39479s(1968).
301. G. Schorre, German Patent 1,238,474; *Chem. Abstr.*, **68**, 39480k(1968).
302. G. Schorre, German Patent 1,244,786; *Chem. Abstr.*, **68**, 49466z(1968).
303. C. H. Schramm, Belgian Patent 616,540; *Chem. Abstr.*, **58**, 12776e(1963).
304. C. H. Schramm, British Patent 970,955; *Chem. Abstr.*, **61**, 14904d(1964).
305. E. Schraufstätter, G. Schmidt-Kastner, and W. Wirth, German Patent 1,078,129; *Chem. Abstr.*, **55**, 22345b(1961).
306. O. E. Schultz and K. K. Gauri, *Arch. Pharm.* (Weinheim), **295**, 146(1962); *Chem. Abstr.*, **57**, 16579f(1962).
307. A. Senning, *Acta Chem. Scand.*, **18**, 269(1964).
308. J. T. Sheehan and H. L. Yale, U.S. Patent 3,157,647.
309. T. Shimamoto, M. Ishikawa, and M. Inoue, Japan Patent 8620/67; *Chem. Abstr.*, **68**, 2815t(1968).
310. A. K. Sijpesteijn, U.S. Patent 3,022,216; *Chem. Abstr.*, **57**, 14236g(1962).
311. A. K. Sijpesteijn, J. Kaslander, and G. J. M. Van Der Kerk, *Biochim. Biophys. Acta*, **62**, 587(1962); *Chem. Abstr.*, **57**, 17186d(1962).
312. A. K. Sijpesteijn, J. E. Rombouts, O. M. van Andel, and J. Dekker, *Mededel. Landbouwhogeschoolen Opzoekingsstas. Staat Gent*, **23**, 824(1958); *Chem. Abstr.*, **54**, 7044e(1960).
313. R. P. Singh and W. Korytnyk, *J. Med. Chem.*, **8**, 116(1965).
314. W. R. Sitterly, *Plant Disease Rep.*, **45**, 200(1961); *Chem. Abstr.*, **55**, 15810b(1961).
315. Société des usines chimiques Rhône-Poulenc, French Patent 1,167,657; *Chem. Abstr.*, **55**, 9436a(1961).
316. E. Spinner, *J. Chem. Soc.*, 1237(1960).
317. E. Spinner, *J. Chem. Soc.*, 3127(1962).
318. R. H. Sprague, J. A. Stewart, and J. M. Lewis, Belgian Patent 646,106; *Chem. Abstr.*, **63**, 10906g(1965).
319. B. A. Starrs, U.S. Patent 3,027,371; *Chem. Abstr.*, **57**, 12442a(1962).
320. B. A. Starrs, U.S. Patent 3,027,372; *Chem. Abstr.*, **57**, 12441i(1962).
321. F. B. Strandskov, J. B. Bockelmann, and V. J. Carroll, U.S. Patent 3,048,488; *Chem. Abstr.*, **57**, 13022b(1962).
322. F. B. Struble and L. S. Morrison, *Plant Disease Rep.*, **45**, 441(1961); *Chem. Abstr.*, **55**, 22689h(1961).
323. L. A. Summers, *Nature*, **215**, 381(1967).

324. P. Sun, Q. Fernando, and H. Freiser, *Anal. Chem.*, **36**, 2485(1964).

325. H. Suter and W. Kuendig, Swiss Patent 368,500; *Chem. Abstr.*, **60**, 1779a(1964).

326. G. J. Sutton, *Aust. J. Chem.*, **19**, 733(1966).

327. Y. Suzuki, *Yakugaku Zasshi*, **81**, 917(1961); *Chem. Abstr.*, **55**, 27305i(1961).

328. Y. Suzuki, *Yakugaku Zasshi*, **81**, 1146(1961); *Chem. Abstr.*, **56**, 3450c(1962).

329. Y. Suzuki, *Yakugaku Zasshi*, **81**, 1151(1961); *Chem. Abstr.*, **56**, 3447a(1962).

330. Takeda Chem. Ind. Ltd., Japanese Patent 25668/67.

331. Z. Talik and E. Plazek, *Bull. Acad. Pol. Sci., Ser. Sci. Chim.*, **8**, 219(1960); *Chem. Abstr.*, **60**, 9241c(1964).

332. T. Talik and Z. Talik, *Rocz. Chem.*, **40**, 1675(1966); *Chem. Abstr.*, **66**, 94889j(1967).

333. K. Thomae, G. M. B. H., British Patent 906,236; *Chem. Abstr.*, **59**, 5641f(1963).

334. P. Tomasik and Z. Skrowaczewska, *Rocz. Chem.*, **41**, 275(1967); *Chem. Abstr.*, **67**, 11404u(1967).

335. M. S. Tomita, U.S. Patent 3,251,849; *Chem. Abstr.*, **65**, 2230a(1966).

336. S. Toyoshima, S. Tanaka, and K. Hashimoto, Japanese Patent 6816/67; *Chem. Abstr.*, **67**, 43685c(1967).

337. T. A. Trofimova and B. G. Boldyrev, *Zh. Org. Khim.*, **2**, 103(1966); *Chem. Abstr.*, **64**, 19468f(1966).

338. W. P. Trompen and H. O. Huisman, *Rec. Trav. Chim. Pays-Bas*, **85**, 175(1966).

339. E. Uhlemann and H. Mueller, *J. Prakt. Chem.*, **30**, 163(1965).

340. E. Uhlig and G. Heinrich, *Z. Anorg. Allg. Chem.*, **330**, 40(1964); *Chem. Abstr.*, **61**, 6614c(1964).

341. D. M. Updegraff, *J. Infec. Dis.*, **114**, 304(1964); *Chem. Abstr.*, **61**, 16476g(1964).

342. O. Vaartaja, *Can. Dept. Agr., Div. Forest Biol. Progr. Rep.*, **15**(2) (1959); *Chem. Abstr.*, **54**, 21590f(1960).

343. R. T. Vanderbilt Company, Inc., Belgian Patent 609,487; *Chem. Abstr.*, **57**, 9964i(1962).

344. E. Vincke, *Arzneim.-Forsch.*, **15**, 821(1965).

345. A. F. Vompe and N. V. Monich, *Dokl. Acad. Nauk SSSR*, **170**, 89(1966); *Chem. Abstr.*, **66**, 5352j(1967).

346. A. V. Voropaeva and Kh. Sh. Chibilyaev, *Zh. Obshch. Khim.*, **34**, 1548(1964); *Chem. Abstr.*, **61**, 5605c(1964).

347. A. V. Voropaeva and I. Kh. Fel'dman, *Khim. Geterotsikl. Soedin., Akad. Nauk Latv. SSR, 1965,* 271; *Chem. Abstr.*, **63**, 6967a(1965).

348. G. Wagner, E. Fickweiler, P. Nuhn and H. Pischel, *Z. Chem.*, **3**, 62(1963); *Chem. Abstr.*, **59**, 7636g(1963).

349. G. Wagner and R. Metzner, *Naturwissenschaften*, **52**, 83(1965); *Chem. Abstr.*, **62**, 16565d(1965).

350. G. Wagner and R. Metzner, *Pharmazie*, **20**, 752(1965); *Chem. Abstr.*, **64**, 19748f(1968).

351. G. Wagner and P. Nuhn, *Arch. Pharm.*, (Weinheim), **297**, 461(1964).

352. G. Wagner and P. Nuhn, *Z. Chem.*, **3**, 64(1963); *Chem. Abstr.*, **59**, 7636c(1963).

353. G. Wagner and H. Pischel, *Arch. Pharm.*, (Weinheim), **296**, 576(1963).

354. G. Wagner and H. Pischel, *Pharmazie*, **19**, 197(1964); *Chem. Abstr.*, **60**, 15972b(1964).

355. G. Wagner, H. Pischel, and R. Schmidt, *Z. Chem.*, **2**, 86(1962); *Chem. Abstr.*, **58**, 5776d(1963).

356. L. H. Werner and G. de Stevens, U.S. Patent 3,124,575; *Chem. Abstr.*, **60**, 15895b(1964).

357. T. H. Wicker, Jr., M. A. McCall, and R. L. McConnell, U.S. Patent 2,964,528; *Chem. Abstr.*, 55, 7748c(1961).
358. T. Wieland and H. Biener, *Chem. Ber.*, 96, 266(1963).
359. C. L. Winek and E. V. Buehler, *Toxicol. Appl. Pharmacol.*, 9, 269(1969).
360. J. W. Wrathall and D. K. Busch, *Inorg. Chem.*, 2, 1171(1963).
361. H. L. Yale and J. Bernstein, U.S. Patent 2,943,086.
362. H. L. Yale and J. Bernstein, U.S. Patent 3,106,561.
363. H. L. Yale, K. Losee, and J. Bernstein, *J. Amer. Chem. Soc.*, 82, 2042(1960).
364. H. L. Yale and J. T. Sheehan, *J. Org. Chem.*, 26, 4315(1961).
365. M. Yokoyama and T. Takahashi, Japanese Patent 2043/62; *Chem. Abstr.*, 58, 7914h(1963).
366. V. G. Zhiryakov and P. I. Abramendo, *Khim. Geterotsikl. Soedin.*, *Akad. Nauk Latv. SSR*, 334(1965); *Chem. Abstr.*, 63, 13231d(1965).
367. E. Ziegler, R. Salvador, and Th. Kappe, *Monatsh.*, 94, 941(1963); *Chem. Abstr.*, 60, 2885d(1964).
368. O. Zima and G. Schorre, U.S. Patent 3,010,966; *Chem. Abstr.*, 56, 13022f(1962).
369. O. Zima and G. Schorre, German Patent 1,186,064; *Chem. Abstr.*, 62, 10417g(1965).
370. A. A. Ziyaev and A. S. Saclykov, *Nauch Tr.*, *Tashkent. Gos. Univ.*, 286, 85(1966); *Chem. Abstr.*, 67, 82057(1967).
371. P. A. Van Zwieten, M. Gerstenfeld, and H. O. Huisman, *Rec. Trav. Chim. Pays-Bas*, 81, 604(1962).
372. P. A. Van Zwieten, J. Meltzer, and H. O. Huisman, *Rec. Trav. Chim. Pays-Bas*, 81, 616(1962).
373. P. A. Van Zwieten, J. A. Van Velthuijsen, and H. O. Huisman, *Rec. Trav. Chim. Pays-Bas*, 80, 1066(1961).
374. J. Dever and I. C. Popoff, U.S. Patent 3,373,167; *Chem. Abstr.*, 69, 43442n (1968).
375. E. Maruszewska-Wieczorkowska, J. Michalski, and A. Skowronska, *Rocz. Chem.*, 30, 1197(1956); *Chem. Abstr.*, 51, 11347a(1957).
376. T. Naito and R. Dohmori, *Chem. Pharm. Bull.* (Tokyo), 3, 38(1955).
377. R. E. Smith, S. Boatman, and C. R. Hauser, *J. Org. Chem.*, 33, 2083(1968).
378. H. Johnston, U.S. Patent, 3,364,223; *Chem. Abstr.*, 69, 27254x(1968).
379. E. G. Novikov and I. N. Tugarinova, USSR Patent 192,796; *Chem. Abstr.*, 69, 27256z(1968).
380. D. G. Markees, V. C. Dewey, and G. W. Kidder, *J. Med. Chem.*, 11, 126(1968).
381. A. E. S. Fairfull and D. A. Peak, *J. Chem. Soc.*, 796(1955).
382. J. C. Howard and J. G. Michels, *J. Org. Chem.*, 25, 829(1960).
383. K. Kenzo, O. Ari, T. Fushimi, and S. Yurugi, Japanese Patent 6725,668; *Chem. Abstr.*, 69, 52014b(1968).
384. B. G. Boggiano. V. Petrow, O. Stephensen, and A. M. Wild, *J. Pharm. Pharmacol.*, 13, 567(1961).
385. R. H. Rigterink, and H. Raymond, U.S. Patent 3,385,859; *Chem. Abstr.*, 69, 59105r(1968).
386. G. N. Shibanov and T. M. Zhigaleva, USSR Patent 211,538; *Chem. Abstr.*, 69, 59108u(1968).
387. Y. Ueno, S. Asakawa, and E. Imoto, *Nippon Kagaku Zasshi*, 89, 101(1968); *Chem. Abstr.*, 69, 59049(1968).
388. Howard Johnston, U.S. Patent 3,371,011; *Chem. Abstr.*, 69, 59109v(1968).
389. Farbenfabriken Bayer A.-G., Netherlands Patent 68 02,176.

390. M. Numata, Y. Oka, H. Hirano, and K. Masuda, Japanese Patent 68 02,710; *Chem. Abstr.*, **69**, 67239(1968).

391. K. Masuda, M. Numata, and Y. Oka, Japanese Patent 6802,711; *Chem. Abstr.*, **69**, 67234x(1968).

392. J. R. Geigy A.-G., British Patent 1,108,975; *Chem. Abstr.*, **69**, 67226w(1968).

393. G. K. Helmkamp and D. C. Owsley, *Quart. Rep. Sulfur Chem.*, **2**, 303(1967); *Chem. Abstr.*, **69**, 67189m(1968).

394. L. S. Kobrina, G. G. Furin, and G. G. Yakobson, *Zh. Obshch. Khim.*, **38**, 514(1968); *Chem. Abstr.*, **69**, 77085p(1968).

395. Merck, A.-G., Belgium Patent 711,087.

396. H. Hirano, K. Masuda, and T. Fushimi, Japanese Patent 6802,712; *Chem. Abstr.*, **69**, 86827j(1968).

397. M. A. Elkaschef and M. N. Nosseir, *J. Amer. Chem. Soc.*, **82**, 4344(1960); see, also, M. A. Elkaschef, M. H. Nosseir, and A. A. Kader, *J. Chem. Soc.*, 440(1963).

398. A. Albert and G. B. Barlin, *J. Chem. Soc.*, 2384(1959).

399. A. R. Katritzky and R. A. Jones, *J. Chem. Soc.*, 2947(1960).

400. (a) W. E. Feeley and E. M. Beavers, *J. Amer. Soc.*, **81**, 4004(1959); (b) H. Tani, *Chem. Pharm. Bull.* (Tokyo), **7**, 930(1959); (c) H. Tani, *J. Pharm. Soc. Japan*, **81**, 141(1961).

401. O. Cervinka, *Collect. Czech. Chem. Commun.*, **27**, 567(1962).

402. (a) V. J. Traynelis and R. F. Martello, *J. Amer. Chem. Soc.*, **82**, 2744(1960); (b) S. Oae, T. Kitao, and Y. Kitaoka, *J. Amer. Chem. Soc.*, **84**, 3359(1962).

403. J. F. Bunnett, and R. E. Zahler, *Chem. Rev.*, **49**, 304(1951).

404. H. L. Yale and F. Sowinski, *J. Amer. Chem. Soc.*, **80**, 1651(1958).

405. R. A. Abramovitch, K. S. Ahmed, and C. S. Giam, *Can. J. Chem.*, **41**, 1752(1963).

406. D. E. Ames and J. L. Archibald, *J. Chem. Soc.*, 1475(1962).

407. F. E. Torba, U.S. Patent 3,609,158; *Chem. Abstr.*, **76**, 3699q (1972).

408. G. H. Keller, L. Bauer, and C. L. Bell, *J. Heterocycl. Chem.*, **5**, 647(1968).

409. R. Lawrence and E. S. Waight, *J. Chem. Soc., B,* 1 (1968).

410 Fujisawa Pharmaceutical Co., Ltd., Japanese Patent 68 04,226; *Chem. Abstr.*, **70**, 11942a(1969).

411. E. G. Novikov and I. N. Tugarinova, *Khim. Geterostikl. Soedin.*, 278 (1968); *Chem. Abstr.*, **70**, 47250s (1969).

412. F. M. Hershenson and L. Bauer, *J. Org. Chem.*, **34**, 655 (1969).

413. F. M. Hershenson and L. Bauer, *J. Org. Chem.*, **34**, 660 (1969).

414. R. S. Egan, F. M. Hershenson, and L. Bauer, *J. Org. Chem.*, **34**, 665 (1969).

415. E. Merck A.-G., Netherlands Patent 68 08,029 (1968).

416. L. H. Klemm, C. E. Klopfenstein, R. Zell, D. R. McCoy, and R. A. Klemm, *J. Org. Chem.*, **34**, 347 (1969).

417. L. H. Klemm, J. Shabatai, D. R. McCoy, and W. K. T. Kiang, *J. Heterocycl. Chem.*, **5**, 883 (1968).

418. R. H. Rigterink, U.S. Patent 3,378,565 (1968); *Chem. Abstr.*, **70**, 3845k (1969).

419. Takeda Chemical Industries, Ltd., Japanese Patent 69 28095 (1969); *Chem. Abstr.*, **72**, 43475n (1970).

420. Takeda Chemical Industries, Ltd., Japanese Patent 69 16654 (1969).

421. Daiichi Seiyaku Co., Ltd., Japanese Patent 69 16655 (1969).

422. B. Paul and W. Korytnyk, *Tetrahedron*, **25**, 1071 (1969).

423. K. Masuda and Y. Hara, Japanese Patent 68 26174 (1968); *Chem. Abstr.*, **70**, 57652p (1969).

424. Fugisawa Pharmaceutical Co., Ltd., Japanese Patent 69 26107 (1969); *Chem. Abstr.*, **72**, 12740r (1970).

425. R. T. Vanderbilt Co., Inc., U.S. Patent, 3,412,033 (1968); *Chem. Abstr.*, **70**, 79413p (1964).
426. Daiichi Pharmaceutical Co., Ltd., Japanese Patent, 68 26296 (1968); *Chem. Abstr.*, **70**, 57667x (1969).
427. B. Loev, U.S. Patent 3,480,636 (1969); *Chem. Abstr.*, **72**, 42793c (1970).
428. G. Schorre, H. Nowak, and O. Saiko, German Patent 1,812,794 (1970); *Chem. Abstr.*, **73**, 77074s (1970).
429. E. Poetsch, German Patent 1,809,467 (1970); *Chem. Abstr.*, **73**, 66443k (1970).
430. H. Sliwa, *Bull. Soc. Chim. Fr.*, 642 (1970).
431. J. Schnayder and Z. Kosiara, German Patent 1,936,743 (1970); *Chem. Abstr.*, **72**, 90300m (1970).
432. R. H. Rigterink, U.S. Patent 3,478,148 (1969); *Chem. Abstr.*, **72**, 90297r (1970).
433. R. A. Abramovitch and E. E. Knaus, *J. Heterocycl. Chem.*, **6**, 989(1969).
434. E. Ager, B. Iddon, and H. Suschitzky, *J. Chem. Soc., C*, 193 (1970).
435. W. Goebel and H. J. Koenig, *Acta Chim.* (Budapest), **63**, 107(1970); *Chem. Abstr.*, **72**, 55188j (1970).
436. A. V. Vorapaeva and N. G. Garbar, *Khim. Geterotsikl. Soedin.*, 677 (1969); *Chem. Abstr.*, **72**, 43378h (1970).
437. H. Mrozik, German Patent 1,810,162 (1969); *Chem. Abstr.*, **72**, 43482n (1970).
438. W. Walter and P. M. Hell, *Ann. Chem.*, **727**, 35 (1969).
439. W. Walter and P. M. Hell, *Ann. Chem.*, **727**, 50 (1969).
440. Pharmacia Aktiebolag, British Patent 1,166,684 (1969); *Chem. Abstr.*, **72**, 3381x (1970).
441. J. S. Kelyman, U.S. Patent 3,475,440 (1969); *Chem. Abstr.*, **72**, 31621u (1970).
442. M. Inoue, H. Ishikawa, T. Shimamoto, and M. Ishidawa, German Patent 1,917,539 (1969); *Chem. Abstr.*, **72**, 66840s (1970).
443. T. Talik and Z. Talik, *Rocz. Chem.*, **43**, 1667 (1969); *Chem. Abstr.*, **72**, 21576u (1970).
444. R. T. Rigterink, South African Patent 68 00893 (1969); *Chem. Abstr.*, **71**, 101721g (1970).
445. S. Umio, T. Kishimoto, I. Ueda, Japanese Patent 70,03,779, (1970); *Chem. Abstr.*, **72**, 121373u(1970).
446. J. A. Zoltewicz and N. Carlo, *J. Org. Chem.*, **34**, 765 (1969).
447. C. D. Crawford, U.S. Patent 3,415,832 (1968); *Chem. Abstr.*, **70**, 57654r (1969).
448. R. A. Abramovitch, F. Helmer, and M. Liveris, *J. Org. Chem.*, **34**, 1730 (1969).
449. R. A. Abramovitch and P. Tomasik, unpublished results.
450. R. A. Abramovitch and A. J. Newman, unpublished results.
451. D. R. Grassett and J. F. Murray, *Anal. Chim. Acta*, **46**, 139 (1969); *Chem. Abstr.*, **71**, 91245p (1969).
452. T. Fushimi, Japanese Patent 69 16,654 (1969); *Chem. Abstr.*, **71**, 91314k (1969).
453. L. Dall'Asta and A. Pedrazzoli, French Patent 1,548,392 (1968); *Chem. Abstr.*, **71**, 91323n(1969).
454. E. Merck, A.-G., British Patent 1,156,769 (1969); *Chem. Abstr.*, **71**, 91321k (1969).
455. R. E. Banks, R. N. Haszeldine, D. R. Karsa, F. E. Rickett, and I. M. Young, *J. Chem. Soc., C*, 1660 (1969).
456. M. Murakami, M. Iwanami, and O. Masaru, Japanese Patent 69,11,369 (1969); *Chem. Abstr.*, **71**, 70497g (1969).
457. J. F. Biellmann and H. J. Callot, *Bull. Soc. Chim. Fr.*, 1299 (1969).
458. J. Delarge, *Pharm. Acta Helv.*, **44**, 637 (1969); *Chem. Abstr.*, **71**, 112761w (1969).
459. H. Franz and G. Buchmann, *Pharmazie*, **24**, 301 (1969); *Chem. Abstr.*, **71**, 11276d (1969).

460. E. Agar, B. Iddon, and H. Suschitzky, *J. Chem. Soc., C*, 1530 (1970).
461. E. Poetsch, German Patent 1,811,973 (1970); *Chem. Abstr.*, 73, 55978n (1970).
462. R. H. Rigterink, U.S. Patent 3,385,859 (1968); *Chem. Abstr.*, 69, 59105r (1968).
463. Z. J. Allan, J. Podstata, and Z. Vrba, *Tetrahedron Lett.*, 55, 4855(1969).
464. H. T. Bucherer and J. Schenkel, *Ber.*, 41, 1346(1908).
465. R. H. Rigterink, U.S. Patent 3,478,148 (1969); *Chem. Abstr.*, 72, 90297r (1970).
466. D. R. Grassetti and J. F. Murray, Jr., *J. Chromatogr.*, 41, 121(1969).
467. D. R. Grassetti, J. F. Murray, Jr., and H. T. Ruan, *Biochem. Pharmacol.*, 18, 603 (1969); D. R. Grassetti, *Nature*, 228, 282(1970).
468. Y. Maki, M. Suzuki, and T. Masugi, *Chem. Pharm. Bull.* (Tokyo), 12, 591(1964).
469. R. H. Mizzoni and H. M. Blatter, U.S. Patent 3,718,654; *Chem. Abstr.*, 78, 136088c(1973).
470. Takeda Chemical Industries Ltd., Japanese Patent, 70 27391.
471. R. A. Damico, Canadian Patent, 841,632.
472. A. Signor, L. Biondi, A. Orio, and A. Marzotta, *Atti Accad. Peloritana Pericolanti, Cl. Sci. Fis., Mat. Natur.*, 50, 85(1970); *Chem. Abstr.*, 76, 140455w(1972).
473. S. Rudolf, W. Goebel, and W. Neumann, *Acta Chim.* (Budapest), 64, 267(1970); *Chem. Abstr.*, 74, 3484z(1971).
474. S. A. Notton, Czechoslovakian Patent, 505,094.
475. D. E. Butler, P. Bass, I. C. Nordin, F. P. Hauck, Jr., and Y. J. L'Italien, *J. Med. Chem.*, 14, 575(1971).
476. C. Sapino, Jr. and P. D. Sleezer, German Patent, 2,116,159; *Chem. Abstr.*, 76, 46094f(1972).
477. T. Kuroda, Japanese Patent 71 39866; *Chem. Abstr.*, 76, 46093e(1972).
478. K. Ueno and H. Omura, Japanese Patent 71 39867; *Chem. Abstr.*, 76, 46089h(1972).
479. H. Shibahare, Japanese Patent 71 39868.
480. Arteriosclerosis Research Foundation, Japanese Patent 71 31592.
481. M. Yokoyama, *Bull. Chem. Soc. Jap.*, 44, 3195(1971).
482. L. G. Levkovskaya and T. S. Safonova, *Khim. Geterotsikl. Soedin.*, 7, 1391(1971); *Chem. Abstr.*, 76, 59403f(1972).
483. E. Ager, B. Iddon, and H. Suschitzky, *J. Chem. Soc., Perkin Trans.*, 1, 133(1972).
484. D. R. Grasselli, U.S. Patent 3,597,160.
485. Bristol-Meyers Co., Belgian Patent 765,396.
486. Abbott Laboratories, Belgian Patent 768,305.
487. D. R. Grassetti, Belgian Patent 766,123.
488. S. A. L'Oreal, Belgian Patent 753,822.
489. R. A. Damico, U.S. Patent 3,583,999.
490. Ciba-Geigy AG, Belgian Patent 758,749.
491. Merck and Co., Inc., Belgian Patent 758,758.
492. Merck and Co., Inc., Belgian Patent 758,759.
493. Bristol-Meyers Co., Belgian Patent 758,764.
494. M. Friedman and J. F. Cavino, U.S. Patent 3,607,072.
495. R. A. Damico, U.S. Patent 3,583,999.
496. H. Tani, M. Otani, and T. Matsumoto, Japanese Patent 72 32075; *Chem. Abstr.*, 77, 164513k (1972).
497. J. G. Blum, French Patent 2,079,649; *Chem. Abstr.*, 77, 61831h(1972).
498. K. F. King, *Diss. Abstr. Int. B*, 32, 3870(1972); *Chem. Abstr.*, 76, 126731b(1972).
499. J. G. Blum, French Patent 2,068,429; *Chem. Abstr.*, 76, 126788a(1972).
500. J. Schnayder and Z. Kosiara, Polish Patent 63264; *Chem. Abstr.*, 76, 126794z(1972).
501. L. G. Levkovskaya and T. S. Safonova, *Khim. Geterotsikl. Soedin.*, 7, 1502(1971); *Chem. Abstr.*, 77, 34267x(1972).

502. R. H. Mizzoni and H. M. Blatter, U.S. Patent 3,671,512.
503. R. G. Pews and F. P. Sorson, U.S. Patent 3,641,004.
504. R. E. Manning, U.S. Patent 3,635,995; *Chem. Abstr.,* 76, 99670x(1972).
505. Procter and Gamble Co., Belgian Patent 782,105.
506. P. S. Grand, German Patent 2,217,738; *Chem. Abstr.,* 78, 47798s(1973).
507. M. Nakanishi, S. Saheki, and K. Iimori, Japanese Patent 72 40052; *Chem. Abstr.,* 77, 164494e(1972).
508. E. V. Krumkalns and W. A. White, German Patent 2,213,958; *Chem. Abstr.,* 77, 164498j(1972).
509. M. Izumi, Japanese Patent 72 20168; *Chem. Abstr.,* 77, 164509p(1972).
510. L. Schroeder and K. Thomas, German Patent 2,103,728; *Chem. Abstr.,* 77, 164519s(1972).
511. G. K. Weisse, A. G. Hovey, E. A. Kober, and E. H. Kober, German Patent 2,052,233; *Chem. Abstr.,* 77, 61821e(1972).
512. F. E. Torba, U.S. Patent 3,682,936; *Chem. Abstr.,* 77, 114257j(1972).
513. J. R. Ferraro, B. B. Murray, and N. J. Wieckowicz, *J. Inorg. Nucl. Chem.,* 34, 231(1972).
514. I. Matsumoto, K. Nakagawa, M. Matsuzaki, and K. Horiuchi, Japanese Patent 72 05771; *Chem. Abstr.,* 78, 97499r(1973).
515. H. Shibahara, Japanese Patent 72 24028; *Chem. Abstr.,* 77, 151952c(1972).
516. K. Matsumoto, K. Nakagawa, and K. Horiuchi, Japanese Patent 72 17778; *Chem. Abstr.,* 77, 151979s(1972).
517. G. G. Moore, J. K. Harrington, and J. F. Gerster, U.S. Patent 3,686,192; *Chem. Abstr.,* 77, 151965(1972).
518. P. B. Domenico, U.S. Patent 3,725,421.
519. J. E. Dunbar and J. W. Zemba, U.S. Patent 3,674,795; *Chem. Abstr.,* 77, 88316f(1972).
520. J. E. Delarge, L. N. J. V. Thunus, C. L. A. Lapiere, and A. H. E. Georges, German Patent 2,155,483; *Chem. Abstr.,* 77, 88325h(1972).
521. E. Heinrich and R. Mueller, German Patent 2,117,753; *Chem. Abstr.,* 78, 29636h(1973).
522. P. B. Domenico, U.S. Patent 3,692,792; *Chem. Abstr.,* 78, 16051p(1973).
523. I. Matsumoto, K. Nakagawa, A. Matsuzaki, K. Horiuchi, and H. Hidaka, Japanese Patent 72 40796; *Chem. Abstr.,* 78, 16045q(1973).
524. H. Hidaka, I. Matsumoto, and J. Yoshizawa, German Patent 2,218,248; *Chem. Abstr.,* 78, 16049u(1973).
525. C. E. Berkoff, N. W. DiTullio, and J. A. Weisbach, German Patent 2,216,576; *Chem. Abstr.,* 78, 16044p(1973).
526. F. Ishikawa, Japanese Patent 72 34709; *Chem. Abstr.,* 78, 16048t(1973).
527. M. Nakanishi, S. Saheki, and K. Iimori, Japanese Patent 72 40057; *Chem. Abstr.,* 78, 4131b(1973).
528. M. I. Druzin, A. V. Chistyakova, M. V. Lyubomilova, M. A. Korshumov, R. G. Kuzovleva, A. S. Chernyak, and A. S. Bobrova, U.S.S.R. Patent 341,798; *Chem. Abstr.,* 78, 4134e(1973).
529. S. D. Moshchitskii, G. A. Zalesskii, Ya. N. Ivashchenko, and L. M. Yagupol'skii, *Khim. Geterotsikl. Soedin.,* 1094(1972).
530. T. Kuroda, Japanese Patent 72 22232; *Chem. Abstr.,* 77, 101391y(1972).
531. K. Yoshisue, Japanese Patent 72 22823; *Chem. Abstr.,* 77, 101398f(1972).
532. T. Fushimi, Japanese Patent 72 22581; *Chem. Abstr.,* 77, 101396d(1972).
533. T. Fushimi, Japanese Patent 72 22230; *Chem. Abstr.,* 77, 101401b(1972).
534. A. Ueda, and K. Fujii, Japanese Patent 72 22590; *Chem. Abstr.,* 77, 101402c(1972).

535. B. E. Witzel, C. P. Dorn, Jr., and T. Y. Shen, U.S. Patent 3,715,358.
536. H. Tilles, and M. E. Brokke, U.S. Patent 3,704,236; *Chem. Abstr.*, 78, 43280u(1973).
537. R. A. Damico, U.S. Patent 3,700,676; *Chem. Abstr.*, 78, 43278z(1973).
538. I. Matsumoto, K. Nakagawa, M. Matsuzaki, and K. Horiuchi, German Patent 2,225,482; *Chem. Abstr.*, 78, 58249q(1973).
539. F. E. Torba, U.S. Patent 3,705,170; *Chem. Abstr.*, 78, 58253m(1973).
540. C. D. S. Tomlin, B. Iddon, and E. Ager, British Patent 1,293,909; *Chem. Abstr.*, 78, 58255p(1973).
541. P. I. Abramenko and V. G. Zhiryakov, *Khim. Geterotsikl. Soedin.*, 1541(1972); *Chem. Abstr.*, 78, 58269w(1973).
542. P. I. Abramenko and V. G. Zhiryakov, *Khim. Geterotsikl. Soedin.*, 1539(1972); *Chem. Abstr.*, 78, 58265s(1973).
543. K.K. Kowa, Japanese Patent 72 32075.
544. L. Schröder and K. Thomas, German Patent 2,103,728.
545. P. B. Domenico, U. S. Patent 3,706,751; *Chem. Abstr.*, 78, 97504p(1973).
546. C. W. Muth, R. S. Darlak, and J. C. Patton, *J. Heterocyl. Chem.*, 9, 1003(1972).
547. W. B. Wright, Jr. and H. J. Brabander, *J. Heterocyl. Chem.*, 9, 1017(1972).
548. R. H. Fish and M. Friedman, *Chem. Commun.*, 812(1972).
549. K. R. Reistad, P. Groth, R. Lie, and K. Undheim, *Chem. Commun.*, 1059(1972).
550. Colgate-Palmolive Co., Belgian Patent 782,264.
551. H. Hidaka, Belgian Patent 776,837.
552. Yoshitomi Pharmaceutical Industries, Ltd., Japanese Patent 72 40057.
553. M. Izumi, Japanese Patent 72 0168.
554. M. Friedman and J. F. Cavias, U.S. Patent 3,607,072.
555. Banyu Pharmaceutical Co., Ltd., Japanese Patent 71 7778.
556. Takeda Chemical Industries, Ltd., Japanese Patent 71 38265-R.
557. Yoshitomi Pharmaceutical Industries, Ltd., Japanese Patent 71 38265-Z.
558. I. Matsumoto and M. Okazawa, Japanese Patent 72 00066; *Chem. Abstr.*, 76, 85711z(1972).
559. Takeda Chemical Industries, Ltd., Japanese Patent 72 00061.
560. P. Tomasik, *Pr. Nauk. Inst. Chem. Technol. Nafty Wegla Politech. Wroclaw*, No. 5, 101(1971); *Chem. Abstr.*, 78, 71855d(1973).
561. Warner-Lambert Co., German Patent 2,147,898.
562. G. Carolla, Japanese Patent 72 18112.
563. Sankyo Chemical Industry, Ltd., Japanese Patent 72 22823.
564. Yoshitomi Pharmaceutical Co., Ltd., Japanese Patent 72 22597.
565. Beecham Group Ltd., British Patent 1,280,671.
566. P. B. Domenico, U.S. Patent 3,634,436; *Chem. Abstr.*, 76, 99526e(1972).
567. I. Matsumoto, K. Nakagawa, and K. Kobu, Japanese Patent 72 39080; *Chem. Abstr.*, 78, 58245k(1973).
568. P. B. Domenico, U.S. Patent 3,651,066.
569. P. Pirson and L. Christiaens, *Bull. Soc. Chim. Fr.*, 704(1973).
570. P. B. Domenico, U.S. Patent 3,635,994; *Chem. Abstr.*, 76, 113078u(1972).
571. F. E. Torba, U.S. Patent 3,711,486; *Chem. Abstr.*, 78, 84271k(1973).
572. Sanyo Chemical Industry Co., Ltd., Japanese Patent 72 50366.
573. W. R. Roderick, U.S. Patent 3,708,581; *Chem. Abstr.*, 78, 71928e(1973).
574. K. Edanami, H. Takahashi, and T. Kuroda, *Jap. J. Pharmacol.*, 117(1972).
575. Yoshitomi Pharmaceutical Industries, Ltd., Japanese Patent 73 06467.

576. L. H. Klemm, W. O. Johnson, D. V. White, U.S. Patent 3,709,894; *Chem. Abstr.*, **78**, 72091p(1973). See also *J. Heterocycl. Chem.*, **7**, 473(1970).
577. R. H. Mizzoni and H. M. Blatter, U.S. Patent 3,718,654.
578. P. B. Domenico, U.S. Patent 3,732,234.
579. S. D. Moshchitskii, L. S. Sologub, Ya. N. Ivashchenko, and L. M. Yagupol'skii, *Khim. Geterotsikl. Soedin.* 1634(1972); *Chem. Abstr.*, **78**, 71864f(1973).
580. G. C. Morrison, J. Shavel, Jr., and W. Cetenko, U.S. Patent 3,725,419.
581. I. Matsumoto and M. Okazawa, Japanese Patent 72 00065; *Chem. Abstr.*, **76**, 85709e(1972).
582. I. Matsumoto, K. Nakagawa, and K. Horiuchi, Japanese Patent 73 13,372; *Chem. Abstr.*, **78**, 136097e(1973).
583. P. B. Domenico, U.S. Patent 3,639,413; *Chem. Abstr.*, **76**, 113084t(1972).
584. P. B. Domenico, U.S. Patent 3,719,682; *Chem. Abstr.*, **78**, 136077y(1973).
585. R. D. Haugwitz, U.S. Patent 3,634,438; *Chem. Abstr.*, **76**, 72413d(1972).
586. P. B. Domenico, U.S. Patent 3,629,281; *Chem. Abstr.*, **76**, 72409g(1972).
587. A. M. Samuilov and G. F. Dregval, *Metody Poluch. Khim. Reaktivov Prep.* 149-51(1969); *Chem. Abstr.*, **76**, 14275z(1972).

CHAPTER XVI

Pyridines and Reduced Pyridines of Pharmacological Interest

R. T. COUTTS and A. F. CASY

Faculty of Pharmacy and Pharmaceutical Sciences,
University of Alberta, Edmonton, Canada

I. Introduction

This chapter deals with pyridines and reduced pyridines that possess some pharmacological properties. It is an impossible task to mention all active compounds because literally many thousands of pyridine derivatives have been prepared and tested pharmacologically. The material presented here, therefore, is selective, and an emphasis has been placed on describing work reported in the more recent literature. Much relevant material is to be found in the patent literature.

II. Pyridines

It is convenient to discuss pyridine compounds according to their pharmacological classifications. The groups considered here are as follows: antitubercular compounds, other antimicrobial agents, antiviral and antitumor agents, central nervous system stimulants, hypocholesteremic agents, niacin and niacinamide, vitamin B_6, diagnostic agents, analgesics, antiinflammatory agents, antihistaminics, cholinesterase inhibitors and reactivators, nicotine and other ganglion stimulants, and pyridine derivatives with miscellaneous pharmacological properties.

Many pharmacologically active pyridines are derivatives of nicotinic acid or, to a lesser extent, of isonicotinic acid. References to these derivatives are found in a number of the pharmacological groups listed above.

1. Antitubercular Pyridine Compounds

Derivatives of pyridinecarboxylic acids possess antitubercular activity. Chorine (1) first reported the tuberculostatic activity of nicotinamide (XVI-1) in 1945. Later, in 1948, McKenzie, Kushner, and co-workers (2, 3) pointed out that the activity of nicotinamide was similar to that of p-aminosalicylic acid. Nicotinamide (niacinamide) is also a vitamin of the B group, and it was thought initially that vitamin activity and tuberculostatic activity were interdependent; that is, the tuberculostactic activity of niacinamide was a function of its vitamin activity. Fox (4), however, refuted this postulate when he showed that

3-aminoisonicotinic acid (**XVI-2**) and its methyl ester were tuberculostatic but had no vitamin activity. He extended his studies to the preparation of derivatives of isonicotinic acid (5). He prepared isonicotinaldehyde thiosemicarbazone (**XVI-3**) for evaluation as an antitubercular compound and at the same time tested isonicotinylhydrazine (**XVI-4**), an intermediate in the synthesis of **XVI-3**. Isonicotinylhydrazine (isoniazid) was found to be a very active drug. It showed no serious toxic effects in animals or humans and, together with p-aminosalicylic acid and streptomycin, is one of the three major drugs used in the chemotherapy

of tuberculosis. Syntheses of isoniazid labeled with ^{14}C in the pyridine ring have been reported (6). Animal studies have shown (7) that the lungs contain significant amounts of the drug even after blood levels have become depleted.

Many simple derivatives of isonicotinylhydrazine have been prepared and tested for antitubercular activity. No compounds possessing the general structure **XVI-5** were more active than isoniazid; most were much less active and some were completely devoid of activity (8–13). Substitution of another ring system for the pyridine nucleus destroyed the antitubercular activity of isoniazid (8, 10, 14).

Claims have been made (15–19) that certain derivatives of isoniazid possessing the general structure **XVI-6** are superior to isoniazid; whereas antitubercular activity is similar to that of isoniazid, toxicity to the host is much reduced. Colwell and Hess (20) refute one of these claims.

The preparation of 5-(4-pyridyl)-1,3,4-oxdiazolone (**XVI-7**) from isoniazid by the action of phosgene has been patented (21). It is claimed (22) that this derivative is a less toxic and more active compound than isoniazid in animal studies. It has also been shown (23) to be active in patients who had developed a resistance to isoniazid, p-aminosalicylic acid, and streptomycin.

XVI-7   XVI-8a H   XVI-9
               XVI-8b Et
               XVI-8c n-Pr

Thioisonicotinamide **(XVI-8a)** was first reported in 1954 by Gardner, Wenis, and Lee (24). The events that led to its synthesis have been reviewed by Rist (25). This compound is much more active than niacinamide, but it is also much more toxic especially in humans. Rather surprisingly, substitution in the α-position of **XVI-8a** gave compounds with enhanced antitubercular activity. 2-Ethyl- and 2-n-propyl-thioisonicotinamide [ethionamide **(XVI-8b)** and pro-thionamide **(XVI-8c)**, respectively] are more active than the parent compound (26-28). Related 2-substituted- and 2,6-disubstituted-thioisonicotinamides have been prepared (29, 30) and are claimed to possess therapeutic value in the treatment of tuberculosis.

In recent years there has been a diminished interest in the synthesis of antitubercular compounds. The approach generally has been to synthesize more hydrazides (31-34), hydrazones (35-37), semicarbazones (36, 37), and thiosemicarbazones (38-40) with a pyridine nucleus. Thiosemicarbazones of structure **XVI-9**, for example, are useful bactericides (40), especially for *Mycobacterium tuberculosis*. Some derivatives of thiourea possessing a pyridyl substituent also have tuberculostatic activity (41).

Sulfapyridine **(XVI-10a)** was the first medicinal sulfonamide used in combatting diseases caused by microorganisms, especially pneumonia. It has been largely replaced by equally effective but less toxic congeners, and now is

R
XVI-10a H
XVI-10b CONHNH$_2$

virtually of historical importance only. It was inevitable that the chemical features of sulfapyridine and isoniazid should be combined. The product **(XVI-10b)** inhibits the *in vitro* growth of *M. tuberculosis* (32).

2. Other Antimicrobial Agents

Ochiai (42) has reviewed the biological properties of aromatic amine oxides, covering the literature to 1964. He described the observations made by Newbold and Spring (43), Cunningham (44), and Shaw (45) that 2-hydroxypyridine-1-oxide [or *N*-hydroxy-2(1*H*)-pyridone] **(XVI-11a)** and its derivatives possess significant antibacterial activity. This observation led to the synthesis of 2-mercaptopyridine-1-oxide [i.e., *N*-hydroxy-2(1*H*)-pyridinethione] **(XVI-11b)** and related compounds (46). These cyclic thiohydroxamic acids were far more potent *in vitro* antibacterial agents than the corresponding pyridones **(XVI-11a)**.

2-Mercaptopyridine-1-oxide is an efficient antifungal agent, and numerous patents have been taken out that describe the use of this compound and its

X
XVI-11a O
XVI-11b S

XVI-12

derivatives as pesticides, fungicides, and preservatives. A large number of heavy metal salts and organic salts of **XVI-11b** and the disulfide **(XVI-12)** and its simple derivatives have been described (42, 47). The zinc salt of 2-mercaptopyridine-1-oxide is the active ingredient in commercially available shampoos used to control dandruff.

Condensation of the sodium salt of 2-mercaptopyridine-1-oxide with various benzyl halides gave a variety of 2-benzylthiopyridine-1-oxides that possessed herbicidal properties (48). Related compounds which contain an *N*-heterocyclic group in place of the benzyl group are described (42) as active antiinfection agents. Esters of 1-hydroxy-2-pyridinethiones are also active bactericides and fungicides (49).

Treatment of 2-ethoxypyridine-1-oxide with allyl bromide yielded 1-(allyloxy)-2(1*H*)-pyridone **(XVI-13)**. Numerous other 1-(2-alkenyloxy)-2(1*H*)-pyridones have been obtained in a similar manner; they are antifungal agents (50). 4-Alkoxy derivatives of **XVI-11a** have been prepared and are bacteriocidal (51).

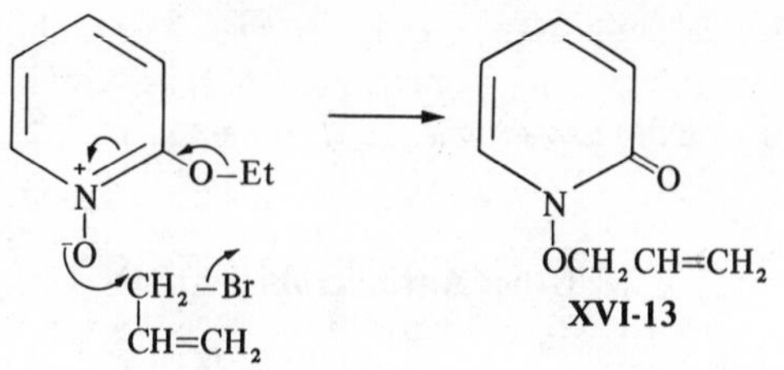

XVI-13

In more recent years, especially in Japan, many hundreds of compounds possessing the 5-nitro-2-furyl group have been synthesized and evaluated for their antifungal and antibacterial properties. Nitrofurantoin **(XVI-14)** and nitrofurazone **(XVI-15)** are two examples which are used clinically. Chemically

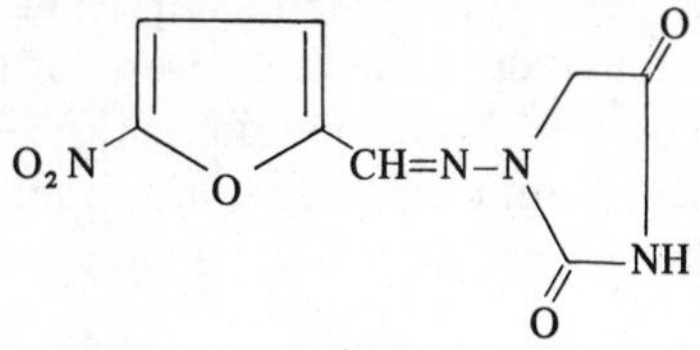

XVI-14 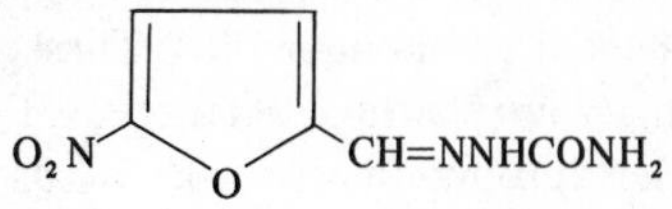   **XVI-15**

related compounds of general structure **XVI-16** in which R is most often H, Me, or Et and the attachment of the pyridine ring is either at the 2 or 4 position are

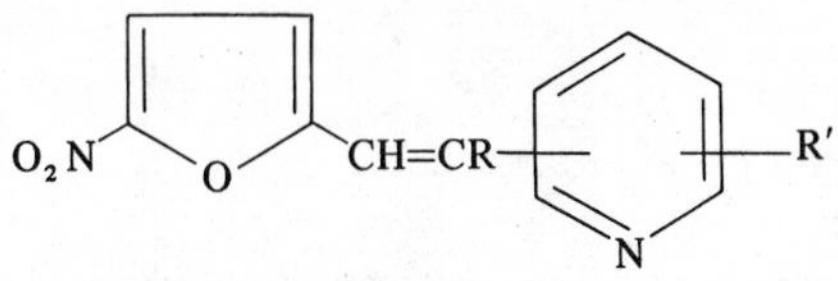

XVI-16

also potent antimicrobial agents (52–57). Their *N*-oxides are also active compounds (55, 58–60).

Previous mention has been made to sulfapyridine **(XVI-10a)**. 2-(*p*-Aminobenzenesulfonamido)pyridine-1-oxide **(XVI-17, R = H)** (sulfapyridine-1-oxide) and its methyl homologs **(XVI-17, R = Me)** were synthesized by Childress and Scudi

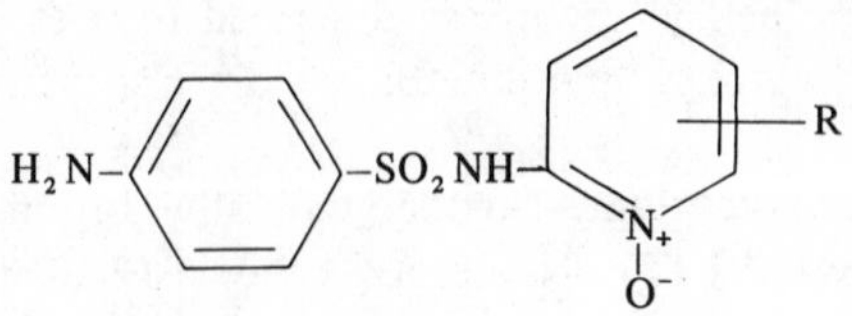

XVI-17

(61) in 1958 and were reported to possess potent antimicrobial activity. The related compound salicylazosulfapyridine (sulfasalazine) **(XVI-18)** is a sulfonamide which is said (62) to have special affinity for connective tissue and is

XVI-18

recommended for the treatment of chronic ulcerative colitis. Its high toxicity compared with other sulfonamides greatly limits its use. Some derivatives of 4-(benzenesulfonyl)pyridine-1-oxide **(XVI-19)** have also been shown (63) to be active antibacterial agents. The monosodium salt of *N*-(sulfinomethyl)isonico-

XVI-19   **XVI-20**

tinoyl) hydrazide, **XVI-20**, is claimed (64) to be very effective against *Bacillus kochi* and *B. hansoni*.

Caerulomycin is an interesting antibiotic that was isolated recently from *Streptomyces caeruleus* cultures (65, 66). Its structure has been shown (66) to be 4-methoxy-2,2′-dipyridyl-6-*syn*-aldoxime **(XVI-21)**. Helpful in the elucidation of its structure was the fact that caerulomycin and certain of its derivatives gave a deep red color with ferrous salts, similar to that given by 2,2′-dipyridyl. The latter compound, 2,2′-dipyridyl, is used clinically as an antidote for heavy metal poisoning. It complexes with many metals (67). 2,2′-Dipyridyl is also undergoing clinical trial as an inhibitor of gastric secretion (67a) (see also p. 480).

Phenazopyridine (2,6-diamino-3-phenylazopyridine) **(XVI-22)** hydrochloride has been used for many years as a urinary antiseptic and analgesic. It is rapidly eliminated by the kidneys, imparting a red color to the urine, and is therefore used in the treatment of urinary tract infections. It is prepared (68, 69) by coupling diazotized aniline with 2,6-diaminopyridine. A number of related compounds have also been prepared and their bacteriostatic properties evaluated (69).

Numerous quaternary ammonium compounds have been prepared and evaluated as antibacterial agents, but only a small number are used clinically for this purpose. Of these, two are pyridine derivatives. They are cetylpyridinium chloride **(XVI-23)** and Emcol-607 **(XVI-24)** and both are used as topical antiseptics (70).

XVI-21

XVI-22

XVI-23

XVI-24

XVI-25

XVI-26

3. Antiviral and Antitumor Agents

Various derivatives of picolinic acid or nicotinic acid have been found to possess antiviral and/or antitumor activity. α-Picolinic acid 2-acetylhydrazide **(XVI-25)** can be used against Walker 256 tumor systems in animals (71). Some anilides of pyridine carboxylic acids that possess an alkylating function also possess some antitumor activity (72). Certain 6-substituted nicotinamides and various alkylating derivatives of nicotinic acid of general structure **XVI-26** have been tested (73, 74) and found to have moderate activity against Walker rat carcinoma and mouse lymphoid leukemia, Ll210.

Some N-(5-alkoxycarbonyl-2-pyridyl)alkanoic acid amides were prepared (75) from 6-aminonicotinic acid and were found to possess antiviral and antitumor activities. Related acylaminonicotinic acids are useful against encephalomyocarditis virus and herpes virus (76), and the guanidides of nicotinic acid, isonicotinic acid, picolinic acid, and 6-methylpicolinic acid are reported (77) to be effective against common cold virus.

Various other N-pyridyl aliphatic amides (78, 79), some β-alkoxy-β-phenyl-ethylpyridinium bromides (80), and numerous pyridine-2-thiones and related compounds (81) are also active compounds.

4. Central Nervous System Stimulants

The isopropyl derivative of isoniazid, called iproniazid **(XVI-27)**, was initially prepared for evaluation as an antitubercular compound and indeed it was found to be more active than isoniazid in humans (82, 83). Unfortunately it also displayed more toxic side reactions. Unlike isoniazid, iproniazid was observed to

XVI-27

XVI-28

produce central stimulation in the patients who were being treated for tuberculosis (84). In addition, the observation was made that iproniazid was a monoamine oxidase (MAO) inhibitor (85). These discoveries led to the clinical examination of iproniazid as a central nervous system (CNS) stimulant in the treatment of mental depression (86). For this it was very effective. Its antidepressant action is thought to be due to its ability to inhibit MAO, which permits a build up in the levels of both serotonin and norepinephrine in the brain. By some unknown mechanism, iproniazid also reduces the frequency and severity of angina pectoris attacks.

Because iproniazid has many toxic side effects, it has now been withdrawn and replaced by other less toxic agents. One of these is the related compound nialamide **(XVI-28)**. Another hydrazide **(XVI-29)** is claimed (87) to be useful for the treatment of angina pectoris and anxiety states. This compound also inhibits MAO and is three to seven times as active in rats as iproniazid. Similarly, the preparation of N'-(2-hydroxyethyl)isonicotinic acid hydrazide (hydroxyethyliso-niazid) **(XVI-30)** has been reported (88). This compound is claimed to be as potent a MAO inhibitor as iproniazid with a more favorable therapeutic index. In contrast, a number of isonicotinoylhydrazones of cyclanones **(XVI-31)** and related hydrazines **(XVI-32)** prepared from isoniazid have been described (89) as possessing CNS depressing activity in mice.

$-CH_2CH_2CONHNHCHMe_2$

XVI-29

$-CONHNHCH_2CH_2OH$

XV-130

$-CONHNH_2$ + O=

R

$-CONHN=$

R

$\xrightarrow[PtO_2]{H_2}$

$-CONHNH-$

R

XVI-31   **XVI-32**

N,N-Diethylnicotinamide (nikethamide) **(XVI-33)** is a compound that is used to counteract respiratory depression. Its moderate CNS stimulant properties have been recognized for many years (90, 91). The dimethyl, dipropyl, and other homologs of **(XVI-33)** are less active and more toxic (92). Their mode of action is unknown.

$CONEt_2$

XVI-33

$COOH$

XVI-34

$\left(-COO \right)_2 AlOH$

XVI-35

5. Hypocholesteremic Agents

High blood cholesterol levels can lead to arteriosclerosis, and yet few drugs are available that can reduce serum cholesterol levels, although considerable research is in progress. Many drugs being investigated are pyridine derivatives.

Altschul (93, 94) found that nicotinic acid **(XVI-34)** in large doses was a hypocholesteremic agent, whereas nicotinamide **(XVI-1)** did not affect serum cholesterol levels in spite of the fact that these two forms of vitamin have the same effect on nutrition. Experience with nicotinic acid in large doses to reduce serum cholesterol levels have been reviewed (95). The acid is thought to inhibit cholesterol biosynthesis (96); it is also capable of inhibiting the release of free fatty acids from adipose tissue (97). Despite the fact that the large doses required make it unpleasant for the patient, nicotinic acid has been subjected to extensive clinical trial. Aluminum nicotinate (**XVI-35**) is reported (95) to be as specific in reducing serum cholesterol and total lipids as is nicotinic acid and to be better tolerated. Nicotinic acid-1-oxide (**XVI-36**) as the free acid or as its magnesium or ethanolamine salt also possesses hypocholesteremic properties (98–100). The ethanolamine salt is claimed (99) to be a water-soluble hypocholesteremic agent with low toxicity.

Recently, some emphasis has been placed on the synthesis and preliminary pharmacological evaluation of various nicotinic acid esters. Findings are reported in the patent literature. Nicotinates of glycolic acid (101), of *p*-chlorophenyliso-propylcarbinol (102), of 2,2,6,6-tetramethylol-1-cyclohexanol (103), of pyri-doxine (104), of 3-(*o*-methoxyphenoxy)-1,2-propanediol (105), and of various triols (106) are all reported to be active compounds. A chlorinated derivative of picolinic acid is also active (107).

The tertiary alcohol, α,α-diphenyl-β-(4-pyridyl)ethanol (**XVI-37**), was synthe-sized as a potential tranquilizer, but a pharmacological screening of this compound revealed that it reduced serum cholesterol levels in the mouse (108). Numerous compounds closely related structurally to **XVI-37** were then synthesized, but none was as active. Enolic ethers, esters, and amino-compounds of general structure (**XVI-38**) and ketones and imines possessing the structure (**XVI-39**) were the subject of a detailed study (109–112). These compounds were claimed to be ovulation inhibitors as well as hypocholesteremic agents. On the other hand 2-(*m*-methoxyphenyl)-1-(1-oxido-2-pyridyl)-1-phenyl-2-propanol

XVI-36　　　　　　XVI-37　　　　　　XVI-38

XVI-39　　　　　　XVI-40

$$R(CH_2)_2O-\underset{\text{XVI-41}}{\boxed{}}-CH(NHR')CH_2-\boxed{}$$

(XVI-40) is described in a patent (113) as a hypocholesteremic agent with little estrogenic side effects. Numerous other pyridine derivatives lower the cholesterol concentration in blood (114–117). Phenethylamine derivatives [**(XVI-41)**, in which R is a dialkylamino- or cyclic amino-group and R' is a heterocyclic or aryl group] are claimed (118) to be potent inhibitors of cholesterol biosynthesis.

6. Niacin and Niacinamide

Pellagra in human beings and canine blacktongue in dogs are the result of the same nutritional deficiency. Liver extracts were known to cure these deficiencies. Fractionation of these liver extracts led eventually to the isolation of nicotinamide (niacinamide) **(XVI-1)** in 1937 (119). Administration of nicotinic acid (niacin) **(XVI-34)** also cured canine blacktongue and human pellagra (119–121). Niacin is widely distributed in nature and present in many foodstuffs.

Four basic methods of synthesis for the preparation of niacin have been described (121a), but only one is used extensively on a commercial scale, the oxidation of substituted pyridines. In particular, 3-picoline may be oxidized to niacin with various oxidizing agents including potassium permanganate, chlorine, air, or by electrochemical oxidation. 5-Ethyl-2-methylpyridine is also commonly employed in the manufacture of nicotinic acid. This starting material is easily made by the condensation of paraldehyde and ammonia and is oxidized with nitric acid at elevated temperature to 2,5-pyridinedicarboxylic acid and then to nicotinic acid.

Niacinamide is prepared by direct amination of niacin with ammonia or urea or by autoclaving methyl nicotinate with ammonia. The preparation of niacin and niacinamide from the common intermediate, 3-cyanopyridine, in the manner shown is an attractive synthetic method, but it is not employed industrially because of the high cost of pyridine.

Niacin is found in two coenzymes, coenzyme I (diphosphopyridine nucleotide, DPN) (**XVI-42a**) and coenzyme II (triphosphopyridine nucleotide, TPN) (**XVI-42b**). These coenzymes participate in many stereospecific hydrogen-transfer enzymatic reactions. Both DPN and TPN reversibly alternate between the quaternary pyridinium ion form (i.e., oxidized form) (**XVI-43**) and the

R
XVI-42a $\overline{H}$
XVI-42b $PO(OH)_2$

tertiary amine (reduced) form (**XVI-44**) as shown in Scheme 1. A more detailed account of the synthesis and the nutritional and therapeutic role of niacin and niacinamide is available (122).

Microorganisms require niacin for growth and reproduction. Some organisms, for example, *Staphylococcus aureus* and *Bacillus diphtheriae*, are capable of synthesizing niacin, but others require preformed vitamin or preformed coenzymes DPN and TPN. In the latter instances, pyridine-3-sulfonic acid (123), pyridine-3-sulfonamide (124), 3-acetylpyridine (125), several halogen-substituted nicotinic acids (126), and 6-aminonicotinamide (127) can act as antagonists of niacin or niacinamide.

Preparations of niacin and niacinamide labeled with ^{14}C in the acid or amide group have been reported (128). These compounds were prepared from 3-bromopyridine, and were used (129) in a study of the metabolism of both vitamins, the results of which were inconclusive.

$$R^1CH_2OH + ROPOCH_2$$

Alcohol

XVI-43

Alcohol dehydrogenase

$$R^1CHO + ROPOCH_2$$

Aldehyde

XVI-44

Scheme 1

$$\xrightarrow[{}^{14}CO_2]{BuLi} \quad {}^{14}COOH \longrightarrow {}^{14}CONH_2$$

7. Vitamin B$_6$

The chemistry and biochemical role of vitamin B$_6$ are the subjects of an excellent review (130). Initially, vitamin B$_6$ was thought to be a single substance, 4,5-di-(hydroxymethyl)-3-hydroxy-2-methylpyridine **(XVI-45a)**, now identified by the trivial name *pyridoxine* or *pyridoxol*, but subsequent work has revealed that pyridoxine alone is not responsible for all the activity of naturally occurring vitamin B$_6$. The related compounds *pyridoxal* **(XVI-46a)**, *pyridoxamine* **(XVI-47a)**, and the 5-phosphates **(XVI-45b, XVI-46b, XVI-47b)** are also active compounds. In fact, pyridoxal phosphate is now recognized as being a coenzyme, called codecarboxylase, which functions in the decarboxylation of a number of α-amino-acids including arginine, aspartic acid, glutamic acid, lysine,

CH₂OH structure:

$$\text{XVI-45a}$$
$$\text{XVI-45b}$$

HO, CH₂OH, CH₂OR, Me, N — **XVI-45a / XVI-45b**

HO, CHO, CH₂OR, Me, N — **XVI-46a / XVI-46b**

HO, CH₂NH₂, CH₂OR, Me, N — **XVI-47a / XVI-47b**

a R = H
b R = PO(OH)₂

ornithine and tyrosine (131, 132). The mechanism according to Mandeles *et al.* (133) is shown in Scheme 2 and is corroborated by Udenfriend (134).

HC=O, HO, CH₂OPO(OH)₂, Me, N

$+ \; H_3\overset{+}{N}CHRCO_2^-$

$\underset{+H_2O}{\overset{-H_2O}{\rightleftharpoons}}$

HC=NCHRCO₂⁻, HO, CH₂OPO(OH)₂, Me, $\overset{+}{N}$, H

$+ CO_2 \; / \!\!/ \; -CO_2$

HC–N=CHR, HO, CH₂OPO(OH)₂, Me, N, H

$\longleftrightarrow$

HC=N–$\overset{-}{C}$HR, HO, CH₂OPO(OH)₂, Me, $\overset{+}{N}$, H

$+H^+ \;\big\|\; -H^+$

HC=NCH₂R, HO, CH₂OPO(OH)₂, Me, $\overset{+}{N}$, H

$\underset{+H_2O}{\overset{-H_2O}{\rightleftharpoons}}$

HC=O, HO, CH₂OPO(OH)₂, Me, N

$+ \; H_3\overset{+}{N}CH_2R$

Scheme 2

Enzymatic transamination reactions also require pyridoxal phosphate (135, 136) or pyridoxal itself (137) as a coenzyme. Thus α-aminoacids are enzymatically oxidized to α-ketoacids. The mechanism is illustrated (Scheme 3). Pyridoxal phosphate also catalyzes the nonenzymatic transamination of α-aminoacids.

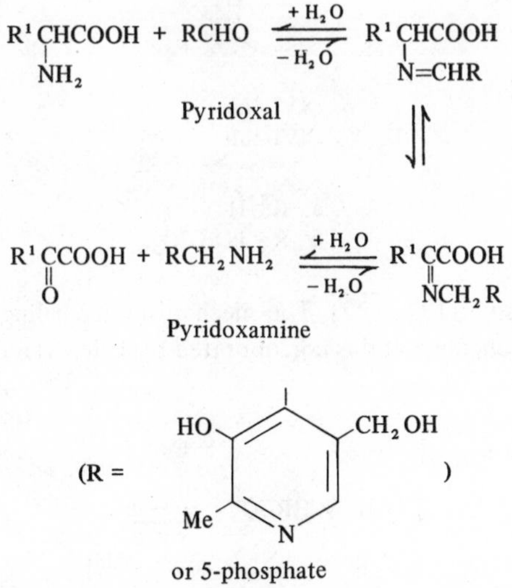

$$R^1CHCOOH + RCHO \xrightleftharpoons[-H_2O]{+H_2O} R^1CHCOOH$$

with NH₂ below R¹CHCOOH on left, Pyridoxal label, and N=CHR on right

$$R^1CCOOH + RCH_2NH_2 \xrightleftharpoons[-H_2O]{+H_2O} R^1CCOOH$$

with O below R¹CCOOH on left (ketone), Pyridoxamine label, and NCH₂R on right

(R =

HO — CH₂OH ring with Me, N

)

or 5-phosphate

Scheme 3

Snell (138, 139) pioneered this work and has reviewed (140) the topic. This nonenzymatic reaction requires the presence of a suitable metal ion (e.g., Cu^{2+}, Fe^{3+}, and Al^{3+}) and is completely reversible. It involves the formation of an intermediate aldimine (**XVI-48**).

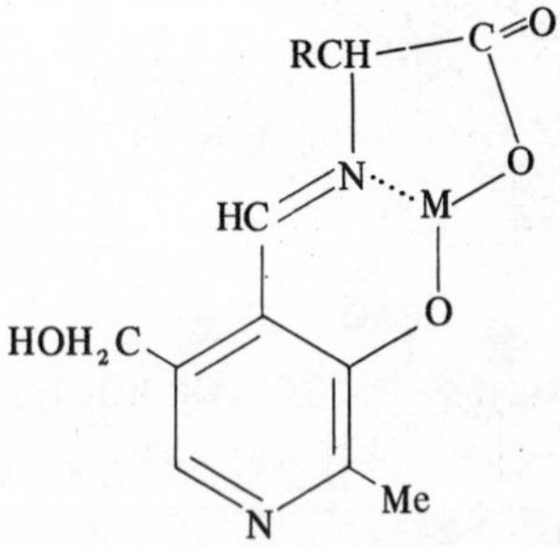

XVI-48 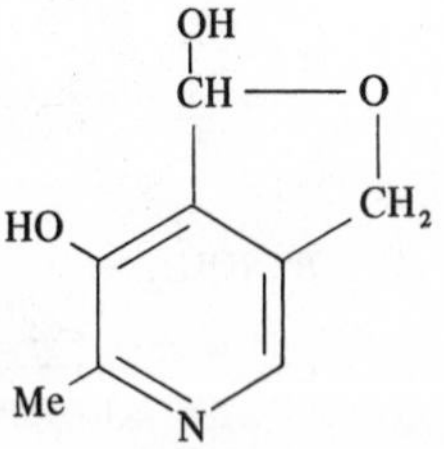   XVI-49

Although pyridoxal and its 5-phosphate are commonly depicted as shown (**XVI-46a**), it is apparent from a study of dissociation constants and ultraviolet and infrared spectra that the hemiacetal form **XVI-49** is the predominant structure in aqueous solution (141).

Concise accounts of some of the methods used to synthesize pyridoxine, pyridoxamine, and pyridoxal are available (122, 142). Numerous other methods (143, 144) have been reported (see also Chapter XIV). Recently, however, an emphasis has been placed on the preparation and evaluation of derivatives of the vitamin B_6 group. Analogs of pyridoxal 5-phosphate in which the methyl group at position 2 was replaced by a hydrogen atom or an ethyl group have been described (145). Both compounds are efficient coenzymes. This observation led to the conclusion that the methyl group of pyridoxal 5-phosphate was not a prerequisite for coenzymic activity of that compound. Pyridoxal and pyridoxamine phosphate derivatives (**XVI-50**, R = CHO or CH_2NH_2) in which the methyl group was replaced by a hydroxymethyl group have been prepared and characterized (146). Schiff bases of general structure **XVI-51** have been described as long-acting vitamin B_6 derivatives; their preparation has been

XVI-50

XVI-51

XVI-52

patented (147). Another Schiff's base **(XVI-52)** of pyridoxal has been characterized as a new phenylalanine metabolite isolated from the brain tissue of phenylketonuric rats (148).

3-Deoxypyridoxine was isolated (149) from the mother liquors obtained during the synthesis of pyridoxine by the method of Harris and Folkers (150). This compound **(XVI-53)** arrested significantly the growth of lymphosarcoma implants in rats maintained on a pyridoxine-deficient diet.

Other chemically modified vitamin B_6 derivatives have been reported. The preparation and properties of an analog of pyridoxal, namely 3-(4-formyl-3-hydroxy-2-methyl-5-pyridyl)propionic acid **(XVI-54)**, and the corresponding

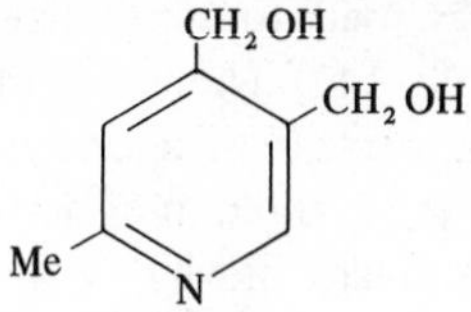

XVI-53

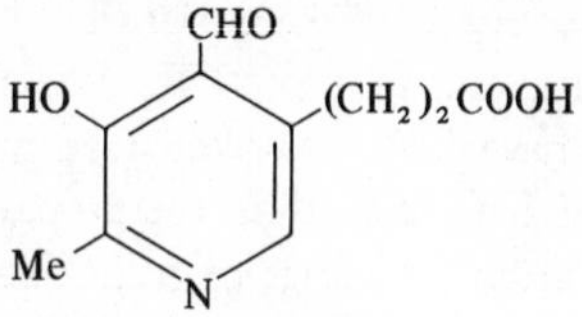

XVI-54

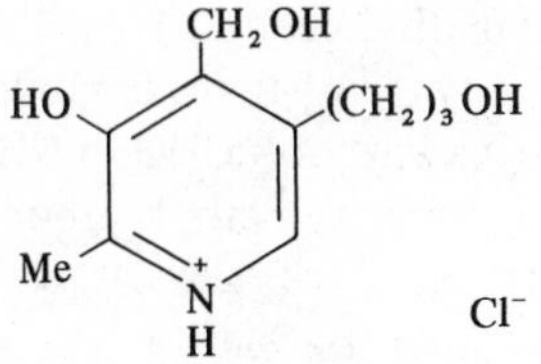

XVI-55

4-aminomethyl compound have been described (151). When the 5-side-chain in pyridoxine was extended by two carbon atoms, very potent antagonists of pyridoxine (e.g., **XVI-55**) were obtained (152). The 5-side-chain was also involved in the preparation of various thioethers and disulfides (e.g., **XVI-56**) of pyridoxine. The former are reported (153, 154) to be useful in the treatment of cerebral disorders.

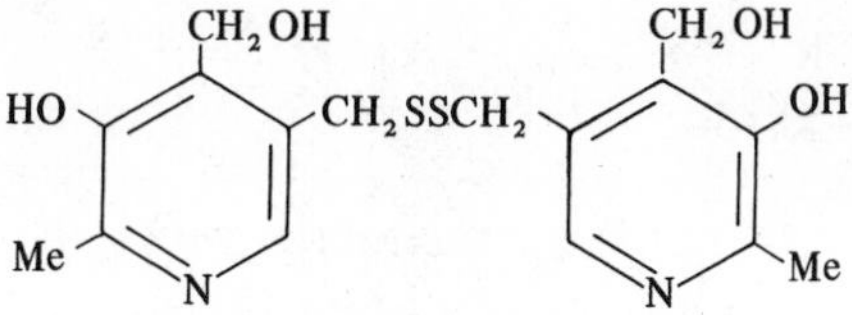

XVI-56

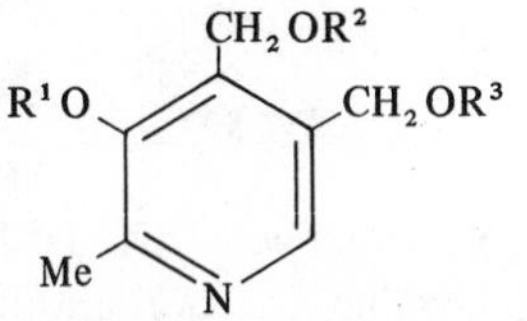

	R^1	R^2	R^3
XVI-57a	H	H	X
XVI-57b	X	X	X
XVI-57c	OCOEt	OCOEt	H
XVI-57d	Me(CH$_2$)$_{10}$CO	MeCO	MeCO

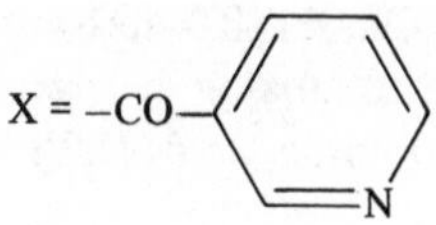

$$X = -CO-$$

XVI-58

Some disulfides exhibit vitamin B_6-like and analgesic action (155), some are useful remedies for epilepsy (156) and others are claimed (157) to possess tranquilizing, emetic, and euphoric properties.

A large number of simple derivatives of pyridoxine have been reported. The quaternary methiodide is useful (158) in the treatment of cardiovascular disease. The glucuronate is a detoxicant; the ingestion of poisons such as morphine or chloral hydrate gave no intoxication when pyridoxine glucuronate was taken simultaneously (159). Pyridoxine mononicotinate (XVI-57a) the related $\alpha^4,5$-O-isopropylidene-α^3-O-nicotinoylpyridoxine (XVI-58) and the trinicotinoyl derivatives of pyridoxine (XVI-57b), and pyridoxamine exhibit (160–162) pharmacological properties of both nicotinic acid and vitamin B_6. The fat-soluble vitamin B_6 derivative (XVI-57c) was prepared (163) by treating pyridoxine with ethyl chlorocarbonate. Another fat-soluble derivative, 4,5-bis(acetoxymethyl)-3-lauroyloxy-2-methylpyridine (XVI-57d) has low toxicity and can be used for the treatment of skin diseases such as sebarrhea (164).

8. Diagnostic Agents

These compounds are used to detect impaired function of body organs or to identify abnormalities in tissues. A few pyridine derivatives are useful for this purpose.

A. X-Ray Contrast Media

Radiopaques or contrast media are substances which are used for visualization of body organs or cavities. For this effect they depend on their ability to absorb x-rays. A review of the radiopaques commonly used in medicine has been published (165).

Derivatives of 3,5-diiodo-4-pyridone are useful compounds. Propyliodone U.S.P., B.P., is propyl 3,5-diiodo-4-pyridone N-acetate (XVI-59a). The acid is prepared from 4-pyridone by iodinating with an iodide/iodate mixture followed by reaction with chloroacetic acid. Propyliodone is used in bronchography. Diodone B.P., the diethanolamine salt of 3,5-diiodo-4-pyridone *N*-acetic acid, is

	n	R
XVI-59a	1	OPr
XVI-59b	1-3	NHR′
XVI-59c	10	OMe

XVI-60   XVI-61

used in pyelography. Newer derivatives of 3,5-diiodo-4-pyridone have been prepared which possess the general formula **XVI-59b** and are reported (166) to be suitable for lymphography and bronchography. Related compounds such as methyl 3,5-diiodo-4-pyridone *N*-undecanoate (**XVI-59c**) and the corresponding acid have been patented (167) as contrast agents useful in angiography. The authors of this patent reported earlier (168) the preparation and biological evaluation as x-ray contrast compounds of a large number of acids and esters of general formula (**XVI-59**).

Sodium iodomethamate (iodoxyl) (**XVI-60**) was one of several organic iodinated compounds prepared by von Lichtenberg (169). It has been used for many years as a radiopaque compound in urography and pyelography.

B.   Glandular Function

2-Methyl-1,2-di-(3-pyridyl)-1-propanone (Metyrapone U.S.P.) (**XVI-61**) is a useful diagnostic agent. This compound possesses the ability to inhibit the enzyme 11-β-hydroxylase which promotes 11-β-hydroxylation during the biosynthesis of cortisol, corticosterone, and aldosterone (170, 171). The drug, therefore, is valuable as a diagnostic aid when hypopituitarism and Cushing's syndrome are suspected. It is used as a test of glandular function.

Certain 2-pyridylindoles also inhibit the biosynthesis of aldosterone from 11-deoxycorticosterone (172). 3-Methyl-2-(3-pyridyl)indole (LXII), as its

XVI-62   XVI-63

methanesulfonate, is reported to be a potent compound.

The biosynthesis of adrenal corticosteroids is also inhibited by various azachalcones (173). Several compounds possessing the general structure **XVI-63** have an inhibiting effect equal to or better than that of Metyrapone.

9. Pyridine Analgesics

Phenyramidol   [2-(β-hydroxy-β-phenylethylamino)pyridine   hydrochloride] **(XVI-64)** was introduced in 1959 as a muscle relaxant with analgesic activity (174). It is used primarily for the relief of pain associated with muscle strain or spasm. Other derivatives of 2-aminopyridine are useful analgesics including

XVI-64   XVI-65

N-acyl-N-(2-pyridyl)-1-piperidino-2-aminopropanes **(XVI-65)** (175) and derivatives of 2-anilino-5-aminopyridine (176).

1-(6-Methyl-2-pyridyl)-2-aminopropane **(XVI-66)** is described (177) as a compound that produces a marked analgesia but that possessed an unfavorable therapeutic index. The related compound **(XVI-67)** is also an analgesic (178). Certain phenyl(pyridyl)carbinol ethers possess analgesic and diuretic activities (179). The simple phenol **(XVI-68)**, that is, the ester of salicylic acid and 3-pyridylcarbinol, is also an analgesic (180). The somewhat related compound, 3-pyridylmethylamonium salicylate **(XVI-69)** is described (181) as being a powerful analgesic and antirheumatic medicament for external use.

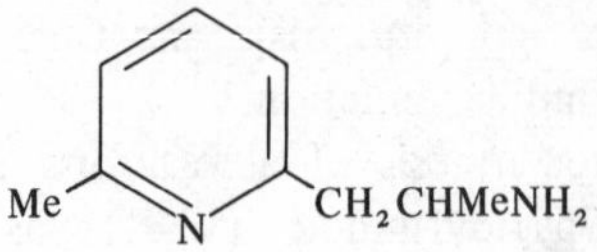

XVI-66 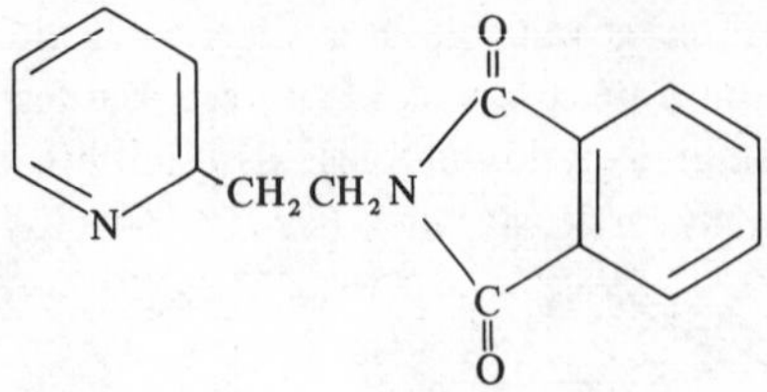   XVI-67

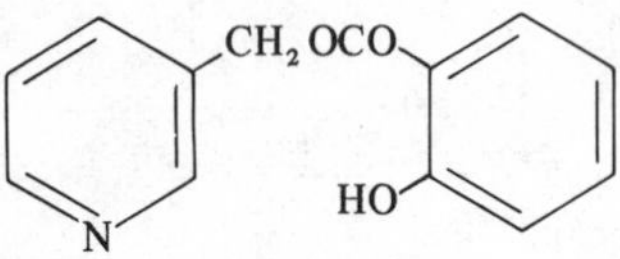

XVI-68 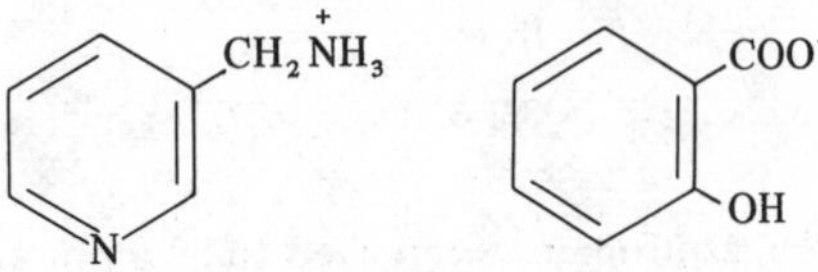   XVI-69

d-Propoxyphene **(XVI-70)** (Darvon®) is a clinically useful analgesic agent that does not possess any appreciable addiction liability (182). It was of interest to deStevens *et al.* (183) and deStevens (184) to prepare some heterocyclic and cyclic analogs of **XVI-70**. Among the compounds they studied was the pyridine analog **(XVI-71)**, which was prepared as shown. Compound **XVI-71** was completely devoid of analgesic effects. The related compound **(XVI-72)**, however, was a more potent analgesic than propoxyphene.

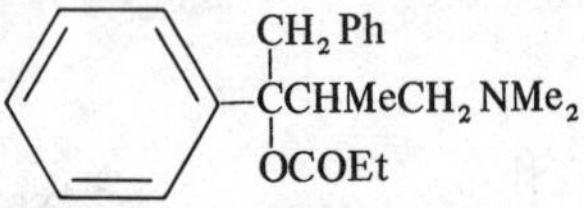

XVI-70 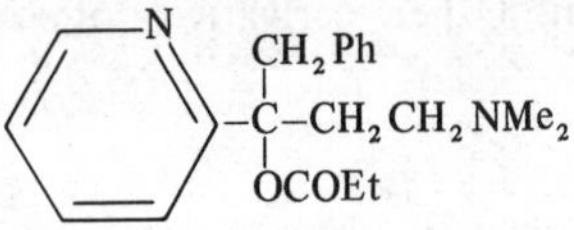   XVI-71

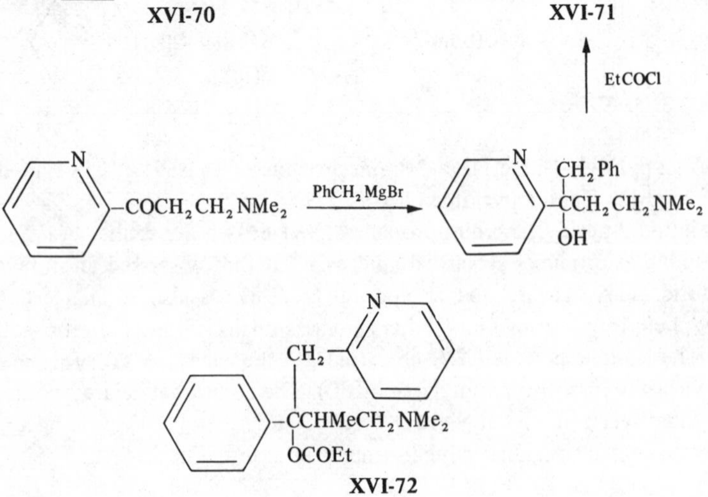

EtCOCl

PhCH₂ MgBr

XVI-72

Compounds **XVI-73**, **XVI-74**, and **XVI-75** were also evaluated. The tetrahydro-naphthalene derivative **(XVI-73)** was five times more potent than morphine, the oxygen analog **(XVI-74)** was only 10% as active an analgesic as **XVI-73**, and the piperidine derivative **(XVI-74)** was inactive. This latter observation indicated that the enhancing analgesic effect of the 2-pyridylmethyl group on a quaternary C atom is not general.

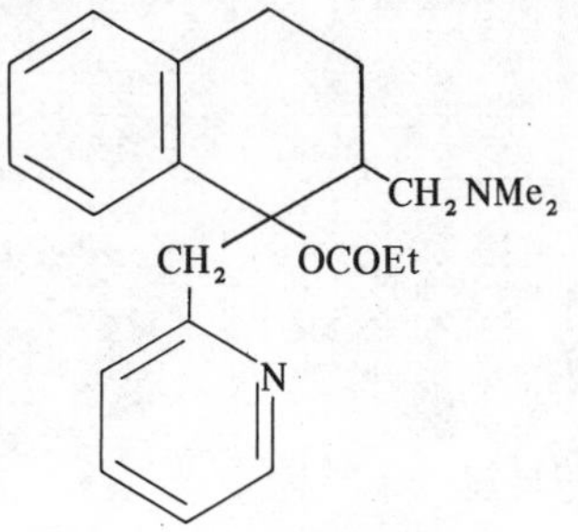

XVI-73

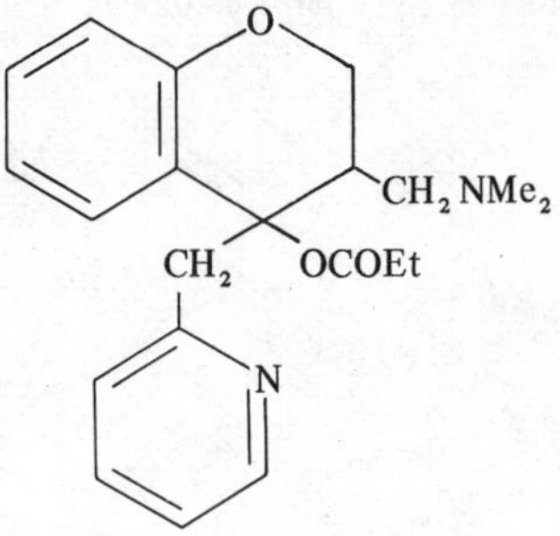

XVI-74

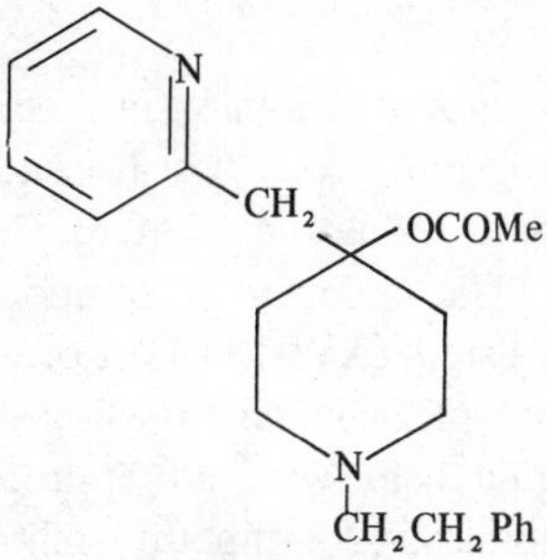

XVI-75

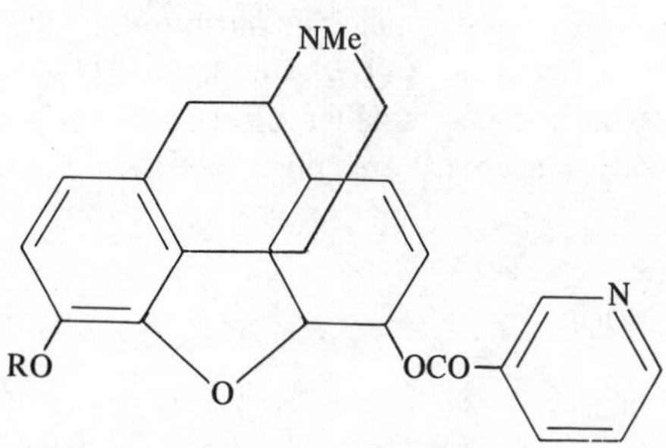

	R
XVI-76a	$-CO-$ pyridyl
XVI-76b	H

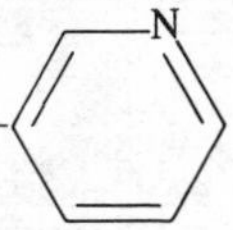

Two complex esters of nicotinic acid are used clinically. They are nicotinyl derivatives of morphine and codeine named nicomorphine (3,6-dinicotinoyl-morphine) **(XVI-76a)** and nicocodine (6-nicotinoylcodeine) **(XVI-76b)**, respectively. The former is a narcotic analgesic (185, 186) and the latter is a narcotic antitussive (187).

The nifenazone molecule **(XVI-77)** combines the features of nicotinamide and phenazone. It possesses analgesic and antiinflammatory properties (185, 188).

XVI-77

10. Antiinflammatory Agents

Nifenazone, mentioned above, is a nicotinic acid derivative that possesses antiinflammatory (antiphlogistic) properties. Recently, some emphasis has been placed on the synthesis and evaluation of other nicotinic acid derivatives with this property (189, 190). The majority of compounds studies are substituted 2-anilinonicotinic acids or esters **(XVI-78)** most of which are described in recent patents (189) and are not yet available for clinical use. They are prepared by interacting 2-chloronicotinic acid with an appropriately substituted aniline. Some compounds, including 2-(2,3-dimethylanilino)nicotinic acid **(XVI-79)**, possess (190) an antiinflammatory activity comparable with that of mefenamic acid **(XVI-80)**, a clinically used analgesic, antipyretic, and antiinflammatory agent. Others of general structure **XVI-78** are claimed (191) to have antiinflammatory properties with a much smaller ulcer-active effect than other non-steroidal antiinflammatory agents. Some pharmaceutical formulations have been suggested.

XVI-78   XVI-79

Me Me

XVI-80

An extremely large number of pyridines, 2-pyridones, and pyridine-1-oxides with various substituents have been prepared and evaluated as antiinflammatory agents. A detailed account is not warranted here. The reader is referred to the recent patent literature.

11. Antihistaminics

Many body tissues contain low concentrations of histamine (XVI-81) in bound form. Physical injury and the action of many chemical agents as well as antigen-antibody reactions can induce the release of free histamine which has potent pharmacologic activity on blood vessels, on gastric secretion, and on bronchial smooth muscle. When the skin is injured the 'triple response of Lewis' (192), that is, localized redness followed by edema and then diffuse redness, is often observed. The damaged area itches. Wasp and nettle stings cause itching and edema because of their histamine content. Histamine is also released in anaphylaxis and in allergic diseases. Further details on histamine are available in most pharmacology textbooks.

$RCH_2CH_2NMe_2$

XVI-81 XVI-82

Like histamine, most antihistaminics are derivatives of ethylamine and can be represented by the general formula XVI-82. Although the nature of the substituent R can vary greatly, it incorporates the pyridine nucleus in a significant number of antihistaminics. Those drugs that are or have been used clinically are now mentioned.

2-{α-[2-(Dimethylamino)ethyl] benzyl}pyridine (*pheniramine*) (XVI-83a), its halogenated derivatives, *brompheniramine* (XVI-83b) and *chlorpheniramine* (XVI-83c), and their dextrorotatory isomers (*dexbrompheniramine* and *dex-chlorpheniramine*) are derivatives of propylamine with antihistaminic properties.

Two potent ethanolamine derivatives are 2-{p-chloro-α-[2-(dimethylamino)-ethoxy]benzyl}pyridine (*carbinoxamine*) (**XVI-84a**) and 2-{α-[2-(dimethyl-amino)ethoxy]-α-methylbenzyl}pyridine (*doxylamine*) (**XVI-84b**).

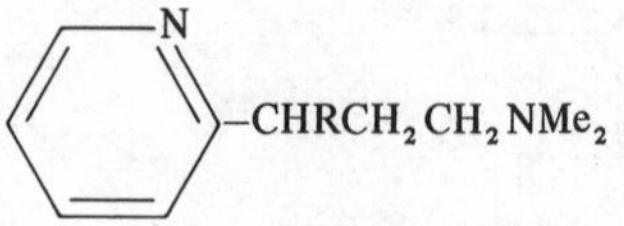

	R
XVI-83a	Ph
XVI-83b	p-PhBr
XVI-83c	p-PhCl

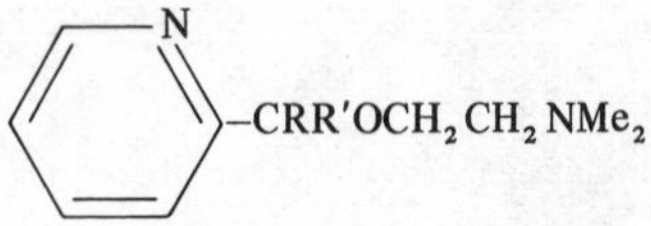

	R	R'
XVI-84a	H	p-PhCl
XVI-84b	Me	Ph

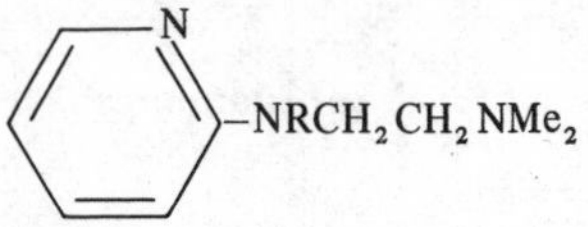

	R
XVI-85a	−CH₂Ph

XVI-85b −CH₂−⟨C₆H₄⟩−Cl

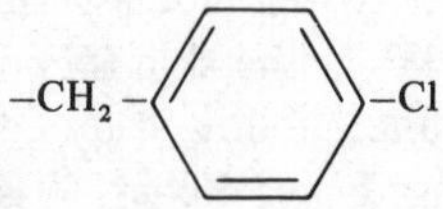

XVI-85c −CH₂−⟨C₆H₄⟩−OMe

XVI-85d −CH₂-thienyl

XVI-85e −CH₂-thienyl

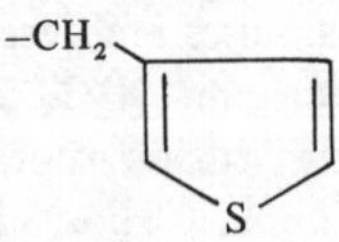

XVI-85f −CH₂-chlorothienyl

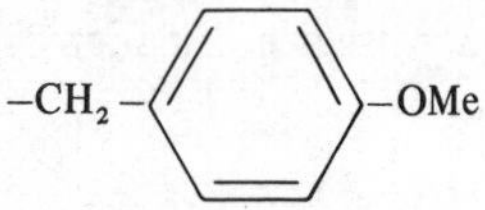

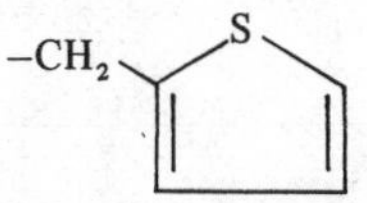

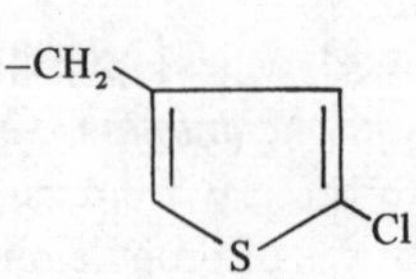

Antihistaminics that are derivatives of ethylenediamine are more common. Under this classification are found *tripelennamine* 2-{benzyl[2-(dimethylamino)-ethyl]amino}pyridine **(XVI-85a)**, and its simple derivatives, *halopyramine* **(XVI-85b)** and *mepyramine (pyrilamine)* **(XVI-85c)**. *Methapyrilene* **(XVI-85d)**, *thenyldiamine* **(XVI-85e)**, and *chlorothen* **(XVI-85f)** are chemically related compounds. *Dimethindene* **(XVI-86)** and *triprolidine* **(XVI-87)** are two other clinically useful antihistaminics that possess a pyridine ring in their structures.

XVI-86   XVI-87

XVI-88

Structure and activity relationships of the antihistamines have been reviewed (193). Antihistamines are more potent in preventing the actions of histamine than in reversing these actions after they have developed. They are used in the treatment of urticaria and hay fever but are of little benefit in the treatment of asthma. Some are used in the prevention of motion sickness, and some are being used in the treatment of Parkinson's disease.

Few new antihistaminic compounds with pyridine substituents have appeared in recent years. One interesting reference (194) describes the preparation of certain ketoximino esters of 2-, 3-, and 4-benzoylpyridine oxime (e.g., **XVI-88**). It was found that all the compounds prepared exhibited antihistaminic activity with almost no anticholinergic action. The propionyl analogs were the most active compounds.

12. Cholinesterase Inhibitors and Reactivators

The autonomic (involuntary) nervous system regulates the functions of all smooth muscle and glands in the body. There are two main divisions to the autonomic nervous system, the *sympathetic* and *parasympathetic* divisions. Of these, the latter is the main adjuster of the internal environment under normal resting conditions. It regulates the heart rate, gastric secretion, intestinal motility and tone, respiratory smooth muscle, iris size, lachrymation, bladder and rectal emptying, sexual function, salivation, and other systems. At the ending of the parasympathetic nerve fiber, a chemical transmitter (acetylcholine) is synthesized and released when the nerve is stimulated. The acetylcholine travels to the effector cell in the smooth muscle or gland with which it combines reversibly thus stimulating the cell. The acetylcholine is then released by the effector cell and hydrolyzed at the nerve ending to choline and acetic acid by the action of the enzyme cholinesterase.

A. *Inhibitors of Cholinesterase*

Certain drugs inhibit the action of cholinesterase; some (e.g., quaternary ammonium anticholinesterases) combine reversibly with the enzyme and therefore have a limited duration of action. Others, for example, the organophosphate anticholinesterases (see below), combine in a virtually irreversible manner with the enzyme and their effects may persist for months. A comprehensive review of anticholinesterase agents was published recently (195). The result of inhibiting the action of cholinesterase is to prolong the action of acetylcholine on the effector cell. The effect is often more prominent on certain structures than on others. Three pyridine derivatives are used clinically for their anticholinesterase properties. They are pyridostigmine bromide (**XVI-89a**), and two related compounds, benzpyrinium bromide (**XVI-89b**), and distigmine bromide (**XVI-90**). All are chemically reminiscent of the potent anticholinesterase, neostigmine bromide (**XVI-91**).

	R
XVI-89a	Me
XVI-89b	CH_2Ph

Me₂NOCO / OCONMe₂ structures

XVI-91   XVI-92

3-(Dimethylcarbamyloxy)-1-methylpyridinium bromide (pyridostigmine bromide) is synthesized (196) by first treating 3-hydroxypyridine with dimethylcarbamyl chloride and then quaternizing the resulting dimethylcarbamate with methyl bromide. Pyridostigmine bromine is used in the treatment of myasthenia gravis (197). 1-Benzyl-3-(dimethylcarbamyloxy)pyridinium bromide (benzpyrinium bromide) is used clinically for its action on the bowel or bladder or to treat delayed menstruation (198, 199). It is synthesized in a manner similar to that employed for the preparation of pyridostigmine bromide (200). *N,N'*-Hexamethylenebis[1-methyl-3-(methylcarbamyloxy)pyridinium bromide] (distigmine bromide) is less widely used clinically. It was one of a series of compounds that possessed the general formula $(CH_2)_n(NRCO_2Z)$ in which R was an alkyl group, n an integer from 2 to 10, and Z a group containing a tertiary nitrogen atom, and which were prepared (201) for testing as cholinesterase inhibitors. Distigmine bromide is used to treat myasthenia gravis.

Another bis-pyridinium bromide derivative has been prepared (202). This compound, 1,2-xylylenebis[1-(3-dimethylcarbamyloxy)pyridinium bromide] (**XVI-92**), is reported (199) to be more active than benzpyrinium bromide.

B. *Reactivators of Inhibited Cholinesterases*

Enzymes such as acetylcholinesterase (AChE), butyrylcholinesterase, and isopropylcholinesterase are inhibited by organophosphorus compounds, especially the phosphofluoridates, for example, TEPP (tetraethyl pyrophosphate), DFP (diisopropyl phosphorofluoridate), DDP (di-*n*-propyl 2,2-dichlorovinyl phosphate), TABUN (ethyl *N,N*-dimethylphosphoramidocyanidate), SARIN (isopropyl methylphosphonofluoridate), and SOMAN (pinacolyl methylphosphonofluoridate) which possess the general formula **XVI-93** in which R' is an alkoxy group, R^2 is an alkoxy or alkyl or alkylamido group, and X is a halogen atom or a cyanide or a phenoxy or a disubstituted phosphoryloxy group. These compounds are *per se* phosphorylating agents of AChE and are potential war nerve gas poisons. A second group of organophosphates consists of compounds that are weak anticholinesterase agents but are converted into highly potent phosphorylating agents *in vivo*. To this group belong some of the organophosphate insecticides used at present; O,O-diethyl-O-(4-nitrophenyl) phosphorothioate (Parathion) is a typical example.

The active center of AChE contains two subsites called anionic and esteratic sites (203, 204). The enzyme reacts with organophosphorus compounds to give phosphorylated AChE, which is extremely stable. Its hydrolysis with water proceeds at a very slow rate. In 1951, Wilson reported (203) that hydroxylamine reactivated phosphorylated AChE considerably faster than did water. Information gained in these studies precipitated a search for more potent reactivators and led to the discovery that certain hydroxamic acids and oximes were active compounds.

The most potent reactivators of phosphorylated AChE are oximes; hydroxamic acids are less active. Excellent reviews on this subject have been published (195, 205). The best reactivators among the hydroxamic acids are the pyridine derivatives *N*-nicotinoylhydroxylamine methiodide **(XVI-94)**, and *N*-picolinoyl-hydroxylamine **(XVI-95)**, and the chemically related compound pyrimidine-2-

XVI-93   XVI-94   XVI-95

XVI-96   XVI-97   XVI-98

hydroxamic acid **(XVI-96)**. The quaternary hydroxamic acids are thought to react in the manner shown (Scheme 4) (206, 207).

Other quaternary hydroxamic acids derived from pyridine, of general structure **XVI-97**, in which R is H or alkyl, and X is $(CH_2)_{1,2}$ or CHMe, have been shown (208) to be active in preventing and reversing some of the physiological effects of cholinesterase inhibition. Wilson and Ginsburg (209) hypothesized that the anionic site on the surface of the enzyme was not affected when AChE was inhibited by organophosphorous compounds and prepared many compounds that possessed a potentially good reactivating group (the oxime group)

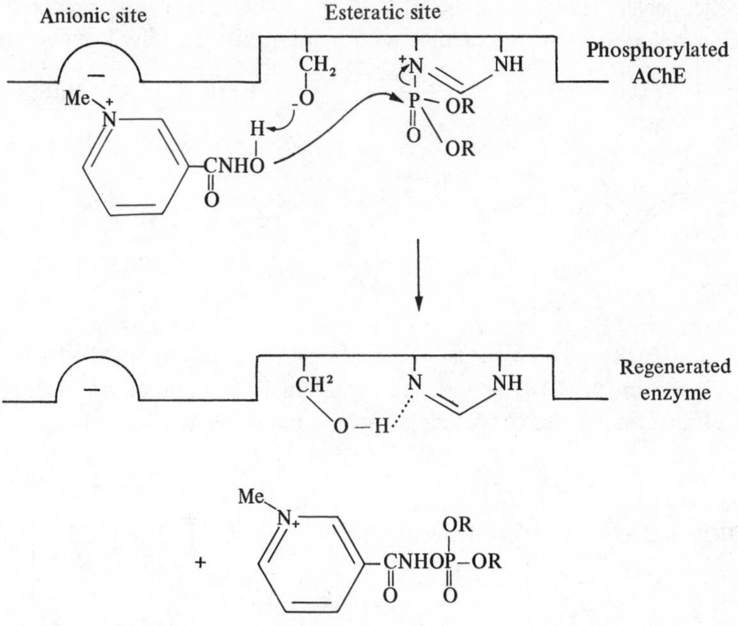

Scheme 4

and a quaternary nitrogen atom. This led to the discovery of the very effective reactivator, pyridine-2-aldoxime methiodide (2-PAM) (XVI-98), also available as the chloride and called pralidoxime chloride. This quaternary salt was 50,000 times faster in action than picolinohydroxamic acid (XVI-95) and almost a million times as active as hydroxylamine when tested against TEPP-inhibited cholinesterase. Pyridine-2-aldoxime methiodide can adopt two configurations, a colorless A form stable only at low temperature and a yellow B form. Some doubts exist (205) as to which of these two forms has the *syn* configuration and which is *anti*.

Various attempts were made to enhance the activity of 2-PAM. Compounds possessing two and three quaternary nitrogen atoms were synthesized and evaluated for their reactivating properties (210–212). One of the most effective compounds was 1,1'-trimethylenebis(4-formylpyridinium bromide)dioxime (TMB-4) (XVI-99).

2-PAM and TMB-4 are currently considered to be the most potent reactivators of organophosphorus-inhibited cholinesterase. An emphasis has been placed recently on the preparation of compounds which are chemically related to 2-PAM or TMB-4. Ethers, thioethers, sulfoxides, and sulfones of general structure XVI-100 (in which Z is O, S, SO, or SO_2) are active compounds (213).

The chemically related diol [**XVI-100**, Z = (CHOH)$_2$] was prepared from pyridine-4-aldoxime by interaction with 1,4-dibromo-2,3-dihydroxybutane and

XVI-99   XVI-100

was found (214) to be capable of preventing inhibition of AChE. Diamines of general structure **XVI-101** and related diquaternary compounds are also reported to be efficient antidotes for organophosphate poisoning (215, 216).

XVI-101   XVI-102

Arecoline analogs of 2-PAM and TMB-4 were prepared (217) by reducing the corresponding pyridinium aldoximes with sodium borohydride. The products were tertiary bases of structure **XVI-102** in which $n = 1$ or 2 and R was Me or (CH$_2$)$_{2'}$. Biological results showed that although these tertiary tetrahydropyridines were less toxic than quaternary aldoximes, they were much less effective reactivators of inhibited cholinesterase.

Various salts of pyridine-2-aldoxime have been reported in the literature. All possess antidotal properties against lethal alkylphosphate intoxication (218–221). The synthesis of pyridine-2-aldoxime-^{14}C methiodide has been achieved (222).

13. Nicotine and Other Ganglion Stimulants

Nicotine [1-methyl-2-(3-pyridyl)pyrrolidine] (**XVI-103**) is an alkaloid obtained from the leaves of the tobacco plant, *Nicotiana tabacum*. It is one of the most toxic compounds known; the lethal dose in man is in the neighborhood of 50 mg (223). Because of this toxicity, nicotine is used mainly as a horticultural pesticide. Pharmacologically, nicotine causes an initial rise in blood pressure as a result of ganglionic stimulation and norepinephrine release, followed by

	R	R'
XVI-104a	Me	H
XVI-104b	Me	OH
XVI-104c	H	H

XVI-103

ganglionic paralysis that causes the blood pressure to fall. Death is the result of respiratory depression. The absolute configuration of nicotine is known. No difference in toxicity between its L and D forms was found (224). Nicotine in small doses is used as a prototype of a ganglion stimulant agent in spite of many disadvantages (225).

The metabolism of nicotine in man and various animal species has been studied (226, 227). Oxidation occurs at position 5 of the pyrrolidine ring and yields cotinine **(XVI-104a)**, which is further metabolized to hydroxycotinine **(XVI-104b)** (the position of the OH group has not been determined), to norcotinine **(XVI-104c)**, and by further oxidation then cleavage of the pyrrolidine ring to γ-(3-pyridyl)-β-oxo-*N*-methylbutyramide **(XVI-105)**.

A large number of pyridine derivatives have been developed from nicotine and evaluated for their ganglionic stimulant properties (228). The compounds investigated were of general structure **XVI-106** in which *n* was 1 or 2, R was a

XVI-105   **XVI-106**   **XVI-107**

pyrrolidine or piperidine ring or a trialkylammonium group, and the point of attachment on the pyridine ring was at the 2, 3, or 4 position. Maximum activity was observed with compounds which possessed a quaternary ammonium group attached to the 3-position of the pyridine ring. The trimethylammonium compounds **(XVI-107)**, for example, had a potency twice that of nicotine. The quaternary methiodide of nicotine **(XVI-108)** is reported (229) to be as potent a ganglion stimulant as nicotine without the secondary blocking action of the latter.

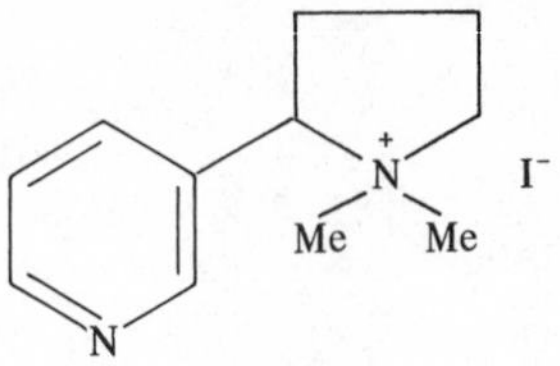

XVI-108

14. Miscellaneous Pyridine Derivatives

There are numerous recent references in the patent literature to pyridine derivatives which possess sedative, hypnotic, tranquilizing, or antidepressant activity (230–239). Structurally novel compounds have been described including 1-pyridyl-2-substituted cyclopropane esters, alcohols, and ketones with general structures **XVI-109** and **XVI-110**, which are claimed (230, 231) to possess both central nervous system activity and autonomic nervous system blocking activity.

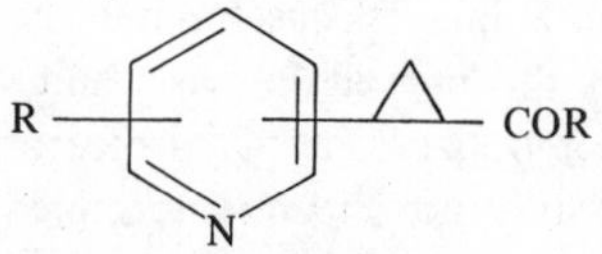

XVI-109 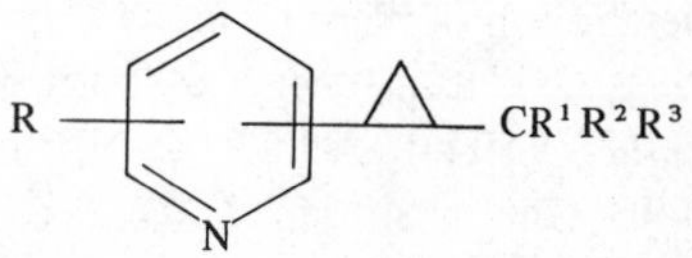   **XVI-110**

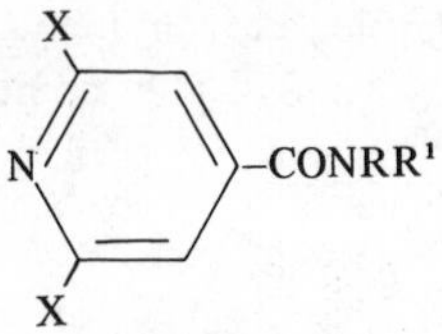

XVI-111 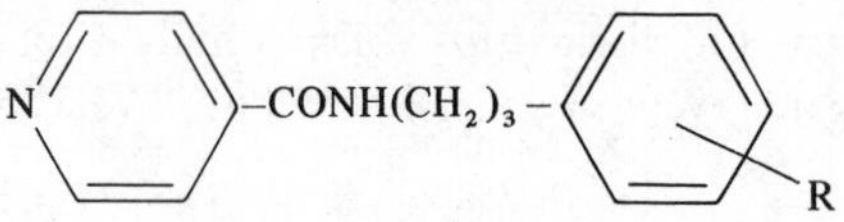   **XVI-112**

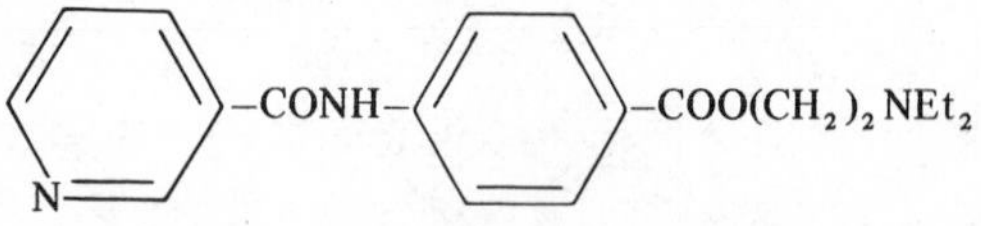

XVI-113

Derivatives of 2,6-dihaloisonicotinamide **(XVI-111)** are said to be highly effective tranquilizers and sedatives (238, 239). Other derivatives of isonicotinamide **(XVI-112** in which R is Cl or OMe) possess the property of lowering body temperature (240). The nicotinamide derivative, 2-diethylaminoethyl 4-(nicotinoylamino)benzoate **(XVI-113)**, is described in a patent (241) as a local anesthetic suitable for use in geriatrics.

As well as possessing vitamin activity, nicotinic acid is a weak peripheral vasodilator. This observation led to the synthesis of related compounds including the methyl, isopropyl, benzyl, and tetrahydrofurfuryl esters (XVI-114a to XVI-114d, respectively) of nicotinic acid (242–245). All possess vasodilating properties. Nicotinyl alcohol (3-pyridylcarbinol) (XVI-115) as its tartrate is a commercially available vasodilator. Its action is of longer duration, causing less

	R
XVI-114a	Me
XVI-114b	CHMe$_2$
XVI-114c	CH$_2$Ph
XVI-114d	CH$_2$ ...

flushing than nicotinic acid, and it is active orally (246, 247). It is prepared by reducing ethyl nicotinate with lithium aluminum hydride (248). There is a continued interest in the preparation of pyridine derivatives with cardiovascular activity. Numerous nicotinic esters (249) and related compounds (250–252) have been reported, mainly in the patent literature, to possess potent vasodilating properties.

Many pyridine derivatives have been prepared in recent years and shown to possess hypotensive properties (253–255), but none are used clinically.

Certain 4-picolyl ketones, carbinols, and related compounds have been prepared and shown (256, 257) to be useful anticonvulsant agents. Some 3- and 5-halogenated pyridine-2-sulfonamides are both anticonvulsants and analgesics (258).

A series of substituted α,β-diphenyl-α-trifluoromethyl-β-(2-pyridyl)ethanols have been synthesized (259). Many of these compounds were potent estrogens.

The literature, especially the patent literature, abounds with references to simple and complex pyridine derivatives that are claimed to possess one or more of the following activities: herbicidal, fungicidal, pesticidal, insecticidal, parasiticidal, anthelmintic, and nematocidal. A monumental effort would be required to summarize and reference the work done in these areas. It is sufficient to mention that the compounds investigated include pyridyl ureas, urethanes, carbamates, aldoximes, ketoximes, alkylamines, thiocarbonates, ethers, and thioethers; various quaternary pyridinium salts; halogenated pyridines; phosphorylated pyridines; nitro-, sulfinyl-, and sulfonyl-pyridines; derivatives of 3-pyridylmethane; N-substituted pyridones; and metal pyridine compounds.

Some pyridine derivatives have been found to be potent irreversible inhibitors of the enzyme, α-chymotrypsin (260). Others possess diuretic properties (261–263). 5-Hydroxy-2-formylpyridine thiosemicarbazone **(XVI-116)** when administered intraperitoneally is claimed (264) to be a highly active antileukemic agent. 2-Picoline derivatives **(XVI-117)** are reported (265) to be antitussive agents, as active against cough as codeine but devoid of codeine's narcotic properties.

$$HO \underset{N}{\underset{\parallel}{\bigcirc}} CH=NNHCSNH_2 \qquad \underset{N}{\underset{\parallel}{\bigcirc}} CH_2NR^1(CH_2)_2NR^2R^3$$

XVI-116   XVI-117

A few derivatives of picoline are useful antispasmodic drugs and can be used to treat intestinal spasm and peptic ulcer. Spasms of the stomach and other smooth muscles are often due to the release at the neuromuscular junction of more acetylcholine than is necessary. The presence of peptic ulcers can bring on spasm, which, together with the action of hydrochloric acid, can cause pain and may perpetuate the ulcer. This condition can be controlled by anticholinergic drugs that reduce smooth muscle spasm and reduce the secretion of hydrochloric acid.

A series of substituted pyridines was synthesized (67a) and found to inhibit gastric secretion without anticholinergic, ganglion blocking, or adrenergic blocking activity. The most active compounds were 3-phenoxypyridine, 4-phenoxypyridine, 2-phenylpyridine-1-oxide, 2-(2-thienyl)pyridine, 3-phenyl-pyridine, 2,2'-dipyridyl, and several of its isomers. Of these compounds, 2,2'-dipyridyl has been chosen for clinical trial; a study on its antisecretory properties and other pharmacological actions was published previously (265a).

All antispasmodic drugs have a basic center and most have some stereochemical resemblance to acetylcholine (266) which they displace from the receptor sites in the smooth muscle. 1-Cyclohexyl-1-phenyl-4-(2-pyridyl)-1-butanol **(XVI-118)** was prepared as illustrated and found to possess antispasmodic properties (267). A series of compounds of general structure **XVI-119**, in which R = Me, Et, or

$$\underset{N}{\bigcirc}-(CH_2)_3COPh + \bigcirc-MgBr \longrightarrow \underset{N}{\bigcirc}-(CH_2)_3\underset{\bigcirc}{\overset{Ph}{\underset{|}{C}}}-OH$$

XVI-118

XVI-119   XVI-120

allyl, was synthesized by treating 3-hydroxy-2-phenylpropionyl chloride with the appropriate amine. The products are described (268) as possessing excellent spasmolytic properties. *N*-Ethyl-2-phenyl-*N*-(4-pyridylmethyl)hydracrylamide (Tropicamide) (**XVI-119**, R = Et) is marketed, however, as a short acting mydriatic drug. Various 4-alkylpyridines and 4-alkyl-2-aminopyridines were evaluated as spasmolytic agents on guinea pig and rabbit intestine (269). The most active compounds possessed a heptyl, nonyl, or 3-(2-methylheptyl) group. The presence of the amino-group was not required for activity. Other picoline derivatives (**XVI-120**) are said to be useful for the treatment of peptic ulcer (270).

Various heterocyclic 2,2-dichloro-*N*-(2-hydroxyethyl)-*N*-substituted acetamides containing 2-, 3-, and 4-pyridyl, and other substituents were prepared by dichloroacetylation of the corresponding heterocyclic *N*-(2-hydroxyethyl)-amines. Several of these compounds possessed high amebicidal properties when evaluated against *Endamoeba histolytica*. 2,2-Dichloro-*N*-(2-hydroxyethyl)-*N*-(3- and 4-pyridylmethyl)acetamides (**XVI-121**) were among the most active compounds (271). The amebicidal activity of 2-diethanolamino-5-nitropyridine (**XVI-122a**) was found to equal that of chiniofon on rats infected with *E. histolytica* (272). This discovery prompted the preparation of a number of

XVI-121

	R	R¹
XVI-122a	CH_2CH_2OH	CH_2CH_2OH
XVI-122b	CH_2CH_2OH	$CH_2CHMeOH$
XVI-122c	CH_2CONH_2	Me
XVI-122d	$CH_2CONHNH_2$	Me

related compounds by condensing 2-chloro-5-nitropyridine with various amines. Only three of these compounds (**XVI-122b, c, d**) were active (273).

In view of the fact that 2,6-diaminopyridine was known to possess antiplasmodial activity (274), a series of 4-alkoxy derivatives (**XVI-123**) of

2,6-diaminopyridine was prepared from the dibutyl ester of chelidamic acid as shown (Scheme 5), and each compound was evaluated against the protozoa

XVI-123

Scheme 5

Tetrahymena pyriformis and *Crithidia fasciculata.* The nature of R in **XVI-123** varied from CH_3 to C_6H_{13}. The longer the side-chain was, the greater was the antiplasmodial activity (275).

Bis(*p*-acetoxyphenyl)-2-pyridylmethane (Bisacodyl) **(XVI-124)** is a purgative (276) which is used clinically for the treatment of constipation and in the preparation of patients for surgery or radiography. It is thought to act directly on nerve endings in the mucosa of the colon and rectum in deacetylated form to increase peristaltic action. It does not affect the uterus or the small intestine (277).

2-(2-Methoxyethyl)pyridine **(XVI-125)** is an interesting pyridine derivative. It is a water-soluble liquid and an effective anthelmintic in mice and other animals when administered subcutaneously! It is prepared from 2-vinylpyridine (278) and is reported (279) to be more active against nematodes than phenothiazine.

XVI-124   **XVI-125**

III. Dihydropyridines

Very few compounds possessing a dihydropyridine ring system have been investigated for their pharmacological properties. The most important group of compounds studied are piperidinediones that can be considered to be dihydrodihydroxypyridines. These compounds possess hypnotic and other properties; their pharmacology and clinical uses have been reviewed (280, 281). *Methyprylon* (**XVI-126**) (282) and *glutethimide* (**XVI-127**) (283) are two popular nonbarbiturate hypnotics which are rapid in onset and of short duration. *Aminoglutethimide* (**XVI-128**) (284) possesses anticonvulsant proper-

XVI-126   XVI-127   XVI-128

XVI-129   XVI-130

XVI-131   XVI-132   XVI-133

ties and was claimed to be effective in the treatment of all types of epilepsy. Because of its toxicity it has been withdrawn. *Phenglutarimide* (**XVI-129**) (285) is the diethylamino derivative of glutethimide and has different pharmacological properties. It is an anticholinergic drug and is used in the treatment of parkinsonism.

The nonbarbiturate hypnotic and sedative, *thalidomide* **(XVI-130)** (286) was widely used until 1961 when it became apparent that it had teratogenic effects when administered to women early in pregnancy. Consequently, it was withdrawn. Thalidomide, however, may possess immunosuppressive activity and is being tried in the treatment of cancer (287).

An interest is still being shown in the preparation of hypnotics related to the ones just described. The preparation of 5-ethyl-3,3-dimethyl-1-isopropyl-2,4-(1*H*, 3*H*)pyridinedione **(XVI-131)** and related compounds have been patented recently (288). The compounds show sedative and hypnotic properties. An earlier reference (289) describes the preparation of 3,3-disubstituted-2-amino-3,4-dihydro-4-oxopyridines **(XVI-132)** in which the substituents were alkyl or alkenyl groups. These compounds are claimed to have intensive hypnotic action.

Bemegride **(XVI-133)** (290) is an interesting compound. It is structurally related to the hypnotics just described and yet it possesses CNS stimulant properties. It counteracts the depressant effect of anesthetics and hypnotics and is used clinically and in veterinary medicine to treat barbiturate intoxication.

Various esters of structure **XVI-134**, in which R is aryl, cycloalkyl, or cycloalkenyl, and R^1 and R^2 are alkyl groups were synthesized as illustrated (291–293). Some *N*-methyl derivatives were also prepared (291). These

$$RCHO + 2R^2COCH_2COOR^1 + NH_3 \longrightarrow$$

XVI-134

compounds possess hypotensive activity and are useful in the treatment of coronary insufficiency and angina pectoris.

Compounds structurally similar to **XVI-134** are capable of inducing changes in porphyrin metabolism. Porphyrin accumulation in chick embryo liver cells and in intact guinea pigs can be induced by 3,5-diethoxycarbonyl-1,4-dihydro-2,4,6-trimethylpyridine (DDC) **(XVI-135a)** and homologs **(XVI-135b, c)**. Two substituted pyridines **(XVI-136a, b)** were also active compounds (294). Replacement of the 3- and 5-ethoxycarbonyl substituents of DDC with acetyl substituents leads to a loss of activity. Analogs of DDC, namely, 3,5-dicyano-1,4-dihydro-2,4,6-trimethylpyridine and the related 2,6-dimethyl-4-*t*-butylpyridine were also inactive (295), thus emphasizing the importance of the ethoxycarbonyl substituents for porphyria-inducing activity in this series.

	R		R
XVI-135a	Me	XVI-136a	Me
XVI-135b	Et	XVI-136b	Et
XVI-135c	n-Pr		

Some dihydropyridines have been shown to be antibacterial agents. 1-Methyl-2-sulfanilimido-1,2-dihydropyridine (XVI-137) is one of the products of treating sulfapyridine with diazomethane. It is an active antibacterial agent against *E. coli* (296). When pyridine-1-oxide is treated with phenylmagnesium bromide, the product obtained is 1-hydroxy-2-phenyl-1,2-dihydropyridine (XVI-138) which is a bactericide (297).

XVI-137   XVI-138

IV. Tetrahydropyridines

Compared to the number of piperidines which have been evaluated pharmacologically, relatively few tetrahydropyridines have been examined. Many of the compounds studies are the tetrahydropyridine counterparts of piperidines known to be pharmacologically active. Those which have received the greatest attention are derivatives of 1,2,5,6- (or 1,2,3,6-) tetrahydropyridine. Many of them are potent hypotensive agents, a limited number are analgesics and some possess both properties.

1. Hypotensive Agents

These are agents that reduce blood pressure. There are numerous mechanisms by which this occurs, for example, ganglion blockade. Pempidine (1,2,2,6,6-pentamethylpiperidine) (XVI-139) is an efficient ganglionic blocking agent which is used to treat hypertension (298, 299). One hundred and fifty congeners of pempidine including derivatives of 1,2,5,6-tetrahydropyridine (**XVI-140**, in which X = H, Me, Et, C≡CH, $CH_2CH_2CH_2NMe_2$; and R = H, Me, NHMe, NH_2) were tested for ganglion blocking activity and many were found to be at least as active as hexamethonium (300, 301). Introduction of the 3,4-double bond had

XVI-139   **XVI-140**   **XVI-141**

no effect on magnitude of activity but duration of activity was diminished and toxicity was increased. 1,2,2,6-Tetramethyl-1,2,3,4-tetrahydropyridine (**XVI-141**) is a short acting ganglion blocking agent (300). Recently, more derivatives of general structure **XVI-140** were reported (302). Amines (**XVI-140**; X = $CH_2CH_2NR_2'$, R = Me) were much stronger ganglion blockers in various animals and on isolated tissues than esters or amides (**XVI-140**; X = CH_2COR', R = Me).

Aralkoxyamides of 4-phenyl-1,2,5,6-tetrahydropyridinoalkanoic acids and related compounds are useful in the treatment of hypertension and peripheral vascular disease (303). The synthesis of one such compound (**XVI-142**) is illustrated. It is of interest that the chemically related acetylene derivative (**XVI-143**) and other N-alkynyl- and N-alkynyloxy-amides of (4-phenyl-1,2,5,6-tetrahydropyridino)alkanoic acids exhibit analgesic activity (304). The two

$$PhCH_2ONH_2 + ClCOCH_2CH_2Cl \longrightarrow PhCH_2ONHCOCH_2CH_2Cl$$

XVI-143   **XVI-142**

related compounds 2-methyl-3-[1-(4-phenylpiperidino)]propionohydroxamic acid (**XVI-144**) (305, 306) and 2-methyl-3-[1-(4-phenyl-1,2,5,6-tetrahydropyridino]propionohydroxamic acid (**XVI-145**) (307) have been studied in some detail. Both are potent hypotensive agents but for different reasons. The former

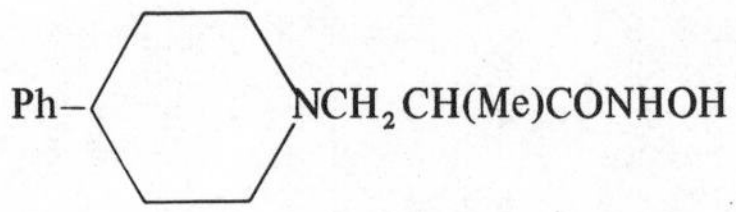

Ph—⟨ ⟩NCH₂CH(Me)CONHOH   Ph—⟨ ⟩NCH₂CH(Me)CONHOH

XVI-144 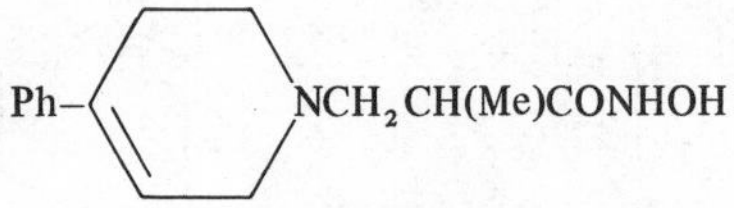   **XVI-145**

is a ganglion blocking agent (306) whereas the tetrahydropyridine derivative is an α-adrenergic blocking agent (307).

N-(2-Guanidinoethyl)-1,2,5,6-tetrahydropyridines **(XVI-146)** are long acting hypotensive agents and are reported (308, 309) to be less toxic than the corresponding piperidino derivatives. N-(2-Guanidinoethyl)-4-methyl-1,2,5,6-tetrahydropyridine **(XVI-147)** is active orally and may be synthesized as shown:

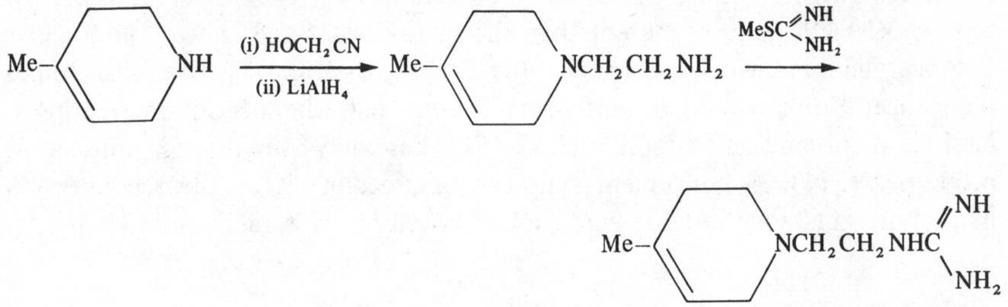

Me—⟨ ⟩NH $\xrightarrow[\text{(ii) LiAlH}_4]{\text{(i) HOCH}_2\text{CN}}$ Me—⟨ ⟩NCH₂CH₂NH₂ $\xrightarrow{\text{MeSC}\overset{\nearrow\text{NH}}{\searrow\text{NH}_2}}$

Me—⟨ ⟩NCH₂CH₂NHC$\overset{\nearrow\text{NH}}{\searrow\text{NH}_2}$

XVI-147

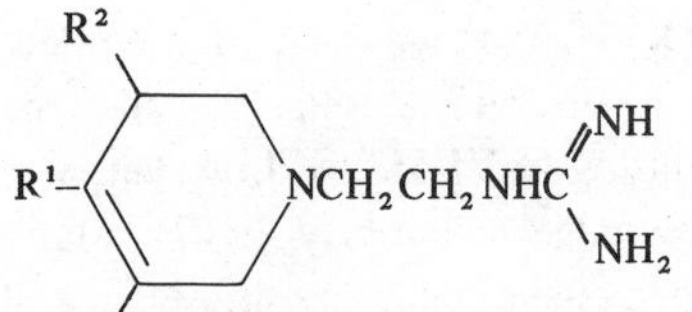

R^1—⟨R^2 R⟩NCH₂CH₂NHC$\overset{\nearrow\text{NH}}{\searrow\text{NH}_2}$

XVI-146

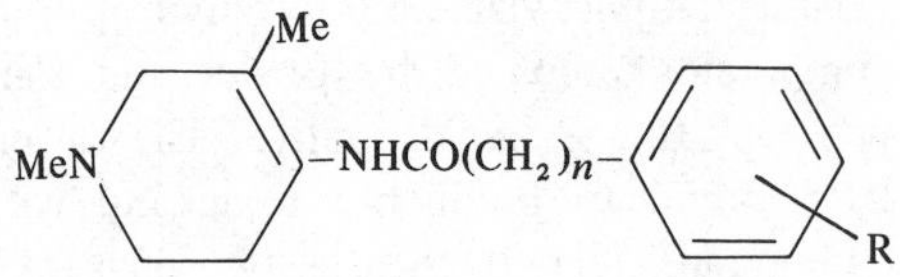

MeN—⟨ Me ⟩—NHCO(CH₂)$_n$—⟨ ⟩R

XVI-148

The somewhat related N-(1,2,5,6-tetrahydro-1,3-dimethyl-4-pyridyl)-substituted phenylalkanoic acid amides **(XVI-148**, in which $n = 1, 2$, or 3) are described in a patent (310) as being useful hypotensive and analgesic agents.

Methyl 1,2,5,6-tetrahydro-1-methylnicotinate (arecoline) **(XVI-149a)** is an alkaloid obtained from the dried ripe seeds of *Areca catechu* (areca nuts, betel nuts). The nondeliquescent hydrobromide is used exclusively in veterinary medicine as an anthelmintic and cathartic because of the strong peristalsis it induces. Arecoline is also a weak hypotensive agent when tested on cats (311). It possesses parasympathetic stimulant (muscarinic) properties approaching those of acetylcholine when evaluated on guinea pig ileum or atrium (312, 313) or rabbit jejunum (314) despite the fact that the alkaloid is a tertiary base and acetylcholine is quaternary. A fully methylated cationic head is usually

XVI-150   XVI-151

	R	pD_2
XVI-149a	Me	7.5
XVI-149b	Et	8.0

considered to be an essential strucţural requirement for muscarinic activity but, surprisingly, the methiodide (XVI-150) has a potency only 1/30th of that of arecoline. No good explanation of this observation has been offered. The ethyl ester (XVI-149b) is more potent than the methyl ester and is even more active than acetylcholine which has a pD_2 value of 7.4 [pD_2 is the negative logarithm of the concentration (ionic in this case) causing half the maximum response]. Methyl nicotinate methiodide (XVI-151) has only very weak muscarinic properties (311). A convenient synthesis of arecoline is by the reduction of methyl nicotinate methiodide with sodium borohydride in methanol (315).

2. Analgesics

1-Butyl-4-phenyl-1,2,5,6-tetrahydropyridine (XVI-152a) was one of the many compounds prepared by Foster and Carman (316) and tested for analgesic potency. It was weakly active. The lower homolog (XVI-152b), however, was later found to be highly toxic and without analgesic activity (317). Many compounds related to the two just mentioned have been synthesized and evaluated. N-Substituted-4-phenyl-1,2,5,6-tetrahydropyridines which do have appreciable analgesic activity were readily obtained (318) in the manner shown

XVI-154

	R
XVI-152a	n-Bu
XVI-152b	Me

for the 2,2-diphenylethyl derivative (XVI-153) in Scheme 6. Related 1-(ω-aroyl-alkyl)-4-alkyl-1,2,5,6-tetrahydropyridine (XVI-154, in which R = alkyl and n = 1-3) were prepared from a 4-alkyl-1,2,3,6-tetrahydropyridine and an appropriate phenone and found to be useful analgesics, mydriatics, hypnotics, and barbiturate potentiators (319).

XVI-153

Scheme 6

In recent years, the J. R. Geigy Company has reported (320, 321) the synthesis of a large number of amides and ketones of general structure XVI-155 and the corresponding alcohols (e.g., XVI-156). These compounds were shown to possess useful pharmaceutical properties, especially oral and parenteral analgesic activity and an inhibiting action on the cough stimulus. 1-[N-Acyl-N-phenylaminoalkyl]-4-phenyl-1,2,5,6-tetrahydropyridines (XVI-157) and related compounds also possess antitussive and analgesic activity in cats and rats (322).

XVI-155　　　　XVI-156　　　　XVI-157

Certain pyridones exhibit analgesic properties. Dihydropyridones (XVI-158, in which R, R^1, and R^2 are H, alkyl, or phenyl) have been shown (323) to have analgesic, antipyretic, and antiphlogistic activity. Similar claims are made (324) for the closely related compound (XVI-159) named in the patent as 1,2,2,4-tetramethyl-5-cyano-6-oxo-1,2,3,6-tetrahydropyridine, which was prepared from methylamine, methyl cyanoacetate, and mesityl oxide.

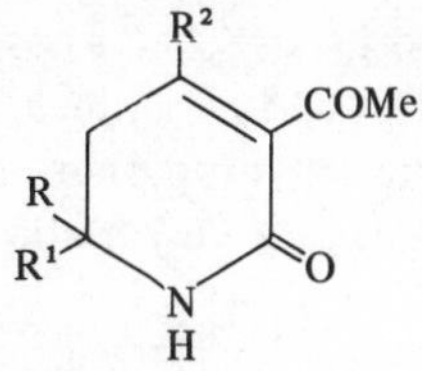

XVI-158 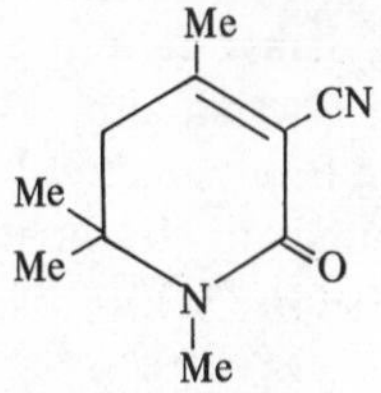   XVI-159

3. Miscellaneous Tetrahydropyridine Derivatives

Derivatives of 5-benzoyl-1-methyl-4-phenyl-1,2,5,6-tetrahydropyridine (XVI-160, in which R = Me, CH$_2$Ph, or allyl, and R^1 and R^2 = H, Cl, F, Me, MeO) were prepared as shown (Scheme 7) and reduced by means of lithium aluminum hydride or sodium borohydride to the corresponding secondary alcohols (XVI-161). The ketones (XVI-160) are said (325) to be useful antipyretic and antiinflammatory agents. The alcohols (XVI-161) possess diuretic properties (326).

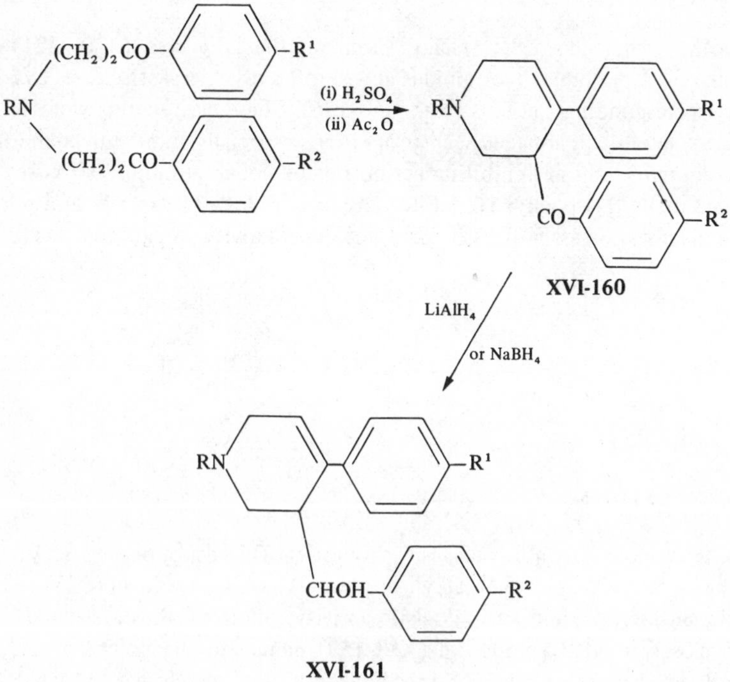

XVI-161

Scheme 7

1-[2-(*p*-Phenoxybenzoyl)ethyl]-4-phenyl-1,2,5,6-tetrahydropyridine **(XVI-162)** and related compounds were synthesized by treating 4-phenyl-1,2,5,6-tetrahydropyridine with the appropriate alkyl halide. These products, and the corresponding alcohols, are claimed to be useful muscle relaxants (327). The diamine **(XVI-163)** and its *p*-chloro derivative are active in inhibiting the growth of protozoa (328). 1-Methyl-4-(*m*-chloro- and *m*-trifluoromethyl-phenyl)-1,2,5,6-tetrahydropyridine **(XVI-164, R = Cl or CF$_3$)** were prepared as illustrated in

XVI-162   XVI-163

XVI-165

Scheme 8. The products are useful veterinary anthelmintics (329). Various amide derivatives of 4-methyl-1,2,5,6-tetrahydropyridine **(XVI-165)** exhibit a strong choleretic activity and have low toxicity (330).

XVI-164

Scheme 8

Incomplete catalytic hydrogenation of 4,4-disubstituted heptanedinitriles **(XVI-166)** gave rise to 2,3-disubstituted-3-(3-aminopropyl)-3,4,5,6-tetrahydropyridines **(XVI-167)** which possessed fungicidal and pesticidal properties (331).

XVI-166   XVI-167

V. Fully Reduced Pyridines

Most of this section comprises an account of compounds containing the fully reduced pyridine ring system, namely piperidine; examples of biologically active compounds with partially reduced pyridine nuclei are rarer. Most of the compounds mentioned do not have complex structures; hence polycyclic molecules containing reduced pyridines such as alkaloids are not described in any detail. Emphasis is placed on drugs and natural products that are, or have been, used clinically.

1. General

Many pharmacologically active compounds are tertiary bases, typified by the natural products atropine and morphine and the synthetic drugs methadone and chlorpromazine. A general principle of drug design is to screen analogs that differ in tertiary base group structure, and in this respect piperidine derivatives are commonly examined. Replacement of dimethylamino (often the basic feature of the prototype drug) by piperidino does not markedly alter the dimensions of the drug molecule because the piperidino group has a compact cyclic structure which is certainly less space demanding than diethylamino and higher alkylamino functions which possess freely rotating alkyl groups. The basic properties of dimethylamino and piperidino analogs are also alike as judged from the pK_a values of dimethylamine (10.77) and piperidine (11.22) (332). It is usually difficult to provide a reason why the piperidino member of a particular series is chosen for development as a clinical agent since many considerations, for example, solubility, absorption, distribution, and metabolism of the drug, and even patent evasion, may be involved. A few examples follow.

In the narcotic analgesic field, dipipanone (XVI-168a) is marketed as an alternative to the dimethylamino analog, methadone (XVI-168b). It is about half

RCHMeCH$_2$CPh$_2$COEt

	R
XVI-168a	N(CH$_2$)$_5$
XVI-168b	NMe$_2$

as potent as methadone both in the mice hot-plate test (333) and clinically in man (334). A 20 to 25 mg dose of dipipanone was found to be as effective as 10 mg of morphine against chronic and postoperative pain with minimal side effects and a similar duration of action and addiction liability (333, 335). Its use as an adjunct to anesthesia in obstetrics and surgery has also been reported (336). In analgesics based on propionanilide the piperidino analog **XVI-169** (phenampromid) is reported to have a potency similar to that of codeine in mice (337)

$$\text{PhNCH(Me)CH}_2\text{N}\overset{\displaystyle\text{COEt}}{\underset{\displaystyle}{|}}$$

XVI-169

while doses of 25 to 50 mg approach but do not equal the effect of 10 mg of morphine upon postoperative pain (338). Comparative data on the dimethylamino derivative is lacking. In the 2-thienyl analogs **(XVI-170)**, the piperidino member is distinctly the more active (339).

XVI-170

X	ED_{50} mg/kg (hot-plate test, mice)
NMe_2	45.2
$N(CH_2)_5$	17.4
(Morphine)	6.8

Potency rankings in the highly potent 2-benzylbenzimidazole analgesics **XVI-171** are unusual in that the piperidino member is more active than the

XVI-171

R	Activity in mice (morphine = 1)
NMe_2	20
$N(CH_2)_5$	100
NEt_2	1000

dimethylamino but less active than the diethylamino derivative (340, 341).

The 1-piperidino group is also a feature of the antihistaminic agent **XVI-172** (342) and the spasmolytic benzhexol **XVI-173** (Artane) (343). The latter compound is distinctly more active than the pyrrolidino analog, procyclidine, as

Ph   H
     \   /
      C=C
PhCH₂   CH₂N⟨ ⟩

     C₆H₁₁
       \
        C(OH)CH₂CH₂N⟨ ⟩
       /
     Ph

XVI-172   XVI-173

an antagonist of guinea-pig ileum contractions induced by acetylcholine (344) but it is to be noted that dimethylamino and pyrrolidino bases are commonly encountered in both classes of drug.

2. The Piperidine Ring as a Molecular Framework

In the following examples a reduced pyridine nucleus forms an essential part of the molecular structure and does not merely endow the drug with basic properties.

A. *Narcotic Analgesics*

This class provides the richest source of examples of pharmacologically active piperidines. The simplest structures are derivatives of 4-phenylpiperidine typified by the original member ethyl 1-methyl-4-phenylpiperidine-4-carboxylate (**XVI-174**), well known as pethidine, meperidine, or Dolantin. It was first introduced in 1939 by Eisleb and Schauman (345) and it remains the most

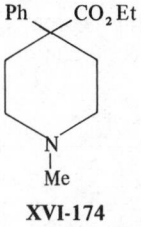

Ph   CO₂Et

N
|
Me

XVI-174

widely used synthetic analgesic. In potency, pethidine is graded between codeine and morphine (50 to 100 mg equivalent to 10 mg morphine in man) (346) and is useful for the management of mild to moderate pain, especially in patients intolerant to opiates. Its toxicity is low, and its action is somewhat shorter than that of morphine. At equivalent dosage, pethidine is at least as depressant as morphine upon respiration and morphine-like side effects such as nausea and vomiting frequently occur. It has found extensive use in the relief of labor pain even though it increases the incidence of delay on the first breath and cry of the newborn infant. Tolerance to the drug develops slowly and its addictive liability is judged to be lower than that of morphine (333). Full accounts of the clinical use of pethidine are available (335, 347).

The unique role of the piperidine fragment in 4-phenylpiperidine analgesics is demonstrated by the substantial reduction in activity which follows replacement of the 6-membered ring system of pethidine and its reversed ester (see below) by a 5-, 7-, or 8-azacycloalkane ring (348). Hot-plate test activities in animals on pethidine analogs are given below.

$$\text{MeN} \underset{[CH_2]_y}{\overset{[CH_2]_x}{\diagup}} \underset{CO_2Et}{\overset{Ph}{\diagdown}}$$

Pyrrolidine (x = 1, y = 2)	Piperidine (x = y = 2)	Azacycloheptane (x = 2, y = 3)
Inactive in mice at 200 mg/kg (349)	ED_{50} 9.9 mg/kg in mice (333); ED_{50} 11.2 mg/kg in rats (350) (pethidine)	ED_{50} 42.6 mg/kg in mice (333); ED_{50} 33.5 mg/kg in rats (350) (ethoheptazine)

Pethidine and its N-substituted analogs are synthesized by reactions shown in the Scheme 9; this process and its modifications have been reviewed (351, 352). Intermediates are phenylacetonitrile and substituted bis(2-chloroethyl)amines.

$$PhCH_2CN \xrightarrow[NaNH_2]{R'N(CH_2CH_2Cl)_2} \quad \xrightarrow[EtOH]{H_2SO_4}$$

Scheme 9

The latter are highly toxic and may be replaced to advantage by the N-p-toluenesulfonyl analog. When R' is benzyl or p-toluenesulfonyl, the end product is readily converted to the corresponding secondary base which may be alkylated to provide a variety of N-substituted pethidine analogs.

A vast number of 4-aryl (generally phenyl) piperidine derivatives have been screened as analgesics and several reviews on the structure-activity data so accumulated have been published (353–356). The structural variation most thoroughly investigated is that of replacement of N-methyl by other groups, notably phenalkyl. These studies probably stem from the observation made in 1956 by Perrine and Eddy (357) that N-phenethylnorpethidine is twice as active as pethidine in mice. Compounds in clinical use developed in this way are XVI-175b (anileridine, Leritine) (358), XVI-175c (piminodine, Alvodine) (359), and XVI-175d (phenoperidine). The last derivative has been used for neuro-

XVI-175a	R = Me
XVI-175b	R = p-NH$_2$C$_6$H$_4$(CH$_2$)$_2$
XVI-175c	R = PhNH(CH$_2$)$_3$
XVI-175d	R = PhCH(OH)(CH$_2$)$_2$
XVI-175e	R = PhCO(CH$_2$)$_2$
XVI-175f	R = H

leptanalgesia, and anesthetic technique in which an analgesic-tranquilizer mixture is given intravenously usually as an adjunct to nitrous oxide (360). Many of these N-substituted derivatives have very high potencies, the Mannich base XVI-175e, for example, (obtained from XVI-175f, formaldehyde, and acetophenone) being over 50 times as active as pethidine in mice (361).

Replacement of the ethoxycarbonyl group of pethidine XVI-175a by a propionyloxy function, giving the so-called reversed ester XVI-176a, results in a potency rise (362). This change, coupled with the introduction of a 3-methyl

XVI-176a	R = H	XVI-177
XVI-176b	R = Me	

substituent into the piperidine ring gives alphaprodine (Nisentyl) (XVI-176b) which has been used clinically (335). Alphaprodine is one of two possible racemic diastereoisomers and is less potent than the second form, betaprodine

(363, 364). The key intermediates in the synthesis of reversed esters of pethidine are 1,3-disubstituted-4-piperidones **XVI-177**; reaction of the piperidone with an aryl-lithium and subsequent acylation of the resultant tertiary alcohol gives the required ester **XVI-176**. Preparation of the ketones involves a Dieckmann cyclization of a diester as penultimate step, the reactions being illustrated for 1,3-dimethyl-4-piperidone in Scheme 10.

$$MeNH_2 \; + \; CH_2{=}CMeCO_2Me \longrightarrow MeNHCH_2CHMeCO_2Me \xrightarrow{CH_2{=}CHCO_2Me}$$

Scheme 10

These reactions (especially the formation of the diester) proceed less smoothly when arylalkylamines are employed (365), and the base-exchange procedure **XVI-178** → **XVI-179**, developed in Russia (366), offers an attractive alternative

XVI-178   **XVI-179**

for making piperidones with N-substituents such as benzyl and phenethyl.

The stereochemistry of the prodines has been investigated extensively (367, 368), and alphaprodine has been assigned a *trans* 3-Me/4-Ph and the beta isomer a *cis* configuration, as depicted in the chair conformations **XVI-180** and **XVI-181**.

α **XVI-180**   β **XVI-181**

Stereoisomeric pairs of the same type, for example, the 3-methyl analogs of pethidine itself **XVI-182** (369), have been examined and, in all cases, the *cis* (β-) member has the superior potency. The optical isomers of alpha- and beta-

XVI-182

R	ED_{50} in mice(hot-plate)
H	4.7 mg/kg
α-Me	3.6 mg/kg
β-Me	0.42 mg/kg

prodine also show marked activity differences (370), the 3*R*, 4*S* (dextro α-) and 3*S*, 4*S* (dextro β-) forms being the more potent.

Isomeric forms of the 1,2,5-trimethyl reversed ester analog **XVI-183** have been studied and activity variations observed (371, 372). The γ-isomer (2-Me *cis* to

XVI-183

4-Ph and *trans* to 5-Me), termed trimeperidine or Promedol, is used clinically in the USSR (335). Synthesis of the precursor ketone **XVI-184** of this series (Scheme 11) is based on vinyl acetylene chemistry (acetylene link showed

XVI-184

Scheme 11

nonlinear for convenience) (373). The ketone **XVI-184** is a mixture of isomers in which the *trans* 2-Me/5-Me form preponderates, as recently established by a PMR study (374).

Other pethidine congeners of interest are the 4-propionyl derivative (**XVI-185**), termed ketobemidone which is still used in Germany and Scandinavia in

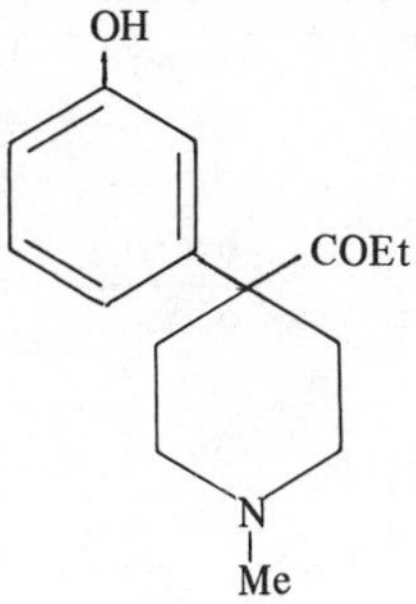

XVI-185

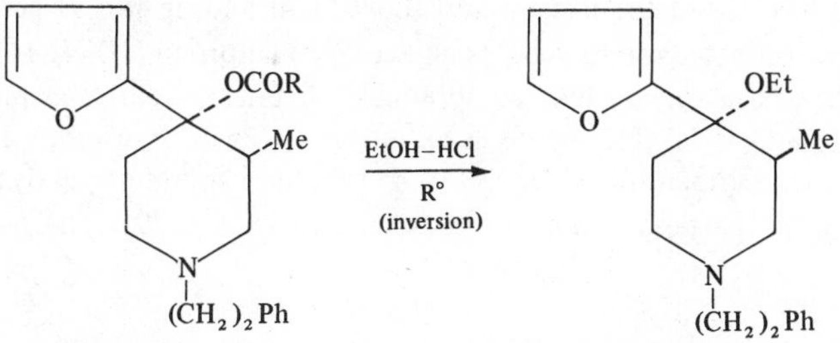

XVI-186   **XVI-187**

spite of its high addiction liability (335), and the 4-(2-furyl) ether (**XVI-187**). The last compound, the product of an unusually facile alkyl-oxygen ethanolysis **XVI-186** → **XVI-187**, is several times more active than pethidine in mice and has a low toxicity (375, 376).

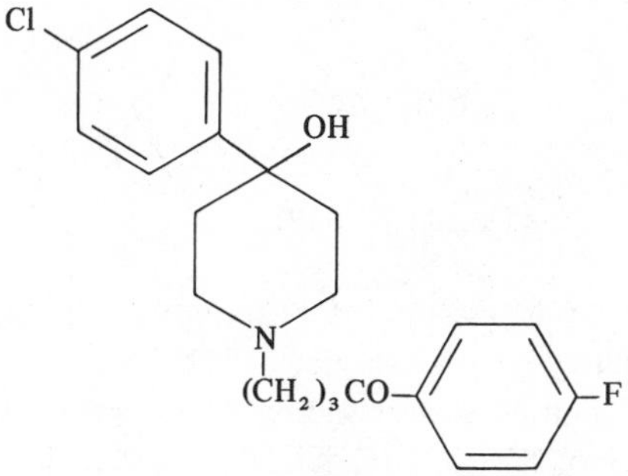

XVI-188

Some 4-arylpiperidin-4-ols, in contrast to their derived esters, are central nervous depressants of the tranquillizing type rather than analgesics, for example, haloperidol (XVI-188) (377).

Reduction of the Schiff base XVI-189 formed between 1-phenethyl-4-piperidone and aniline, and acylation of the resultant diacidic base, gives the analgesic Fentanyl (XVI-190) (378). This compound, although based on piperidine,

PhN
(i) LAH
(ii) (EtCO)₂O
PhNCOEt H
N
(CH₂)₂Ph
XVI-189

N
(CH₂)₂Ph
XVI-190

cannot be classified as a pethidine analog because it has an acylated anilino substituent in place of the usual 4-aryl substituent and, in this respect, it bears some relationship to the acyclic analgesic phenampromid (XVI-169) already mentioned. It is a very potent compound (200 times as active as morphine in mice) (379) and is in clinical use as an analgesic (380). Innovan, a mixture of Fentanyl and Droperidol (XVI-191), a tetrahydropyridine derivative with tranquilizing properties) is used in neuroleptanalgesia (379). The N-phenethyl

H
N
O=
N

N
(CH₂)₃CO—⟨benzene⟩—F

XVI-191

substituent of Fentanyl is essential for activity, the N-methyl analog, for example, being completely inactive in mice (381).

Normally, tertiary alcohols corresponding to reversed esters of pethidine are inactive. The free alcohol XVI-192 (R = OH), however, is highly potent in rats

Ph R

(CH$_2$)$_2$N(COEt)Ph

XVI-192

Ph OH

CH$_2$CHMeN(COEt)Ph

XVI-193

(50 times pethidine) and its activity is, in fact, reduced on esterification (382). Even higher potency levels are reached in branched chain congeners; for example, **XVI-193** is 150 times as active as morphine in mice (383).

The derivatives **XVI-192** and **XVI-193** are more appropriately classified with acyclic basic anilides such as phenampromid **(XVI-169)** than with the pethidine class, and on this basis the entire 4-phenylpiperidin-4-ol moiety appears to serve as the basic unit with *N*-phenyl rather than *N*-piperidylphenyl as the prime aromatic feature of the molecule.

Janssen has linked the cyanide precursor of normethadone to norpethidine to produce diphenoxylate **(XVI-194)**; this compound is not an analgesic but has the constipating action of morphine and is used as an antidiarrhoeal agent (384).

Ph CN

C

Ph CH$_2$CH$_2$N Ph

CO$_2$Et

XVI-194

The related 4-aminocarboxamide **XVI-197**, Pirinitramide, obtained from *N*-benzyl-4-piperidone **(XVI-195)** *via* the cyano-amine **XVI-196** is an analgesic, however,

O

N

CH$_2$Ph

XVI-195

KCN
Piperidine →

NC N

N

CH$_2$Ph

XVI-196

Several
Steps →

$\overset{O}{\overset{\|}{H_2NC}}$ N

N

CH$_2$CH$_2$CPh$_2$CN

XVI-197

and is twice as active as morphine in mice (385). Pirinitramide has been used in the clinic (386). Several spiranes, obtained by condensing 4-anilino analogs of **XVI-197** with formamide are active in the mouse hot-plate test (e.g., **XVI-198**

XVI-198

has ED_{50} = 0.49 mg/kg) and behave as powerful chlorpromazine-like sedatives in animals (387).

The 4-phenylpiperidine entity also forms part of some polycyclic narcotic analgesics, notably the alkaloid morphine **(XVI-199)** itself, the chief active ingredient of opium. This structural aspect of morphine is illustrated with

XVI-199   XVI-200

greater clarity in the partial conformational representation **XVI-200**. The fact that both morphine and pethidine have a common 4-phenylpiperidine unit is not necessarily significant, however, since the relative orientation of the phenyl and heterocyclic rings in preferred conformations of pethidine and its congeners (the 4-phenyl group will adopt an equatorial conformation) is radically different from that obtaining in the rigid morphine skeleton (*cf.*, **XVI-180, XVI-181**, and **XVI-200**).

Our knowledge of the structure and absolute stereochemistry of morphine is complete (388–390) and the alkaloid has been synthesized (391, 392). Numerous modifications of morphine have been made (335, 350, 393) but discussion of these is limited to those directly involving the piperidine ring (ring D), since detailed considerations of fused ring systems are outside the scope of this chapter.

Removal of the *N*-methyl group of morphine gives normorphine; this compound, a probable metabolite of morphine, is less potent when given by normal routes of administration, but is of equal potency intracisternally in mice

(354). Several alkylated derivatives of normorphine have been examined, the most significant being the *N*-allyl compound, nalorphine **(XVI-201)**. This substance antagonizes a wide spectrum of morphine effects and is an antidote for morphine poisoning (394). Although it lacks analgesic properties in

XVI-201   XVI-202

laboratory animals, it is a potent, essentially nonaddicting analgesic in man, and this property has led to the development of several clinically useful analgesics based on morphine antagonists (395). Nalorphine itself cannot be used clinically because it has undesirable psychotomimetic side-effects. Naloxone, the *N*-allyl analog of oxymorphone **(XVI-202)** is one of the most potent morphine-antagonists yet examined (19X nalorphine in rats), and it gives no analgesic response in the phenylquinone writhing test, a procedure that detects analgesic properties in other morphine antagonists (396). Harris (397) regards Naloxone as the most nearly pure narcotic antagonist so far tested.

Morphinan **(XVI-203)** and the later developed benzomorphan **(XVI-204)** analgesics are progressively simpler versions of morphine; all possess the

XVI-203   XVI-204

a   R = Me
b   R = CH$_2$CH$_2$Ph
c   R = CH$_2$CH=CMe$_2$
d   R = CH$_2$—◁

4-phenylpiperidine moiety. The synthesis of both groups is based on methods developed by Grewe and others (398), and is illustrated for the benzomorphans in Scheme 12. The best known morphinan is (−)-N-methyl-3-hydroxymorphinan (levorphanol, Dromoran, **XVI-203a**). Levorphanol is highly effective by oral

Scheme 12

administration and is more potent than morphine (2 to 3 mg of levorphanol are equal to 10 mg of morphine in man) (399); its addiction liability is, however, at least as great as that of morphine (400). The corresponding (+)-isomer (dextrorphan) is devoid of analgesic activity. Structure-activity relationships in the morphinans (and the benzomorphans) mirror those of morphine. Thus, methylation of the phenolic group of (−)- **XVI-203a** results in a large decrease in activity, while replacement of N-methyl by N-allyl gives a potent morphine antagonist levallorphan (401). Activity is retained in levo-3-hydroxy-N-methyl-isomorphinan (**XVI-205**), an isomer of levorphanol in which rings C and D are *cis* fused (402). Benzomorphan isomers having the 5,9-dimethyl groups *cis* (α-) and *trans* (β-) with respect to the hydroaromatic ring have been isolated, the former being related sterically to morphine and the morphinans, and the latter to isomorphinan. The α-N-methyl analogue (**XVI-204a**), metazocine, is a potent analgesic [the racemate is almost as active as morphine in mice (403)], but the clinically important benzomorphans are the N-phenethyl (phenazocine, Prinadol,

XVI-204b) and N-3,3-dimethylallyl (pentazocine, Talwin, **XVI-204c**) derivatives. Pentazocine is the most promising clinical analgesic yet developed from analgesic antagonists, which is itself a feeble antagonist of the effects of morphine on the tail-flick reaction (404), but in man is an effective analgesic in a wide variety of pain situations as is evident from the numerous clinical reports now available (356). On the average a 30 to 40 mg intramuscular dose is equal to

XVI-205

10 mg of morphine, although higher doses have been found necessary in cancer patients (405). Most significantly, the drug has a very low addiction liability (406); it was released by the FDA in 1967 and is not covered by the Harrison Narcotic Act.

The corresponding N-cyclopropylmethyl analog **XVI-204d**, cyclazocine, is about twice as active a morphine antagonist as nalorphine and is a potent analgesic in man (407). It has addiction liability, but the abstinence syndrome that follows its withdrawal from addicts is not as severe as that seen after morphine (408).

The analgesic potency of several 6,7-benzomorphan derivatives is influenced both by the relative configurations of the 5,9-dialkyl substituents (β-diastereoisomers are more potent than the α-forms) and by absolute configuration within a particular enantiomorphic pair; in all cases examined, the activity of both α- and β-racemates largely resides in the levo antipode (409, 410). In related derivatives that are analgesic antagonists, activity differences between ($\pm$)-diastereoisomers are insignificant, but pronounced potency variations among enantiomers are still found (411).

Reviews on both morphinan (412) and benzomorphan analgesics (409) are available, and Martin has reviewed related antagonists (395).

B. Postganglionic Parasympathetic Agonists and Antagonists

This section refers chiefly to compounds that mimic or antagonize the muscarinic properties of acetylcholine (ACh). Nicotine and related compounds are mentioned elsewhere (p. 476). Reference is also made to the CNS stimulant and other properties of these compounds.

A few simple piperidine derivatives have ACh-like activities but potencies are generally low (413). Some data on two ring-nitrogen analogs of ACh, namely XVI-206 (414) and XVI-207 (415), are given below.

	XVI-206	XVI-207
Activity (ACh = 1)	0.03 (Cat blood pressure)	0.001 (Rabbit ileum)

The quinuclidine derivative XVI-208a, an analog of the 3-piperidyl ester XVI-207 in which the 1 and 4 positions are bridged by a bimethylene chain, is a potent cholinergic agent with a predominant muscarinic action (416). It is

XVI-208a   R = Me
XVI-208b   R = CHPh$_2$

known as Aceclidine in the USSR and used in the treatment of glaucoma and postoperative atonia of the bowels and bladder. The methiodide is about 200 times less active than the tertiary base XVI-208a itself; this unexpected base-quaternary salt potency ranking also occurs in the arecoline series as mentioned earlier. The parent alcohol of Aceclidine has been resolved (417) and the derived enantiomorphic acetate methiodides examined. The S-(−)-isomer had significant, albeit weak, muscarinic properties and was a good substrate for acetylcholinesterase (AChE), while the R-(+)- form virtually lacked muscarinic properties and was inactive as an AChE substrate (418). The report that the levo isomer has the same configuration of (+)-β-methylACh, a potent muscarinic agent and good AChE substrate, has recently been amended.

It is significant that the diphenylacetate of 3-quinuclidinol (XVI-208b) is a potent antagonist of ACh and that most of the activity of the racemic material resides in the levo isomer, this being about twice as active as atropine (417).

Piperidine derivatives with a 2,6-bimethylene bridge form an important part of this section, the so-formed bicyclic nucleus **XVI-209** forming the basis of the tropane alkaloids (419). Atropine **XVI-210**, an active principle of belladonna,

hyoscyamus, and other solanaceous plants, is the best known cholinergic antagonist and has long been used in medicine for its spasmolytic and mydriatic properties. Chemically it is an ester formed from the aminoalcohol tropine and racemic tropic acid. The related alkaloid hyoscyamine is formed from tropine and $S(-)$-tropic acid. Apart from the problem of the conformation of the piperidine ring, shown as a chair in **XVI-210**, this molecule offers two points of stereochemical interest: the dependence of activity on the configuration of the tropic acid residue, and on the relative orientations of the basic center and the 3-acyloxy function. Regarding the first, a variety of comparisons of atropine and hyoscyamine have shown that most of the activity of the racemic product resides in the levo isomer (420). Long et al. (421) compared (−)-hyoscyamine sulfate and the (+)-camphor-10-sulfonate and found the latter to be virtually inactive, while Bovet and Bovet-Nitti (422) stated that (−)-hyoscyamine was 10 to 100 times more active than the dextro form. Similarly, (−)-hyoscine, an atropine analog in which the dimethylene bridge carries an epoxide function, is twice as active as the racemic form (420). From one early report on tropyl-ψ-tropine (*cis*-N/3-OCOR) itself (423) and a more recent study of the tropyl and ψ-tropyl esters of benzoic acid (424), it may be concluded that a *trans* orientation of the basic center and the 3-acyloxy function is optimal for activity in atropine and related compounds. *Cis*-isomers, although of reduced potency, are not devoid of activity, except, apparently, at sites in the eye.

For the preferred conformation of tropine (chair or boat piperidine ring) PMR spectra strongly support the chair form for tropine base in $CDCl_3$ and the hydrochloride in D_2O (425, 426). PMR evidence shows that the same conformation is preferred for atropine (427).

The marked specificity of action of hyoscyamine isomers is also noted in tropyl α-methyltropate, the (−)/(+) potency ratio being 50 (428).

The benzyhydryl ether of tropine (**XVI-211**, benzotropine) also has strong anticholinergic properties and is used for the treatment of Parkinson's disease

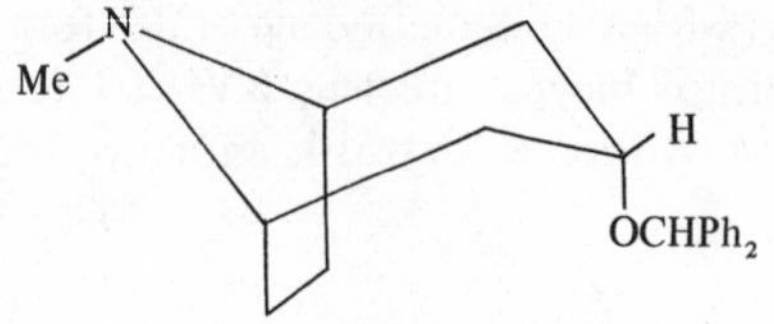

XVI-211

(429). No potent ACh-like agonists have been reported based on tropine, the acetate methiodides **XVI-212a** and **b** having weak muscarinic properties, and **XVI-212c** negligible ones (430, 431).

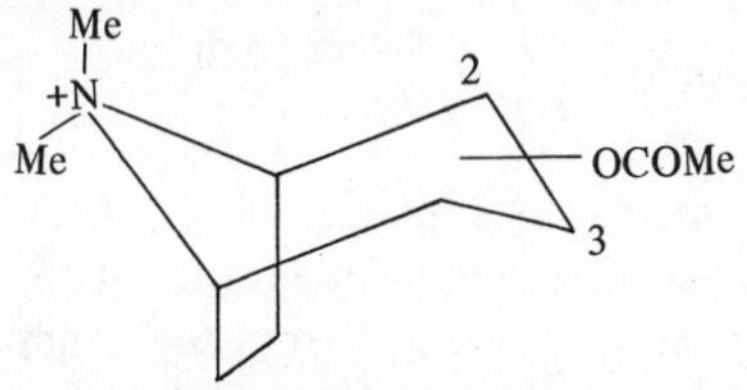

XVI-212a   3-OCOMe   *trans* to N
XVI-212b   2-OCOMe   *trans* to N
XVI-212c   2-OCOMe   *cis* to N

Certain 3-piperidyl benzilates have high anticholinergic properties, the derivative **XVI-213**, for example, being somewhat more potent than atropine (432),

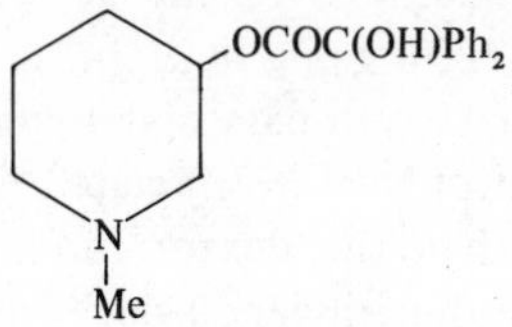

XVI-213

while the 4-substituted analog has a similar order of potency (433). The 3-piperidyl esters also elicit powerful psychotomimetic and antidepressant effects in man and this finding has led to the development of several analogs such as Ditran **(XVI-214)** that are used in the treatment of mental disorders; the

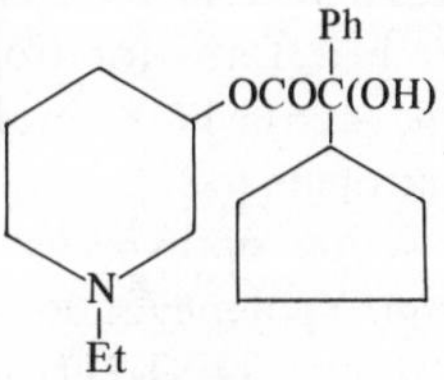

XVI-214

threshold doses for producing an increase in spontaneous motor activity in rats were 0.05 and 1.2 mg/kg for Ditran and atropine, respectively (432).

Other piperidine derivatives of interest in these respects are pipradol (**XVI-215**) (434), methylphenidate (**XVI-216**) (435), and diphenamil (**XVI-217**)

XVI-215   **XVI-216**   **XVI-217**

(436); the first two are used as central stimulants and the last as an anticholinergic agent. The configurations of enantiomorphic forms of **XVI-215** and **XVI-216** (these isomers exhibit significant differences in activity) have recently been reported (437, 438). A series of 1-benzyl-4-[(2,6-dioxo-3-phenyl)-3-piperidyl]piperidines are potent, orally active anticholinergics. Some of these derivatives, for example, benzetimide (**XVI-218**), have central anticholinergic potencies comparable with that of benztropine (439).

XVI-218

XVI-219   **XVI-220a** *e*-3-OH
                    XVI-220b *a*-3-OH

C. *Local Anesthetics*

Tropane derivatives are also prominent in this pharmacological group. (−)-Cocaine, the active principle of coca leaves and the best known local anesthetic, is 2*R*-methoxycarbonyl-3*S*-benzoyltropine **(XVI-219)** (419, 440); ecgonine **(XVI-220a)**, its principal product of hydrolysis, is a 2-carboxylic acid related to pseudotropine. Pseudococaine, the C-2 epimer of cocaine (derived from pseudoecgonine, the product of equilibrating ecgonine with base), and racemic forms of allococaine and allopseudococaine, C-2 epimers of alloecgonine **XVI-220b** (derived from the reduction product of 2-methoxycarbonyltropinone), have all been reported (441). The relative configurations of these isomers are based on chemical and PMR studies (441–443). Pharmacological activity variations among cocaine and its isomers are not pronounced. Thus, Gottlieb (444) found that (–)-cocaine, (±)-cocaine, (+)-pseudococaine, and (±)-pseudococaine all had about the same order of local anesthetic activity when tested on the frog's sciatic plexus and the dog's cornea. α-Cocaine **(XVI-221)**, an analog in which the C-3 carbon atom carries both the benzoyl and carbomethoxy functions, is inactive on the tongue and rabbits' cornea (unlike cocaine, it cannot penetrate mucous membranes) but is quite potent in tests where the drug is either absorbed from the stomach or injected intradermally (445). Its synthesis from

XVI-221

tropinone and potassium cyanide is stereospecific and leads to an isomer of the configuration shown (446). Norcocaine **(XVI-219, NMe replaced by NH)** is at least as active as cocaine, but cocaine methiodide, ecgonine, methyl ecgonine, and benzoylecgonine are all inactive (420).

Many substituted piperidines were examined as cocaine surrogates following Willstätter's elucidation of the general structure of the alkaloid (447), and the derivatives α-eucaine **(XVI-222)** and β-eucaine **(XVI-222, Benzamine)** enjoyed

XVI-222

XVI-223

some clinical application. Both substances produce effects on the guinea-pig's cornea but β-eucaine is the more active, although not as active as cocaine (448). β-Eucaine and its C-4 epimer (iso-β-eucaine) have been resolved, and all isomers have similar activities on the rabbit's cornea. In the frog's sciatic test enantiomorphic forms of β-eucaine proved to be equiactive and stronger than (+)- and (−)-iso-β-eucaine. Recently, Perks and Russell (449) have shown that β-eucaine has *cis* N/OCOPh geometry as in cocaine itself. Piperocaine (**XVI-224**),

Me

PhCO$_2$CH$_2$CH$_2$CH$_2$N⟨ ⟩   H$_2$N—⟨ ⟩—CO$_2$CH$_2$CH$_2$NEt$_2$

XVI-224   **XVI-225**

another example (450), is modelled upon procaine (**XVI-225**) rather than cocaine.

References

1. V. Chorine, *C. R. Acad. Sci., Paris, Sec. C.*, **220**, 150 (1945).
2. S. Kushner, H. Dalalian, R. T. Cassell, J. L. Sanjurjo, D. McKenzie, and Y. Subbarow, *J. Org. Chem.*, **13**, 834 (1948).
3. D. McKenzie, L. Malone, S. Kushner, J. J. Oleson, and Y. Subbarow, *J. Lab. Clin. Med.*, **33**, 1249 (1948).
4. H. H. Fox, *J. Org. Chem.*, **17**, 542, 547 (1952).
5. H. H. Fox, *J. Org. Chem.*, **15**, 555 (1952).
6. L. Pichat, C. Baret, and M. Audinot, *Bull. Soc. Chim. Fr.*, **21**, 88 (1954).
7. L. J. Roth and R. W. Manthel, *Proc. Soc. Exptl. Biol. Med.*, **81**, 566 (1952).
8. H. H. Fox and J. T. Gibas, *J. Org. Chem.*, **17**, 1653 (1952).
9. H. H. Fox and J. T. Gibas, *J. Org. Chem.*, **18**, 983 (1953).
10. H. H. Fox and J. T. Gibas, *J. Org. Chem.*, **18**, 994 (1953).
11. H. H. Fox and J. T. Gibas, *J. Org. Chem.*, **18**, 1375 (1953).
12. H. H. Fox and J. T. Gibas, *J. Org. Chem.*, **20**, 60 (1955).
13. H. H. Fox and J. T. Gibas, *J. Org. Chem.*, **21**, 356 (1956).
14. H. H. Fox and J. T. Gibas, *J. Org. Chem.*, **18**, 990 (1953).
15. G. Brouet, B. N. Halpern, J. Marche, and J. Mallet, *Presse Méd.*, **61**, 863 (1953).
16. E. M. Bavin, E. Kay, and D. E. Seymour, *Lancet*, **2**, 337 (1954).
17. E. M. Bavin, B. James, E. Kay, R. Lazare, and D. E. Seymour, *J. Pharm. Pharmacol.*, **7**, 1032 (1955).
18. S. D. Rubbo, J. Edgar, and G. Vaughan, *Amer. Rev. Tuberc. Pulmonary Diseases*, **76**, 331 (1957); *Chem. Abstr.*, **53**, 3472i (1959).
19. S. D. Rubbo and G. Vaughan, *Amer. Rev. Tuberc. Pulmonary Diseases*, **76**, 346 (1957).
20. C. A. Colwell and A. R. Hess, *Amer. Rev. Tuberc. Pulmonary Diseases*, **73**, 892 (1956); *Chem. Abstr.*, **51**, 3016f (1957).
21. J. A. Aeschlimann, U.S. Patent 2,665,279 (1954); *Chem. Abstr.*, **49**, 2521g (1955).
22. A. E. Wilder-Smith, *Science*, **119**, 514 (1954).
23. H. Brodhage, *Lancet*, **1**, 570 (1955).
24. T. S. Gardner, E. Wenis, and J. Lee, *J. Org. Chem.*, **19**, 753 (1954).

25. N. Rist, *Advanc. Tuberc. Res.*, **10**, 69 (1960).
26. F. Grumbach, N. Rist, D. Libermann, M. Moyeux, S. Cals, and S. Calvel, *C. R. Acad. Sci., Paris, Ser. C.*, **242**, 2187 (1956).
27. H. Noufflard-Guy-Loé and S. Berteaux, *Presse Méd.*, **71**, 201 (1963).
28. G. Brouet, J. Chevallier, and P. Nevot, *Presse Méd.*, **71**, 201 (1963).
29. O. Fuchs, V. Senkariuk, A. Nemes, A. Lazar, and T. Somogyi, Austrian Patent 241,461 (1965); *Chem. Abstr.*, **63**, 11518d (1965).
30. K. Palat, M. Celadnik, E. Novacek, E. Matuskova, and M. Pavlas, *Cesk. Farm.*, **14**, 502 (1965); *Chem. Abstr.*, **65**, 20091e (1966).
31. J. Izdebski, *Rocz. Chem.*, **39**, 717 (1965); *Chem. Abstr.*, **64**, 3465e (1966).
32. C. Casagrande, M. Canova, and G. Ferrari, *Boll. Chim. Farm.*, **104**, 424 (1965); *Chem. Abstr.*, **64**, 5037 (1966).
33. K. Palat, L. Novacek, and M. Celadnik, *Collect. Czech. Chem. Commun.*, **32**, 1191 (1967).
34. R. Valette, French Patent 1,468,738 (1967); *Chem. Abstr.*, **68**, 12866d (1968).
35. A. L. Mndzhoyan, V. G. Afrikyan, R. S. Oganesyan, A. O. Shakhmuradova, L. D. Zhuruli, S. G. Karagezyan, and V. G. Sarafyan, *Arm. Khim. Zh.*, **21**, 340 (1968); *Chem. Abstr.*, **70**, 11530u (1969).
36. F. Knotz, *Sci. Pharm.*, **37**, 35 (1969); *Chem. Abstr.*, **71**, 3226e (1969).
37. F. Fujikawa, R. Hirao, T. Shiota, M. Natio, and S. Tsukuma, *Yakugaku Zasshi*, **87**, 1493 (1967); *Chem. Abstr.*, **69**, 43751n (1968).
38. D. Eilhauer, East German Patent 30,865 (1965); *Chem. Abstr.*, **64**, 3494e (1966).
39. VEB-Leuna-Werke ("Walter Ulbright"), French Patent M3325 (1965); *Chem. Abstr.*, **64**, 3496c (1966).
40. J. Kinugawa and S. Takei, Japanese Patent 68 00,517 (1968); *Chem. Abstr.*, **69**, 59112r (1968).
41. E. Winkelmann, H. Hilmer, and W. H. Wagner, German Patent 1,232,964 (1967); *Chem. Abstr.*, **67**, 21832w (1967).
42. E. Ochiai, "Aromatic Amine Oxides," Elsevier, New York, 1967, p. 424.
43. G. T. Newbold and F. S. Spring, *J. Chem. Soc.*, 1684 (1948).
44. K. G. Cunningham, G. T. Newbold, F. S. Spring, and J. Stark, *J. Chem. Soc.*, 2091 (1949).
45. E. Shaw, *J. Amer. Chem. Soc.*, **71**, 67 (1949).
46. E. Shaw, J. Bernstein, K. Losee, and W. A. Lott, *J. Amer. Chem. Soc.*, **72**, 4362 (1950).
47. G. F. Ottmann and I. M. Voynick, U.S. Patent 3,347,863 (1967); *Chem. Abstr.*, **68**, 87164g (1968).
48. J. J. D'Amico, U.S. Patent 3,155,671 (1964); *Chem. Abstr.*, **62**, 4015a (1965).
49. C. O. Obenland, U.S. Patent 3,310,568 (1967); *Chem. Abstr.*, **67**, 82111g (1967).
50. L. A. Paquette, U.S. Patent 3,213,101 (1965); *Chem. Abstr.*, **64**, 2063b (1966).
51. S. Minakami and E. Hirai, Japanese Patent 68 09,225 (1968); *Chem. Abstr.*, **69**, 106563a (1968).
52. K. Harada and S. Emoto, *Chem. Pharm. Bull.* (Tokyo), **13**, 389 (1965).
53. A. Fujita, M. Nakata, S. Minami, and H. Takamatsu, *Yakugaku Zasshi*, **86**, 1014 (1966); *Chem. Abstr.*, **66**, 758885p (1967).
54. Dainippon Pharmaceutical Co. Ltd. British Patent 1,053,730 (1967); *Chem. Abstr.*, **66**, 115605f (1967).
55. S. Minami, A. Fujita, and J. Matsumoto, U.S. Patent 3,414,567 (1968); *Chem. Abstr.*, **70**, 68164g (1969).
56. C. F. Boehringer and Son, French Patent 90,356 (1967); *Chem. Abstr.*, **70**, 87584k (1969).

57. H. Takamatsu, S. Minami, A. Fujita, K. Fujimoto, M. Shimizu, and Y. Takase, Japanese Patent 69 00,541 (1969); *Chem. Abstr.*, **70**, 87596r (1969).
58. S. Minami, A. Fujita, K. Yamamoto, K. Fujimoto, and Y. Takase, Japanese Patent 1178 (1967); *Chem. Abstr.*, **66**, 94912m (1967).
59. S. Minami, Japanese Patent 10,987 (1967); *Chem. Abstr.*, **68**, 21838y (1968).
60. C. F. Boehringer and Son, British Patent 1,128,117 (1968); *Chem. Abstr.*, **70**, 47308s (1969).
61. S. J. Childress and J. V. Scudi, *J. Org. Chem.*, **23**, 67 (1958).
62. E. E. Smissman, in "Textbook of Organic Medicinal and Pharmaceutical Chemistry," C. O. Wilson, O. Gisvold, and R. F. Doerge (Eds.), 5th ed., Lippincott, Philadelphia, 1966, p. 268.
63. S. P. Acharya and K. S. Nargund, *J. Sci. Ind. Res.*, **21B**, 451 (1962).
64. A. German and T. B. Loc, French Patent 1,443,053 (1966); *Chem. Abstr.*, **66**, 37781u (1967).
65. A. Funk and P. V. Divekar, *Can. J. Microbiol.*, **5**, 317 (1959).
66. P. V. Divekar, G. Read, and L. C. Vining, *Can. J. Chem.*, **45**, 1215 (1967).
67. A. Vivante, *Arch. Intern. Pharmacodynamie*, **105**, 241 (1956).
67a. D. E. Butler, P. Bass, I. C. Nordin, F. P. Hauck, and Y. J. L'Italien, *J. Med. Chem.*, **14**, 575 (1971).
68. A. E. Chichibabin and O. A. Zeide, *J. Russ. Phys.-Chem. Soc.*, **46**, 1216 (1914).
69. R. N. Shreve, N. W. Swaney, and E. H. Riechers, *J. Amer. Chem. Soc.*, **65**, 2241 (1943).
70. A. Taub, in "Textbook of Organic Medicinal and Pharmaceutical Chemistry," C. O. Wilson, O. Gisvold, and R. F. Doerge (Eds.), 5th ed., Lippincott, Philadelphia, 1966, pp. 286, 287.
71. C. Pfizer and Company, British Patent 1,065,938 (1967); *Chem. Abstr.*, **68**, 39478r (1968).
72. G. B. Afanas'eva, G. N. Platonova, B. S. Tanaseichuk, L. A. Cechulina, and I. Y. Postovskii, *Khim.-Farm. Zh.*, **2**, 32 (1968); *Chem. Abstr.*, **70**, 96576m (1969).
73. W. C. J. Ross, *Biochem. Pharmacol.*, **16**, 675 (1967).
74. W. C. J. Ross, *J. Med. Chem.*, **10**, 257 (1967).
75. J. R. Geigy, A.-G., French Patent 1,510,321 (1968); *Chem. Abstr.*, **70**, 77803s (1969).
76. A. Stempel and L. H. Sternbach, U.S. Patent 3,415,835 (1968); *Chem. Abstr.*, **70**, 57657u (1969).
77. T. Ueda, Japanese Patent 3,742 (1966); *Chem. Abstr.*, **64**, 693d (1966).
78. J. R. Geigy, A.-G., French Patent 1,510,319 (1968); *Chem. Abstr.*, **70**, 68185q (1969).
79. J. R. Geigy A.-G., Netherland Patent 6,509,930 (1966); *Chem. Abstr.*, **64**, 19571 (1966).
80. Société d'Etudes de Recherches et d'Applications Scientifiques et Médicales, French Patent 1,518,209 (1968); *Chem. Abstr.*, **70**, 106391p (1969).
81. K. Undheim, V. Nordal, and K. Tjonneland, South African Patent 67 02,789 (1968); *Chem. Abstr.*, **71**, 30366g (1969).
82. I. Cheifitz, C. Paulin, H. Tuatay, and E. H. Rubin, *Diseases of the Chest*, **25**, 390 (1954).
83. D. M. Bosworth, *N. Y. State J. Med.*, **56**, 1281 (1956).
84. I. J. Selikoff, E. H. Robitzek, and G. G. Ornstein, *Ann. Rev. Tuberc.*, **67**, 212 (1953).
85. E. A. Zeller, J. Barsky, J. R. Fouts, W. F. Kirchheimer, and L. S. Van Orden, *Experientia*, **8**, 349 (1952).

86. H. P. Loomer, J. C. Saunders, and N. S. Kline, *Psychiat. Res. Rep.*, **8**, 129 (1958).
87. Roussel-UCLAF, French Patent M3268 (1965); *Chem. Abstr.*, **63**, 16314 (1965).
88. Laboratories Houde, French Patent M2861 (1964); *Chem. Abstr.*, **62**, 9113 (1964).
89. F. Sporatore and G. Paglietti, *Boll. Chim. Farm.*, **106**, 12 (1967).
90. E. S. Faust, *Schweiz. Med. Wochschr.*, **54**, 229 (1924).
91. E. S. Faust, *Lancet*, **208**, 1336 (1925).
92. F. Uhlman, *Z. Ges. Exptl. Med.*, **43**, 556 (1924).
93. R. Altschul, *Circulation*, **14**, 494 (1956).
94. R. Altschul and A. Hoffer, *Circulation*, **16**, 499 (1957).
95. J. P. Miale, *Curr. Therap. Res.*, **7**, 392 (1965).
96. W. B. Parsons, *Circulation*, **24**, 1099 (1961).
97. M. F. Oliver, *Mod. Trends Pharmacol. Thera.*, **1**, 221 (1967).
98. Société Amilloise de Produits Chimiques, French Patent M1256 (1962); *Chem. Abstr.*, **57**, 16766h (1962).
99. Société Amilloise de Produits Chimiques, French Patent Cam 158 (1967); *Chem. Abstr.*, **71**, 3283w (1969).
100. M. DiRosa, F. Sarluca, and L. Sorrentino, *Arch. Ital. Sci. Farmacol.*, **15**, 62 (1965); *Chem. Abstr.*, **66**, 17885c (1967).
101. Société Chimique de L'Yvette, French Patent M2541 (1964); *Chem. Abstr.*, **62**, 1634 (1965).
102. J. Nordmann and H. Swierkot, French Patent M3454 (1965); *Chem. Abstr.*, **65**, 13667 (1966).
103. T. Irikura, S. Sato, Y. Abe, and K. Kasuga, Japanese Patent 19063 (1966); *Chem. Abstr.*, **66**, 37779z (1967).
104. N. Sugimoto and S. Imada, British Patent 1,070,120 (1967); *Chem. Abstr.*, **68**, 78148c (1968).
105. H. Fujimura, K. Okamoto, M. Tetsuo, and T. Abe, Japanese Patent 12,355 (1968); *Chem. Abstr.*, **70**, 3847n (1969).
106. Yoshitomi Pharmaceutical Industries, Netherland Patent 6,514,807 (1966); *Chem. Abstr.*, **65**, 10571 (1966).
107. B. C. Fischback, U.S. Patent 3,426,031 (1969); *Chem. Abstr.*, **70**, 68163f (1969).
108. H. B. Wright, D. A. Dunnigan, and U. Biermacher, *J. Med. Chem.*, **7**, 113 (1964).
109. H. A. DeWald, R. D. Westland, and J. D. Dice, U.S. Patent 3,157,666 (1964); *Chem. Abstr.*, **62**, 4011 (1965).
110. H. A. DeWald, U.S. Patent 3,157,664 (1964); *Chem. Abstr.*, **62**, 4013 (1965).
111. H. A. DeWald, U.S. Patent 3,156,694 (1964); *Chem. Abstr.*, **62**, 4014 (1965).
112. W. T. Ely, U.S. Patent 3,306,896 (1967); *Chem. Abstr.*, **67**, 73528w (1967).
113. J. R. Dice and D. R. Westland, U.S. Patent 3,128,281 (1964); *Chem. Abstr.*, **60**, 15842 (1964).
114. J. P. English, F. L. Bach, and S. Gordon, U.S. Patent 3,330,832 (1967); *Chem. Abstr.*, **68**, 49453t (1968).
115. M. Nakanishi, K. Arimura, and T. Muro, Japanese Patent 14,466 (1968); *Chem. Abstr.*, **70**, 57653q (1969).
116. J. H. Biel and E. J. Warawa, U.S. Patent 3,409,629 (1968); *Chem. Abstr.*, **70**, 57655s (1969).
117. J. H. Biel and E. J. Warawa, U.S. Patent 3,413,298 (1968); *Chem. Abstr.*, **70**, 68188t (1969).
118. E. D. Amstutz, F. P. Palopoli, C. H. Tilford, and R. F. Shuman, British Patent 990,340 (1965); *Chem. Abstr.*, **63**, 6978 (1965).
119. C. A. Elvehjem, R. J. Madden, F. M. Strong, and D. W. Woolley, *J. Amer. Chem. Soc.*, **59**, 1767 (1937).

120. P. J. Fouts, O. M. Helmer, S. Lepkovsky, and T. H. Jukes, *Proc. Soc. Exptl. Biol. Med.*, **37**, 405 (1937).

121. D. T. Smith, J. M. Ruffin, and S. G. Smith, *J. Amer. Med. Assoc.*, **109**, 2054 (1937).

121a. C. del Rio-Estrada and H. W. Dougherty, in "Kirk-Othmer Encyclopedia of Chemical Technology," 2nd ed., Vol. 21, A. Standen (Ed.), Interscience, New York, 1970, p. 509.

122. A. F. Wagner and K. Folkers, in "Medicinal Chemistry," A. Burger, (Ed.), 2nd ed., Interscience, New York, 1960, p. 144.

123. H. McIlwain, *Brit. J. Exptl. Pathol.*, **21**, 136 (1940).

124. H. McIlwain, *Nature*, **146**, 653 (1940).

125. E. Auhagen, *Z. Physiol. Chem.*, **274**, 48 (1942).

126. D. E. Hughes, *Biochem. J.*, **57**, 485 (1954).

127. W. J. Johnson and J. D. McColl, *Science*, **122**, 834 (1955).

128. E. Leifer, J. L. Roth, J. R. Hogness, and W. H. Langham, *J. Biol. Chem.*, **176**, 249 (1948).

129. B. C. Johnston and P. H. Lin, *J. Amer. Chem. Soc.*, **75**, 2971, 2974 (1953).

130. Ref. 122, pp. 238–257.

131. I. C. Gunsalus, W. D. Bellamy, and W. W. Umbreit, *J. Biol. Chem.*, **155**, 685 (1944).

132. W. D. Bellamy, W. W. Umbreit, and I. C. Gunsalus, *J. Biol. Chem.*, **160**, 461 (1945).

133. S. Mandeles, R. Koppelman, and M. E. Hanke, *J. Biol. Chem.*, **209**, 449 (1952).

134. S. Udenfriend, in "Chemical and Biological Aspects of Pyridoxal Catalysis" (Proceedings of a symposium of the International Union of Biochemistry, Rome, October 1962), E. E. Snell, P. M. Fasella, A. Braunstein, and A. Rossi Fanelli, (Eds.), Pergamon Press, London, 1963, p. 267.

135. A. E. Braunstein, *Enzymologica*, **7**, 25 (1934).

136. P. P. Cohen, *Biochem. J.*, **33**, 1478 (1939).

137. H. Wada and E. E. Snell, *J. Biol. Chem.*, **237**, 127, 133 (1962).

138. E. E. Snell, *J. Biol. Chem.*, **157**, 491 (1945).

139. E. E. Snell, *J. Amer. Chem. Soc.*, **67**, 194 (1945).

140. E. E. Snell, in "Chemical and Biological Aspects of Pyridoxal Catalysis" (Proceedings of a symposium of the International Union of Biochemistry, Rome, October 1962), E. E. Snell, P. M. Fasella, A. Braunstein, and A. Rossi Fanelli, (Eds.), Pergamon Press, London, 1963, p. 1.

141. K. Nakamoto and A. E. Martell, *J. Amer. Chem. Soc.*, **81**, 5863 (1959).

142. G. M. Dyson and P. May, "Chemistry of Synthetic Drugs," Longmans, London, 1959, p. 392.

143. T. Miki and T. Matsuo, *Yakugaku Zasshi*, **87**, 323 (1967); *Chem. Abstr.*, **67**, 32549k (1967).

144. M. V. Balyakina, Z. N. Zhukova, and E. S. Zhdanovich, *Zh. Prikl. Khim.* (Leningrad), **41**, 2324 (1968); *Chem. Abstr.*, **70**, 68074c (1969).

145. P. F. Muhlradt, Y. Morino, and E. E. Snell, *J. Med. Chem.*, **10**, 341 (1967).

146. Y. Nakai, F. Masugi, N. Ohishi, and S. Fukui, *Bitamin*, **38**, 189 (1968).

147. K. Okumura, Japanese Patent 68 02,716 (1968); *Chem. Abstr.*, **69**, 59107t (1968).

148. Y. H. Loo, *J. Neurochem.*, **14**, 813 (1967).

149. S. A. Harris, E. E. Harris, E. R. Peterson, and E. F. Rogers, *J. Med. Chem.*, **10**, 261 (1967).

150. S. A. Harris and K. Folkers, *Science*, **89**, 347 (1939); *J. Amer. Chem. Soc.*, **61**, 1245, 3307 (1939).

151. I. Chuzo and D. E. Metzler, *J. Heterocycl. Chem.*, **4**, 319 (1967).

152. W. Korytnyk, *J. Med. Chem.*, **8**, 112 (1965).

153. G. Schorre, German Patent 1,238,473 (1967); *Chem. Abstr.*, **68**, 39479s (1968).
154. G. Schorre, German Patent 1,238,474 (1967); *Chem. Abstr.*, **68**, 39480k (1968).
155. K. Hikawa, O. Aki, T. Fushimi, and S. Yurugi, Japanese Patent 67 25,668 (1967); *Chem. Abstr.*, **69**, 52014b (1968).
156. M. Murakami and M. Iwanami, Japanese Patent 19,067 (1966); *Chem. Abstr.*, **66**, 37776w (1967).
157. G. Schorre, German Patent 1,193,049 (1965); *Chem. Abstr.*, **63**, 2961 (1965).
158. M. Richard S.A., French Patent M3115 (1965); *Chem. Abstr.*, **63**, 1772 (1965).
159. Sophymex S.A., French Patent M4028 (1966); *Chem. Abstr.*, **67**, 100017n (1967).
160. S. Imado, Japanese Patent 9341 (1967); *Chem. Abstr.*, **68**, 68893c (1968).
161. N. Sugimoto and S. Imado, Japanese Patent 9348 (1967); *Chem. Abstr.*, **68**, 68892b (1968).
162. Tanabe Seiyaku Company, British Patent 1,101,369 (1968); *Chem. Abstr.*, **69**, 10374p (1968).
163. N. Sugimoto and S. Imado, Japanese Patent 11,832 (1967); *Chem. Abstr.*, **68**, 78146a (1968).
164. H. Ziegler, H. Inion, and F. Binon, German Patent 1,228,261 (1966); *Chem. Abstr.*, **66**, 28661c (1967).
165. N. C. Chenoy, *Pharm. J.*, **194**, 663 (1965).
166. E. Habicht and R. Zubiani, Belgian Patent 654,331 (1965); *Chem. Abstr.*, **64**, 17553 (1966).
167. J. Kejha, O. Radek, V. Jelinek, and O. Nemecek, Czech Patent 126,496 (1968); *Chem. Abstr.*, **70**, 28825 (1969).
168. J. Kejha, O. Radek, V. Jelinek, and O. Nemecek, *Cesk. Farm.*, **16**, 92 (1967); *Chem. Abstr.*, **67**, 108535a (1967).
169. A. von Lichtenberg, *Brit. J. Urol.*, **3**, 119 (1931).
170. M. J. Allen and W. L. Bencze, U.S. Patent 2,966,493, (1960); *Chem. Abstr.*, **55**, 7442 (1961).
171. W. S. Coppage, *J. Clin. Invest.*, **38**, 2101 (1959).
172. H. A. Pfenninger, South African Patent 67 7,171 (1968); *Chem. Abstr.*, **70**, 77795r (1969).
173. J. Durinda, J. Kolena, L. Szucs, L. Krasnec, and J. Heger, *Cesk. Farm.*, **16**, 14 (1967); *Sci. Pharm. Proc., 25th*, **1**, 113 (1965).
174. A. P. Gray and D. E. Heitmeier, *J. Amer. Chem. Soc.*, **81**, 4347 (1959); A. P. Gray, D. E. Heitmeier, and E. E. Spinner, *J. Amer. Chem. Soc.*, **81**, 4351 (1959).
175. Farbenfabriken Bayer A.-G, Netherland Patent 6,610,362 (1967); *Chem. Abstr.*, **67**, 32594w (1967).
176. K. Thiele, South African Patent 67 06,852 (1968); *Chem. Abstr.*, **70**, 68180j (1969).
177. A. Burger and G. E. Ullyot, *J. Org. Chem.*, **12**, 342 (1947).
178. J. R. Lewis, *Arch. Intern. Pharmacodynamie*, **88**, 142 (1951).
179. W. T. Nauta, U.S. Patent 3,354,168 (1968); *Chem. Abstr.*, **68**, 104988y (1968).
180. Aktiebolag Bofors, Netherland Patent 6,608,905 (1966); *Chem. Abstr.*, **68**, 29614t (1968).
181. Far Eastern Detailers Ltd., Belgian Patent 661,609 (1965); *Chem. Abstr.*, **64**, 19570 (1966).
182. A. Pohland, H. R. Sullivan, and R. E. McMahon, *J. Amer. Chem. Soc.*, **79**, 1442 (1957).
183. G. de Stevens, A. Halamandaris, P. Strachan, E. Donoghue, L. Dorfman, and C. F. Huebner, *J. Med. Chem.*, **6**, 357 (1963).

184. G. de Stevens, in "Kirk-Othmer Encyclopedia of Chemical Technology," 2nd ed., Vol. 2, A. Standen (Ed.), Interscience, New York, 1963, p. 379.

185. A. Pongratz and K. L. Zirm, *Monatsh. Chem.*, **88**, 330 (1957).

186. K. L. Zirm and A. Pongratz, *Arzneimittel-Forsch.*, **10**, 137 (1960).

187. R. Cahen, A. Boucherle, J. Nadand, J. Chariot, S. Taurand, and N. Clerembeaux, *Arch. Intern. Pharmacodyn.*, **143**, 466 (1963).

188. B. Zorn and F. Schmidt, *Pharmazie*, **12**, 396 (1957).

189. Patents, see *Chem. Abstr.*, **64**, 712 (1966); **68**, 49457x, 59439g (1968); **70**, 96640c, 106387s (1969); **71**, 13023a (1969).

190. D. Evans, K. S. Hallwood, C. H. Cashin, and H. Jackson, *J. Med. Chem.*, **10**, 428 (1967).

191. Patents, see *Chem. Abstr.*, **71**, 61220j, 61222m, 61232q (1969).

192. T. Lewis and R. T. Grant, *Heart*, **11**, 209 (1924).

193. R. F. Doerge, in "Textbook of Organic Medicinal and Pharmaceutical Chemistry," 5th ed., C. O. Wilson, O. Gisvold, and R. F. Doerge (Eds.), Lippincott, Philadelphia, 1966, p. 617.

194. S. L. Lee, B. B. Williams, and M. M. Kochhar, *J. Pharm. Sci.*, **56**, 1354 (1967).

195. G. B. Koelle (Ed.), *Handbuch der Experimentellen Pharmacologie, vol. 15, Cholinesterase and Anticholinesterase Agents*, Springer-Verlag, Berlin, 1963.

196. R. Urban, U.S. Patent 2,572,579 (1951), *Chem. Abstr.*, **46**, 4576 (1952).

197. J. E. Tether, *J. Amer. Med. Assoc.*, **160**, 156 (1956).

198. E. N. Bookrajian and W. Truter, *Amer. J. Med. Sci.*, **224**, 386 (1952).

199. M. Weinstein and M. Roberts, *J. Amer. Med. Assoc.*, **153**, 268 (1953).

200. H. M. Wuest and E. H. Sakal, *J. Amer. Chem. Soc.*, **73**, 1210 (1951).

201. O. Schmid, U.S. Patent 2,789,981 (1957); *Chem. Abstr.*, **52**, 2089 (1958).

202. E. H. Sakal, U.S. Patent 2,662,890 (1953); *Chem. Abstr.*, **49**, 1108 (1955).

203. I. B. Wilson, *J. Biol. Chem.*, **190**, 111 (1951); **197**, 215 (1952); **199**, 113 (1952); *Biochem. Biophys. Acta*, **7**, 520 (1951).

204. I. B. Wilson and F. Bergmann, *J. Biol. Chem.*, **185**, 479 (1950); **186**, 683 (1950).

205. R. I. Ellin and J. H. Wills, *J. Pharm. Sci.*, **53**, 995 (1964); **53**, 1143 (1964).

206. I. B. Wilson, *J. Amer. Chem. Soc.*, **77**, 2385 (1955).

207. A. H. Beckett, *Proc. Intern. Pharmacol. Meeting, 1st, Stockholm*, **7**, 1961.

208. D. G. Coe, *J. Org. Chem.*, **24**, 882 (1959).

209. I. B. Wilson and S. Ginsberg, *Biochem. Biophys. Acta*, **18**, 168 (1955).

210. E. J. Poziomek, B. E. Hackley, and G. M. Steinberg, *J. Org. Chem.*, **23**, 714 (1957).

211. F. Hobbiger, D. G. O'Sullivan, and P. W. Sadler, *Nature*, **182**, 1498 (1958).

212. T. A. Loomis, M. J. Welsh, and G. T. Miller, *Toxicol. Appl. Pharmacol.*, **5**, 588 (1963).

213. F. Hauschild, R. Schmiedel, and W. D. Wiezorek, East German Patent 38,036 (1965); *Chem. Abstr.*, **63**, 13225 (1965).

214. H. Engelhard, Belgian Patent 638,533 (1964); *Chem. Abstr.*, **62**, 9114 (1965).

215. Y. Ashani, and S. Cohen, *Israel J. Chem.*, **52**, 59 (1967).

216. T. Nishimura, C. Yamazaki, and T. Ishiura, *Bull. Chem. Soc. Jap.*, **40**, 2434 (1967).

217. J. N. Wells, J. N. Davisson, T. Boime, D. R. Haubrich, and G. K. W. Yim, *J. Pharm. Sci.*, **56**, 1190 (1967).

218. W. B. McDowell, U.S. Patent 3,155,674 (1964); *Chem. Abstr.*, **62**, 4012 (1965).

219. Olin Mathieson Chemical Corp., Belgian Patent 640,141 (1964); *Chem. Abstr.*, **62**, 14637 (1965).

220. R. B. Margerison and J. A. Nelson, U.S. Patent 3,385,860 (1968); *Chem. Abstr.*, **69**, 77122y (1968).

221. I. Bulete and I. Popa-Zeletin, *Rev. Chim.* (Bucharest), **18**, 176 (1967); *Chem. Abstr.*, **67**, 82066w (1967).

222. L. Clark and L. J. Roth, *J. Pharm. Sci.*, **52**, 96 (1963).

223. E. G. C. Clarke, "Isolation and Identification of Drugs," The Pharmaceutical Press, London, 1969, p. 441.

224. R. B. Barlow, *Biochem. Soc. Symp.*, **19**, 46 (1960).

225. L. Gyermek, in "Drugs Affecting the Peripheral Nervous System," Vol. 1, A. Burger (Ed.), Marcel Dekker, New York, 1967, p. 180.

226. E. R. Bowman and H. McKennis, *J. Pharmacol. Exptl. Therap.*, **135**, 306 (1962).

227. E. Hansson and C. G. Schmiterlow, *J. Pharmacol. Exptl. Therap.*, **137**, 91 (1962).

228. R. B. Barlow and J. T. Hamilton, *Brit. J. Pharmacol.*, **18**, 510 (1962).

229. R. B. Barlow and N. A. Dobson, *J. Pharm. Pharmacol.*, **7**, 27 (1955).

230. Neisler Laboratories, Belgian Patent 649,145 (1964); *Chem. Abstr.*, **64**, 8151 (1966).

231. A. P. Gray and H. Kraus, British Patent 1,038,728 (1966); *Chem. Abstr.*, **65**, 15344 (1966).

232. F. Hoffman LaRoche and Company, Netherland Patent 6,508,725 (1966); *Chem. Abstr.*, **64**, 15855 (1966).

233. E. I. duPont de Nemours and Company, Netherland Patent 6,607,687 (1966); *Chem. Abstr.*, **67**, 90693p (1967).

234. W. F. Minor, U.S. Patent 3,352,878 (1967); *Chem. Abstr.*, **68**, 105007q (1968).

235. Société Amilloise de Produits Chimiques, French Patent M4,898 (1967); *Chem. Abstr.*, **69**, 106565c (1968).

236. E. B. Sigg, U.S. Patent 3,423,510 (1969); *Chem. Abstr.*, **70**, 87578m (1969).

237. O. Hankovszky, K. Hideg, G. Mehes, L. Decsi, and M. Varszegi, Hungarian Patent 155,659 (1969); *Chem. Abstr.*, **70**, 115014v (1969).

238. B. W. Horrom, U.S. Patent 3,400,132 (1968); *Chem. Abstr.*, **69**, 106569g (1968).

239. B. W. Horrom, U.S. Patent 3,398,155 (1968); *Chem. Abstr.*, **70**, 3843h (1969).

240. H. Leditschke, G. Vogel, and L. Ther, German Patent 1,238,920 (1967); *Chem. Abstr.*, **68**, 78141v (1968).

241. A. Pongratz and K. L. Zirm, German Patent 1,226,102 (1966); *Chem. Abstr.*, **66**, 28669m (1967).

242. Z. J. Vejdelek and M. Protiva, *Chem. Listy*, **45**, 448 (1951).

243. Z. J. Vejdelek and M. Protiva, *Chem. Listy*, **46**, 423 (1952).

244. L. Mombaerts, *Pharm. Acta Helv.*, **32**, 217 (1957).

245. M. Oka, *Acta Med. Scand.*, **145**, 358 (1953).

246. K. Fromherz and H. Spiegelberg, *Helv. Physiol. Pharmacol. Acta*, **6**, 42 (1948).

247. G. S. Roback and A. C. Ivy, *Circulation*, **6**, 90 (1952).

248. A. Cohen, British Patent 631,078 (1949); *Chem. Abstr.*, **44**, 5397 (1950).

249. Patents, see *Chem. Abstr.*, **63**, 6980 (1965); **65**, 20107 (1966); **66**, 104902z (1967); **67**, 90684m (1967); **68**, 95703x (1968); **69**, 52008c (1968); **70**, 87581g (1969).

250. Patents, see *Chem. Abstr.*, **67**, 54039w (1967); **70**, 37653m, 87582h (1969).

251. J. Koo, *J. Pharm. Sci.*, **53**, 1329 (1964).

252. N. I. Kudryashova and N. V. Khromov-Borisov, *Zh. Org. Khim.*, **3**, 1117 (1967).

253. Patents, see *Chem. Abstr.*, **63**, 1773, 9923, 18041 (1965); **64**, 3496 (1966); **67**, 82106j (1967); **68**, 12867e, 49459z, 68890z (1968); **70**, 47306q, 96643f, 106395t (1969).

254. B. Gogolimska, *Acta Polon. Pharm.*, **21**, 343 (1964).

255. K. Thiele, A. Gross, K. Posselt, and W. Schuler, *Chim. Ther.*, **2**, 366 (1967).

256. Hoffman La Roche and Company, Netherland Patent 6,511,532 (1966); *Chem. Abstr.*, **65**, 3847 (1966).

257. B. Brust, R. I. Fryer, and L. H. Sternbach, Belgian Patent 668,701 (1966); *Chem. Abstr.*, **65**, 5446 (1966).
258. F. Aktieselskabet, Danish Patent 107,422 (1967); *Chem. Abstr.*, **69**, 10376r (1968).
259. J. R. Dice, L. Scheinman, and K. W. Berrodin, *J. Med. Chem.*, **9**, 176 (1966).
260. B. R. Baker and J. A. Hurlbut, *J. Med. Chem.*, **12**, 221 (1969).
261. J. B. Bicking, U.S. Patent 3,276,958 (1966); *Chem. Abstr.*, **65**, 20105 (1966).
262. N. V. Koninklijke Pharmaceutische Fabrieken, Netherland Patent 6,511,354 (1966); *Chem. Abstr.*, **65**, 10569 (1966).
263. M. M. Robison, A. A. Renzi, and W. E. Barrett, *Experientia*, **23**, 513 (1967).
264. E. J. Blanz and F. A. French, *Cancer Res.*, **28**, 2419 (1968).
265. Takeda Chemical Industries Ltd., French Patent 1,511,398 (1968); *Chem. Abstr.*, **70**, 77813v (1969).
265a. P. Bass, R. A. Purdon, M. A. Patterson, and D. E. Butler, *J. Pharmacol. Exptl. Therap.*, **152**, 104 (1966).
266. B. V. Rama Sastry, in "Medicinal Chemistry, Part II," 3rd ed., A. Burger (Ed.), Wiley, New York, 1970, p. 1544.
267. J. Bernstein, U.S. Patent 2,727,899 (1955); *Chem. Abstr.*, **50**, 10796 (1956).
268. G. Rey-Bellet and H. Spiegelberg, British Patent 728,579 (1955); *Chem. Abstr.*, **50**, 5773 (1956); U.S. Patent 2,726,245 (1955); *Chem. Abstr.*, **50**, 8745 (1956).
269. A. Wolf and E. von Haxthausen, *Arzneimittel-Forsch.*, **8**, 618 (1958).
270. R. I. Meltzer, U.S. Patent 3,313,822 (1967); *Chem. Abstr.*, **67**, 54038v (1967).
271. E. F. Elslager, E. L. Benton, F. W. Short, and F. H. Tendick, *J. Amer. Chem. Soc.*, **78**, 3453 (1956).
272. R. A. Neal and P. Vincent, *Brit. J. Pharmacol.*, **10**, 434 (1955).
273. A. R. Brown, F. C. Copp, and A. R. Elphick, *J. Chem. Soc.*, 1544 (1957).
274. J. Bernstein, B. Stearns, E. Shaw, and W. A. Lott, *J. Amer. Chem. Soc.*, **69**, 1151 (1947).
275. D. G. Markees, V. C. Dewey, and G. W. Kidder, *Arch. Biochem. Biophys.*, **86**, 179 (1960).
276. K. Thomae, British Patent 730,243 (1955); *Chem. Abstr.*, **50**, 6515f (1956).
277. G. Ferlemann and W. Vogt, *Arch. Exptl. Pathol. Pharmakol.*, **250**, 479 (1965) and the references therein.
278. P. Arnall and N. Greenhalgh, British Patent 889,748 (1962); *Chem. Abstr.*, **56**, 15490 (1962).
279. A. W. J. Broome and N. Greenhalgh, *Nature*, **189**, 59 (1961).
280. L. Lasagna, *Med. Clin. N. Amer.*, **41**, 359 (1957).
281. L. Lasagna, "The Effect of Pharmacological Agents on the Nervous System," Williams and Wilkins Co., Baltimore, 1959.
282. A. H. Lutz and O. Schnider, *Helv. Chim. Acta*, **39**, 81 (1956); U.S. Patent 2,680,116 (1954); *Chem. Abstr.*, **49**, 7603g (1955).
283. E. Tagmann, E. Sury, and K. Hoffmann, *Helv. Chim. Acta*, **35**, 1541 (1952); U.S. Patent 2,673,205 (1954); *Chem. Abstr.*, **49**, 6318d (1955).
284. K. Hoffmann and E. Urech, U.S. Patent 2,848,455 (1958); *Chem. Abstr.*, **53**, 7103i (1959).
285. E. Tagmann, E. Sury, and K. Hoffmann, *Helv. Chim. Acta*, **35**, 1235 (1952); **37**, 185 (1954); K. Hoffmann and E. Tagmann, U.S. Patent 2,664,424 (1953); *Chem. Abstr.*, **49**, 10382c (1955).
286. Chemie Gruenenthal, British Patent 768,821 (1957); *Chem. Abstr.*, **51**, 15595h (1957).
287. P. Dukor and F. M. Dietrich, *Lancet*, **1**, 569 (1967).

288. K. C. Brannock, U.S. Patent 3,432,506 (1969); *Chem. Abstr.*, **70**, 106389u (1969).
289. A. H. Lutz and O. Schnider, *Chimia*, **12**, 291 (1958).
290. W. S. Benica and C. O. Wilson, *J. Amer. Pharm. Assoc. (Sci. Ed.)*, **39**, 451 (1950).
291. B. Loev, U.S. Patent 3,325,505 (1967); *Chem. Abstr.*, **68**, 49455v (1968).
292. F. Bossert and W. Vater, British Patent 1,129,158 (1968); *Chem. Abstr.*, **70**, 19940y (1969).
293. F. Bossert and W. Vater, South African Patent 68 01,482 (1968); *Chem. Abstr.*, **70**, 96641d (1969).
294. G. S. Marks, E. G. Hunter, U. K. Terner, and D. Schneck, *Biochem. Pharmacol.*, **14**, 1077 (1965).
295. D. W. Schneck, W. J. Racz, G. H. Hirsch, G. L. Bubbar, and G. S. Marks, *Biochem. Pharmacol.*, **17**, 1385 (1968).
296. R. G. Shepherd, A. C. Bratton, and K. C. Blanchard, *J. Amer. Chem. Soc.*, **64**, 2532 (1942).
297. T. Kato, H. Yamanaka, and T. Teshigawara, Japanese Patent 17,458 (1965); *Chem. Abstr.*, **63**, 18042 (1965).
298. G. E. Less, W. R. Wragg, S. J. Corne, N. D. Edge, and H. W. Reading, *Nature*, **181**, 1717 (1958).
299. A. Spinks and E. H. P. Young, *Nature*, **181**, 1397 (1958).
300. L. Bretherick, G. E. Lee, E. Lunt, W. R. Wragg, and N. D. Edge, *Nature*, **184**, 1707 (1959).
301. L. Bretherick and W. R. Wragg, British Patent 866,681 (1961); *Chem. Abstr.*, **55**, 24798 (1961).
302. I. M. Sharapov, E. I. Levkoeva, E. S. Nikitskaya, and V. S. Usovskaya, *Khim-Farm. Zh.*, **2**, 29 (1968); *Chem. Abstr.*, **70**, 57584t (1969).
303. J. H. Biel, U.S. Patent 3,221,017 (1965); *Chem. Abstr.*, **64**, 5053 (1966).
304. J. H. Biel and H. B. Hopps, U.S. Patent 3,221,019 (1965); *Chem. Abstr.*, **64**, 5053 (1966).
305. R. T. Coutts, K. K. Midha, and K. Prasad, *J. Med. Chem.*, **12**, 940 (1969).
306. K. K. Midha, R. T. Coutts, J. W. Hubbard, and K. Prasad, *Eur. J. Pharmacol.*, **11**, 48 (1970).
307. D. F. Biggs, R. T. Coutts, J. W. Hubbard, K. K. Midha, and K. Prasad, to be published.
308. Farbenfabriken Bayer A.-G., British Patent 985,354 (1965); *Chem. Abstr.*, **63**, 586 (1965).
309. Farbenfabriken Bayer A.-G., French Patent M3016 (1965); *Chem. Abstr.*, **62**, 16206 (1965).
310. F. Hoffman LaRoche and Company, Netherland Patent 6,407,463 (1965); *Chem. Abstr.*, **62**, 16207 (1965).
311. R. Hunt and R. R. Renshaw, *J. Pharmacol. Exptl. Therap.*, **35**, 75 (1929).
312. A. S. V. Burgen, *J. Pharm. Pharmacol.*, **16**, 638 (1964).
313. A. Christiansen, H. Lüllmann, and E. Mutschler, *Eur. J. Pharmacol.*, **1**, 81 (1967), and the references contained therein.
314. J. M. van Rossum, *Arch. Intern. Pharmacodyn.*, **140**, 592 (1962).
315. R. E. Lyle, E. F. Perlowski, H. J. Troscianiec, and G. G. Lyle, *J. Org. Chem.*, **20**, 1761 (1955).
316. R. H. K. Foster, and A. J. Carman, *J. Pharmacol. Exptl. Therap.*, **91**, 195 (1947).
317. J. Buchi and M. Prost, *Ann. Pharm. Franc.*, **12**, 241 (1954).
318. M. Rajsner, E. Adlerova, and M. Protiva, *Collect. Czech. Chem. Commun.*, **28**, 1031 (1963).

319. P. A. J. Janssen, U.S. Patent 3,030,372 (1962); *Chem. Abstr., 59*, 2780 (1963).
320. J. R. Geigy A.-G., Netherland Patent 6,408,219 (1965); *Chem. Abstr., 63*, 586 (1965); 6,600,519 (1966); *Chem. Abstr., 65*, 16948g (1966); 6,600,522 (1966), *Chem. Abstr., 65*, 16948b (1966); 6,600,524 (1966), *Chem. Abstr., 65*, 16947g (1966).
321. H. H. Kuehnis, H. Ryf, and R. Denss, Swiss Patent 447,164 (1968); *Chem. Abstr., 69*, 52019g (1968).
322. P. M. Carabateas, U.S. Patent 3,226,392 (1965); *Chem. Abstr., 64*, 8150 (1966).
323. H. Kuehnis, R. Denss, and C. H. Eugster, Swiss Patent 417,591 (1967); *Chem. Abstr., 68*, 39482n (1968).
324. G. Ulbricht, East German Patent 39,144 (1965); *Chem. Abstr., 63*, 18037 (1965).
325. M. D. Draper, U.S. Patent 3,272,838 (1966); *Chem. Abstr., 65*, 18563 (1966).
326. M. D. Draper, U.S. Patent 3,272,837 (1966); *Chem. Abstr., 65*, 18563 (1966).
327. J. H. Biel and H. B. Hopps, U.S. Patent 3,426,036 (1969); *Chem. Abstr., 70*, 87583j (1969).
328. A. S. Tomcufcik, P. F. Fabio, and A. M. Hoffmann, British Patent 1,085,066 (1967); *Chem. Abstr., 68*, 95699a (1968).
329. H. Beschke and W. A. Schuler, U.S. Patent 3,284,457 (1966); *Chem. Abstr., 66*, 28672g (1967).
330. H. Oediger and A. Oberdorf, British Patent 1,137,020 (1968); *Chem. Abstr., 70*, 68181k (1969).
331. R. P. Welcher and L. C. Mead, U.S. Patent 3,409,627 (1968); *Chem. Abstr., 70*, 47304n (1969).
332. A. Albert and E. P. Sarjeant, "Ionization Constants of Acids and Bases," Methuen, London, 1962.
333. N. B. Eddy, H. Halbach, and O. J. Braenden, *Bull. World Health Org., 14*, 353 (1956).
334. P. W. Nathan, *Brit. Med. J.,* ii, 903 (1952).
335. N. B. Eddy, H. Halbach, and O. J. Braenden, *Bull. World Health Org., 17*, 569 (1957).
336. D. J. Coleman, J. Levin, and P. O. Jones, *Brit. Med. J.,* i, 1092 (1957).
337. W. B. Wright, H. J. Brabander, and R. A. Hardy, *J. Amer. Chem. Soc., 81*, 1518 (1959); *J. Org. Chem., 26*, 476, 485 (1961).
338. T. J. Dekornfeld and L. Lasagna, *Anesthesiology, 21*, 159 (1960).
339. N. Sugimoto, K. Okumura, N. Shigematsu, and G. Hayashi, *Chem. Pharm. Bull.* Tokyo, 10, 1061 (1962).
340. F. Gross and H. Turrian, *Experientia, 13*, 401 (1957).
341. A. Hunger, J. Kebrle, A. Rossi, and K. Hoffmann, *Helv. Chim. Acta, 43*, 1727 (1960), and references cited therein.
342. A. F. Casy and R. R. Ison, *J. Pharm. Pharmacol., 22*, 270 (1970).
343. R. W. Cunningham, B. K. Harned, M. C. Clark, R. R. Cosgrove, N. S. Daughtery, C. H. Hine, R. E. Vessey, and N. N. Yuda, *J. Pharmacol. Exptl. Therap., 96, 151* (1949).
344. W. M. Duffin and A. F. Green, *Brit. J. Pharmacol., 10*, 383 (1955).
345. O. Eisleb and O. Schaumann, *Dtsch. Med. Wschr., 65*, 967 (1939).
346. L. Lasagna and H. K. Beecher, *J. Pharmacol. Exptl. Therap., 112*, 306 (1954).
347. A. K. Reynolds and L. O. Randall, "Morphine and Allied Drugs," University of Toronto Press, Toronto, 1957, p. 269.
348. A. F. Casy, H. Birnbaum, G. H. Hall, and B. J. Everitt, *J. Pharm. Pharmacol., 17*, 157 (1965).

349. A. D. Macdonald, G. Woolfe, F. Bergel, A. L. Morrison, and H. Rinderknecht, *Brit. J. Pharmacol.*, 1, 4 (1946).
350. O. J. Braenden, N. B. Eddy, and H. Halbach, *Bull. World Health Org.*, 13, 937 (1955).
351. O. J. Braenden and P. O. Wolff, *Bull. World Health Org.*, 10, 1003 (1954).
352. F. Bergel and A. L. Morrison, *Quart. Rev.*, 2, 349 (1948).
353. A. H. Beckett and A. F. Casy, *Bull. Narcot.*, 9, 37 (1957).
354. A. H. Beckett and A. F. Casy, in "Progress in Medicinal Chemistry," Vol. 2, G. P. Ellis and G. B. West (Eds.), Butterworths, London, 1962, p. 43 and references cited therein.
355. R. A. Hardy and M. G. Howell, in "Analgetics," G. DeStevens (Ed.), Academic Press, New York, 1965, p. 179.
356. A. F. Casy, in "Progress in Medicinal Chemistry," Vol. 7, Part 2, G. P. Ellis and G. B. West (Eds.), Butterworths, London, 1970, p. 229.
357. T. D. Perrine and N. B. Eddy, *J. Org. Chem.*, 21, 125 (1956).
358. N. B. Eddy, L. E. Lee, and L. S. Harris, *Bull. Narcot.*, 11, 3 (1959).
359. T. J. Dekornfeld and L. Lasagna, *J. Chron. Dis.*, 12, 252 (1960).
360. E. Nilsson and P. A. J. Janssen, *Acta. Anaesth. Scand.*, 5, 73 (1961).
361. P. A. J. Janssen, A. H. M. Jageneau, P. J. A. Demoen, C. van de Westeringh, A. H. M. Raeymaekers, M. S. J. Wouters, S. Sanczuk, B. F. K. Hermans, and J. L. M. Loomans, *J. Med. Pharm. Chem.*, 1, 105 (1959).
362. P. A. J. Janssen and N. B. Eddy, *J. Med. Pharm. Chem.*, 2, 31 (1960).
363. A. Ziering and J. Lee, *J. Org. Chem.*, 12, 911 (1947).
364. A. H. Beckett, A. F. Casy, G. Kirk, and J. Walker, *J. Pharm. Pharmacol.*, 9, 939 (1957).
365. C. R. Ganellin and R. G. W. Spickett, *J. Med. Chem.*, 8, 619 (1965).
366. E. A. Mistryukov, N. I. Aronova, and V. F. Kucherov, *Izv. Akad. Nauk. SSSR, Otd. Khim. Nauk.*, 932 (1961).
367. A. H. Beckett, A. F. Casy, and N. J. Harper, *Chem. Ind.* (London), 19 (1959).
368. A. F. Casy, *Tetrahedron*, 22, 2711 (1966); *J. Med. Chem.*, 11, 188 (1968).
369. A. F. Casy, L. G. Chatten, and K. K. Khullar, *J. Chem. Soc.*, C, 2491 (1969).
370. P. S. Portoghese and D. L. Larson, *J. Pharm. Sci.*, 57, 711 (1968).
371. I. N. Nazarov, N. S. Prostakov, and N. I. Shvetsov, *J. Gen. Chem. USSR.*, 26, 2798 (1956).
372. N. S. Prostakov, B. E. Zaitsev, N. M. Mikhailova, and N. N. Mikheeva, *J. Gen. Chem. USSR.*, 34, 463 (1964); A. F. Casy and K. McErlane, *J. Pharm. Pharmacol.*, 23, 68 (1971).
373. N. S. Prostakov and N. N. Mikheeva, *Russian Chem. Rev.*, 31, 556 (1962) (translation pagination).
374. M. M. A. Hassan and A. F. Casy, *Org. Magn. Res.*, 2, 197 (1970).
375. A. F. Casy, A. H. Beckett, and N. A. Armstrong, *Tetrahedron*, 16, 85 (1961).
376. A. F. Casy, A. H. Beckett, G. H. Hall, and D. K. Vallance, *J. Med. Pharm. Chem.*, 4, 535 (1961).
377. P. A. J. Janssen, C. van de Westeringh, A. H. M. Jageneau, P. J. A. Demoen, B. K. F. Hermans, C. H. P. van Daele, H. L. Schellekens, C. A. M. van der Eycken, and C. J. E. Niemegeers, *J. Med. Pharm. Chem.*, 1, 281 (1959).
378. P. A. J. Janssen, U.S. Patent 3,141,823 (1964); *Chem. Abstr.*, 62, 14634e (1964).
379. J. K. Gardocki and J. Yelnosky, *Toxicol. Appl. Pharmacol.*, 6, 48 (1964); J. Yelnosky and J. K. Gardocki, *Toxicol. Appl. Pharmacol.*, 6, 63 (1964).
380. J. S. Finch and T. J. Dekornfeld, *J. Clin. Pharmacol.*, 7, 46 (1967).

381. A. F. Casy, M. M. A. Hassan, A. B. Simmonds, and D. Staniforth, *J. Pharm. Pharmacol.*, **21**, 434 (1969).
382. P. M. Carabateas, W. F. Wetterau, and L. Grumbach, *J. Med. Chem.*, **6**, 355 (1963).
383. D. E. Fancher, S. Hayao, W. G. Strychker, and L. F. Sancilio, *J. Med. Chem.*, **7**, 721 (1964).
384. P. A. J. Janssen, A. H. Jageneau, and J. Huygens, *J. Med. Pharm. Chem.*, **1**, 299 (1959).
385. C. van de Westeringh, P. V. Daele, B. Hermans, C. V. van der Eycken, J. Boey, and P. A. J. Janssen, *J. Med. Chem.*, **7**, 619 (1964).
386. A. Saarne, *Acta Anaesth. Scand.*, **13**, 11 (1969).
387. G. E. Deneau, J. E. Villarreal, and M. H. Seevers, 28th Meeting, Committee on Problems of Drug Dependence, National Research Council, 1966, Addendum 2.
388. K. W. Bentley, "The Chemistry of the Morphine Alkaloids," Oxford University Press, Oxford, 1954.
389. J. Kalvoda, P. Buchschacher, and O. Jeger, *Helv. Chim. Acta*, **38**, 1847 (1955).
390. M. Mackay and D. C. Hodgkin, *J. Chem. Soc.*, 3261 (1955).
391. M. Gates and G. Tschudi, *J. Amer. Chem. Soc.*, **74**, 1109 (1952); **78**, 1380 (1956).
392. D. Elad and D. Ginsburg, *J. Chem. Soc.*, 3052 (1954).
393. L. F. Small, N. B. Eddy, E. Mosettig, and C. K. Himmelsbach, *Public Health Rep. Washington*, Supplement No. 138 (1938); *Chem. Abstr.*, **33**, 5998 (1939).
394. L. A. Woods, *Pharmacol. Rev.*, **8**, 175 (1956).
395. W. R. Martin, *Pharmacol. Rev.*, **19**, 463 (1967).
396. H. Blumberg, P. S. Wolf, and H. B. Dayton, *Proc. Soc. Exptl. Biol. Med.*, **118**, 763 (1965).
397. L. S. Harris and W. L. Dewey, in "Annual Reports in Medicinal Chemistry," 1966, p. 33.
398. R. Grewe, A. Mondon, and E. Nolte, *Ann. Chem.*, **564**, 161 (1949).
399. R. D. Hunt and F. F. Foldes, *New Engl. J. Med.*, **248**, 803 (1953).
400. H. Isbell and H. F. Fraser, *J. Pharmacol. Exptl. Therap.*, **107**, 524 (1953).
401. K. Fromhertz and B. Pellmont, *Experientia*, **8**, 394 (1952).
402. M. Gates and W. G. Webb, *J. Amer. Chem. Soc.*, **80**, 1186 (1958).
403. N. B. Eddy, J. G. Murphy, and E. L. May, *J. Org. Chem.*, **22**, 1370 (1957).
404. L. S. Harris and A. K. Pierson, *J. Pharmacol. Exptl. Therap.*, **143**, 141 (1964).
405. W. T. Beaver, S. L. Wallenstein, R. W. Houde, and A. Rogers, *Clin. Pharmacol. Therap.*, **7**, 740 (1966).
406. H. F. Fraser and D. E. Rosenberg, *J. Pharmacol. Exptl. Therap.*, **143**, 149 (1964).
407. L. S. Harris, A. K. Pierson, J. R. Dembinski, and W. L. Dewey, *Arch. Int. Pharmacodyn.*, **165**, 112 (1967).
408. W. R. Martin, C. W. Gorodetsky, and T. K. McClane, *Clin. Pharmacol. Therap.*, **7**, 455 (1966).
409. N. B. Eddy and E. L. May, "Synthetic Analgesics, Part IIB, 6, 7-Benzomorphans," Pergamon, Oxford, 1966.
410. J. Pearl and L. S. Harris, *J. Pharmacol. Exptl. Therap.*, **154**, 319 (1966).
411. B. F. Tullar, L. S. Harris, R. L. Perry, A. K. Pierson, A. E. Soria, W. F. Wetterau, and N. F. Albertson, *J. Med. Chem.*, **10**, 383 (1967).
412. J. Hellerbach, O. Schnider, H. Besendorf, and B. Pellmont, "Synthetic Analgesics, Part IIA, Morphinans," Pergamon, Oxford, 1966.
413. H. L. Friedman, in "Drugs Affecting the Peripheral Nervous System," Vol. 1, A. Burger (Ed.), Marcel Dekker, New York, 1967, p. 79.
414. R. R. Renshaw, M. Zift, B. Brodie, and N. Kornblum, *J. Amer. Chem. Soc.*, **61**, 638 (1939).

415. F. W. Schueler, *Arch. Intern. Pharmacodyn.*, **93**, 417 (1953); *J. Amer. Pharm. Assoc. Sci. Ed.*, **45**, 197 (1956).
416. M. D. Mashkovsky and C. A. Zaitseva, *Arzneimitt. Forsch.*, **18**, 320 (1968).
417. L. H. Sternbach and S. Kaiser, *J. Amer. Chem. Soc.*, **74**, 2219 (1952).
418. J. B. Robinson, B. Belleau, and B. Cox, *J. Med. Chem.*, **12**, 848 (1969); B. Belleau and P. Pauling, *J. Med. Chem.*, **13**, 737 (1970).
419. G. Fodor, in "The Alkaloids," Vol. VI, R. H. F. Manske (Ed.), Academic Press, New York, 1960, p. 145.
420. R. B. Barlow, "Introduction to Chemical Pharmacology," 2nd ed., Methuen London, 1964, p. 214.
421. J. P. Long, F. P. Luduena, B. F. Tullar, and A. M. Lands, *J. Pharmacol. Exptl. Therap.*, **117**, 29 (1956).
422. D. Bovet and F. Bovet-Nitti, "Structure Activité Médicaments Système Nerveux Vegetatif," Karger, Basel, 1948.
423. C. Kieberman and L. Limpach, *Ber.*, **25**, 927 (1892).
424. L. Gyermek, *Nature*, **171**, 788 (1953).
425. C.-Y. Chen and R. J. W. LeFèvre, *J. Chem. Soc.*, 3473 (1965).
426. R. J. Bishop, G. Fodor, A. R. Katritzky, F. Soti, L. E. Sutton, and F. J. Swimbourne, *J. Chem. Soc., C*, 74, (1966).
427. A. F. Casy and W. Jeffery, unpublished results.
428. G. Maffi and G. Bianchi, *Nature*, **185**, 844 (1960).
429. L. György, A. K. Pfeifer, M. Dóda, É. Galambos, J. Molnár, G. Kraiss, and K. Nádor, *Arzneimitt. Forsch.*, **18**, 517 (1968).
430. L. Gyermek and K. Nádor, *J. Pharm. Pharmacol.*, **9**, 209 (1957).
431. S. Archer, A. M. Lands, and T. R. Lewis, *J. Med. Pharm. Chem.*, **5**, 423 (1962).
432. J. H. Biel, P. A. Nuhfer, W. K. Hoya, H. A. Leiser, and L. G. Abood, *Ann. N. Y. Acad. Sci.*, **96**, 251 (1957).
433. L. G. Abood, A. Ostfeld, and J. H. Biel, *Arch. Intern. Pharmacodyn.*, **120**, 186 (1959).
434. J. Tripod, E. Surry, and K. Hoffmann, *Experientia*, **10**, 261 (1954).
435. R. Meier, F. Gross, and J. Tripod, *Klin. Wochschr.*, **32**, 445 (1954).
436. N. Sperber, F. J. Villani, M. Sherlock, and D. Papa, *J. Amer. Chem. Soc.*, **73**, 5010 (1951).
437. P. Portoghese, T. L. Pazdernik, W. L. Kuhn, G. Hite, and A. Shafi'ee, *J. Med. Chem.*, **11**, 12 (1968).
438. A. Shafi'ee and G. Hite, *J. Med. Chem.*, **12**, 266 (1969).
439. B. Hermons, P. van Daele, C. van de Westeringh, C. van der Eycken, J. Boey, J. Dockx, and P. A. J. Janssen, *J. Med. Chem.*, **11**, 797, (1968).
440. E. Hardegger and H. Ott, *Helv. Chim. Acta*, **38**, 312 (1955).
441. A. Sinnema, L. Maat, A. J. van der Gugten, and H. Beyerman, *Rec. Trav. Chim. Pays-Bas*, **87**, 1027 (1968).
442. S. P. Findlay, *J. Amer. Chem. Soc.*, **76**, 2855 (1954).
443. O. Kovacs, G. Fodor, and J. Weisz, *Helv. Chim. Acta*, **37**, 892 (1954).
444. R. Gottlieb, *Arch. Exp. Path. Pharmak.*, **97**, 113 (1923).
445. R. Foster, H. R. Ing, and V. Varagic, *Brit. J. Pharmacol.*, **10**, 436 (1955).
446. A. Heusner, *Z. Naturforsch.*, **12B**, 602 (1957).
447. R. Willstätter, *Ber.*, **29**, 1575, 2216 (1896).
448. G. Vinci, *Virchows Arch. Path. Anat.*, **145**, 78 (1896); **149**, 217 (1897).
449. F. Perks and P. J. Russell, *J. Pharm. Pharmacol.*, **19**, 318 (1967).
450. S. M. McElvain, *J. Amer. Chem. Soc.*, **49**, 2835 (1927).

Subject Index

Cumulative Author Index

Numbers in **boldface** type refer to part numbers for Volume 14. Numbers in parentheses are reference numbers and show that an author's work is referred to although his name is not mentioned in the text. Numbers in *italics* indicate the pages on which the full references appear.

Abashian, **1**: 190(106), *290*

Abblard, J., **1**: 12(103), 21(103), 33(103), *122*

Abbolito, C., **1**: 64(384), *129*; **3**: 829(769), 830(769), 831(769), 888(769), 910(769), 911(769), *1163*

Abbott, **4**: 20(369), 21(462), *110, 113*

Abbott, E. H., **3**: 1001(108), *1146*

Abbott Laboratories, **2**: 545(51), 560(51), *616*; **3**: 111(549,550), 193(549,550), 194 (550), 212(550), 217(549,550), 218(550), *249*

Abbott, S. D., **1**: 48(318b), 145(51), 147 (51), 156(51), *127, 177*; **2**:281(106), 306 (247), 311(106), 414(66a), *384, 388, 477a*

Abdel-Ghaffar, **1**: 187(67), *288*

Abdel-Kader, A., **3**: 646(118), 793(118), 802(118), *1147*; **4**: 195(97), 230(97), 234(97), 270(97), 271(97), *430*

Abe, **4**: 13(194,196), *105*

Abe, A., **4**: 118(7), *183*

Abe, N., **2**: 333(524), 334(524,546), 339 (546), *396, 397*; **3**: 107(534), 189(534), 215(534), 517(306), *249, 594*; **4**: 118(7), *183*

Abe, Y., **4**: 455(103,105), *514*

Abeles, R. H., **1**: 155(123,124), *179*

Abidova, S., **2**: 311(326), *391*

Abood, L. G., **4**: 508(532,533), *524*

Abott, G. G., **1**: 335(16a), *415*

Abou-Zeid, Y. M., **3**: 19(51), 83(433), *39, 245*

Abraham, M. E., **2**: 309(280), *389*

Abramendo, P. I., **4**: 191(366), 259(366), 277(541,542), 278(366), 284(541),

437, 442

Abramenko, **1**: *304*

Abramovitch, **1**: 197(156a), 198(156a), *291*; **4**: 6(82-84,86), 9(119), 12(83,84,86, 189), 15(82), 25(119), 26(82,86,119), 29 (83,84,86), 31(83,84,86), 60(82), 61(82), *102, 103, 105*

Abramovitch, R. A., **1**: 9(59,71), 12(77), 16(141), 19(161), 20(72), 21(172), 23 (183), 31(210), 33(172), 35(71), 37(172), 40(161), 42(266,271,272), 43(271,280-282), 44(289), 45(183), 46(183,296,304-307), 47(183,296,306), 48(172, 313,314, 315,317), 49(172,314), 50(172,314,326, 327), 51(172,326,327-331), 53(329,335, 330), 54(341), 55(354), 56(161,329,331), 57(172,358), 62(172), 65(398), 66(331), 71(398), 72(172,183), 73(431), 75(446), 76(447,448), 77(446), 78(451), 79(452, 453), 81(453), 82(453), 84(141,450,454), 101(536a), 105(544), 106(296,547), 144 (24), 145(52), 156(150), 313(16a), 338 (131), 345(183), 353(131), *120-124, 126-133, 177, 180, 413, 417*; **2**: 14(140), 20 (147b), 37(220), 38(220), 48(220), 51, 52(299a), 55-57(299a), 66, 67(140), 70 (362), 75(374), 80(299a), 96(6), 103(6), 105(6), 108(445), 111(6), 114(6), 125 (472), 127(6), 128(6), 146(520), 148(520), 149(522), 151(536), 152(536a,536b), 153 (536a,b), 156(536a), 157-160(537a), 161 (537b), 164(299a,c), 165(140), 166(299a, c), 171(553a,554b), 172(553a), 173(553a, 554b, 555a), 177(422), 179(522), 197 (140), 200-201(362), 205(602), 207(602), 223(6,147b), 224(147b), 236-237(66),

543

Bach, F. L., Jr., **3**: 57(162,163), 59(162, 163), *237;* **4**: 456(114), *514*
Bachelet, J. P., **1**: 312(14), *412*
Bachman, G. B., **2**: 313(350,351), *392;* **3**: 117(580), 196(580), 198(580), 211(580), *250;* **4**: 149(50), 156(50), *184*
Baclawski, L. M., **2**: 526(198), *620*
Bacot, **1**: 188(58), *288*
Badger, G. M., **1**: 85(459), 87(459), *131;* **2**: 63(329), 98(329), 100(329), 200(329), 203(329), 218(621), 374(768), *251, 260, 404*
Badgett, B., **2**: 330(503), *396*
Badische Anilin und Soda-Fabrek AG., **3**: 248(247), *593*
Bag, S. P., **3**: 281(160), *327*
Bagot, **1**: 242(454), *301*
Bahr, F., **1**: 104(543), 334(115), 370(115), *133, 415*
Bailar, J. C., Jr., **3**: 847(823), *1164*
Bailey, **1**: 247(506), 254(506), *302*
Bailey, A. S., **1**: 145(46), *177;* **2**: 290(848), 437(291), *406, 484;* **3**: 886(928), *1167*
Bailey, D. M., **2**: 13(126b,128), *246*
Bailey, J., **4**: 256(76), 261(76), *428*
Bain, B. M., **2**: 63(328), 130(328,484), 140 (484), 141(328,484), 170(328), *251, 256;* **3**: 131(611), 302(258), 307(258), 311 (258), 715(361), 716(361), 718(361), 720 (361), 866(361), *251, 330, 1153*
Baines, A., **2**: 373(754), *403*
Baiocci, L., **3**: 7(37), 16(37), 19(37), *39*
Baizer, **1**: 242(453), *301;* **4**: 12(153), *104*
Baizer, M. M., **2**: 277(75,76), 297(76), 322 (75,435), 341(75,76), *383, 394;* **3**: 117 (577), 191(577), 258(2), 260(16), 313(2), 340(90), 403(90), 524(90), *250, 322, 588*
Bak, B., **2**: 446(179), *480*
Baker, **1**: 187(110), 197(206), 201(206), *290, 293*
Baker, A. D., **1**: 18(156), *123*
Baker, B. R., **1**: 402(523-527), 427, 330 (500), 334(500), 337-340(500), *396;* **4**: 480(260), *519*
Baker, J. P., **3**: 761(588), *1159*
Baker, J. W., **4**: 249(7), 257(7), 419(7), *426*
Baker, R., **4**: 256(263), 261(263), *434*
Baker, R. H., **1**: 154(99), *178*
Baker, W., **2**: 7(49), 11(49), 130(49), 463,

464(375), 466,467-469(375), 473(374), *244, 486;* **3**: 349(130), 459(130), 528 (130), 533(130), 716(371), *589, 1153*
Bako, E., **2**: 361,362(727,728), 371(727, 728), *402*
Balaban, **1**: 234(437), 236(426,435), 237 (435,436), *300*
Balaban, A. T., **1**: 328(78-80), 329(82), *414,* **2**: 25(177), 26(175b), 27(177), 293 (448), 324(448), *247, 394*
Balandin, A. A., **2**: 307(257-259), 308(258), 334(540), 340(553), *389, 397*
Balassa, M., **3**: 183(688), *254*
Balasubramanian, A., **1**: 357(260), *133, 420*
Balasubramanian, K. K., **2**: 66(334), 67 (334), 71(334), 202(334), *252*
Baldas, J., **1**: 14(127), 15(127), *122*
Baldeschwieler, J. D., **1**: 13(119), *122;* **2**: 303(193), *386*
Baldry, D. A. T., **4**: *426*
Baliah, **1**: 284(669), *307*
Baliah, V., **4**: 387(9), *426*
Ball, **1**: 209(326), 219(326), *296*
Ball, J. S., **2**: 291(148), *385*
Ball, S. J., **4**: 238(10), 257(10), 355(10), *426*
Balme, M., **3**: *247*
Balsiger, R. W., **3**: 760(582), *1158;* **4**: 263 (11), *426*
Balta, E., **3**: 90(458), *246*
Baltrusis, R., **3**: 81(425,426), *245*
Baltz, H., **2**: 307(260), *389*
Balyakina, **4**: 18(250,251,257,258,264), 68 (264), 83(257,258), 84(258), 90(250), *106*
Balyakina, B. V., **3**: 103(511), 221(511), *248;* **4**: 461(144), *515*
Balyatinskaya, L. N., **2**: 590(113), 593(113), *618*
Balykina, **4**: 8(99), 23(99), 19(313), 71 (313), *102, 107, 108*
Bamberger, **1**: 312(15), *412*
Bambury, R. E., **1**: 409(601,602), *430*
Bamford, **1**: 276(597), *305*
Ban, **1**: 279(629), *306*
Ban, H., **3**: 49(90), *235*
Ban, Y., **1**: 55(349), 346(189), *128, 418;* **3**: 46(27), 129(27), 751(314), 752(314), 788 (680,683,684), *232, 1152, 1161*
Banashek, **4**: 17(222), *106*

289), 284(288), 319(289), 328(288), *435*

Collignon, N., **2**: 282(119), *384*

Collins, **1**: 197(185), 200(185), *292*

Collins, D. J., **3**: 44(9), 137-138(9), 144(9), 181(9), *232*

Collins, I., **2**: 432(460), 442(460), *488;* **3**: 885(952), 907(952), *1168*

Collins, R. F., **3**: 931(1023), *1170*

Colombini, **4**: 19(286), 68(286), *108*

Colonge, **1**: 286(676), *307*

Colonna, M., **2**: 18(151), 20(151), 44(267), 45(275), *242, 246, 250*

Colowick, S. P., **1**: 144(27,28), 152(27,28), 365(319), 367(336), *177, 421, 422*

Colwell, C. A., **4**: 447, *511*

Combé, W. P., **2**: 414(68), 415(81), 420 (107), 447(107), *477, 478*

Comrie, A. M., **4**: 199(61), 151(61), 168 (61), 270(61), 299-300(61), 303-304(61), 319(61), 339(61), *428*

Cone, **1**: 187(67), *288*

Connolly, J. D., **2**: 13(126b), *246*

Conover, L. H., **2**: 210-211(611), 220(629), *259-260;* **4**: 205(62), 256(62,63), 277 (62), 280-281(62), 285(62), 286(63), *428*

Conroy, **1**: 223(359), *298*

Cook, **1**: 197(156), 198(156), 209(326), 219(326), 220(334), *291, 296, 297*

Cook, C. E., **3**: 332(16), 357(16), 495(16), 546(16), 560(16), *586*

Cook, D., **3**: 652(144), 655(144), 847(818, 819), 920(818), 921, *1147, 1167;* **4**: *428*

Cook, D. J., **3**: 744(539-541), 749(559), 800-802(559), 832(779), 886(559), 919 (540,541), 920(540), *1157, 1158, 1163*

Cook, G. L., **2**: 291(148), 304(183), *385, 386*

Cook, J. D., **1**: 68(418), *130;* **2**: 439(295), 441-442(438), 451(223,224), 452(224, 235,237,238), 490(6), 491(6,17,19,208), 492(208,209), 493(19), 500(17,208), 503 (109), 506(151), 507(101,151), 508(151), 509(233), 510(48,233), 512(151,209), 513(17,101,151), 514(17,151), 515(6,17, 102), 516(48), 517(6,102,226), 520(151), 553-554(151), 560(48), 580(6,17,101, 102,226,233), *482, 484, 487, 615-621*

Cook, J. M., **1**: 164(194), *181*

Cook, K. C., **3**: 79(409), *245*

Cook, M. J., **1**: 109-110(576), *134*

Cook, N. C., **1**: 150(72), 160(72), *178;* **2**: 570(79), 598-599(79), 601(79), *617*

Cooke, J., **2**: 452(232,233), *482*

Cooks, R. G., **2**: 44(272a), *250*

Coombes, R. G., **2**: 611(138,139), 614(138, 139), *618, 619*

Cooney, **1**: 223(370), *298*

Cooper, **1**: 277(606), 278(610), 281(646), 282(648-650,654,656,657), 283(657, 662), *305, 306*

Cooper, C. W., **4**: 256(176), 261(176), *432*

Cooper, J. C., **1**: 49(319), 53(336), 145(52), 149(53), *127, 128, 177;* **2**: 281(103), 570 (78), 573(78), *384, 617*

Cooper, J. W., **1**: 145(53), 377(374), *177, 423;* **2**: 283(125), 554(181), *385, 620*

Coper, **1**: 186(5), *286*

Copp, F. C., **4**: 481(273), *519*

Coppage, W. S., **4**: 464(171), *516*

Coppens, G., **2**: 29(184), *247*

Coppinger, G. M., **3**: 897(1013), *1170*

Copps, D. B., **3**: 175(675), *253*

Corbally, R. P., **2**: 306(251), 458(326), *388, 484*

Corder, C. N., **1**: 312(13), 403(13), *412*

Cordes, E. H., **1**: 59(368), 145(32), 342 (166a), 347(201-203), 404(554), 404 (554), *128, 177, 417, 418, 428*

Cordier, P., **3**: 729(444), *1155*

Corey, E. J., **1**: 94(497), *132;* **3**: 827(764), 919(764), *1163*

Corley, R. C., **3**: 332(16), 357(16), 495(16), 546(16), 560(16), *586*

Corne, S. J., **4**: 486(298), *520*

Cornea-Ivan, V., **3**: 90(458), *246*

Cornforth, J. W., **1**: 354(239), *419*

Corr, D. H., **1**: 385(407), *424*

Corral, R. A., **2**: 415(73), *477*

Corran, **1**: 232(415), 252(478), 268(561), 270(568), *299, 301, 304*

Corraz, A. J., **3**: 231(719), *255*

Cort, L. A., **1**: 160(169), *180*

Cory, M., **1**: 402(527), *427*

Costa, G., **1**: 119(631), *135*

Costello, C. E., **1**: 326(71), *414*

Costin, R., **2**: 184(576), *258*

Costolow, J. J., **2**: 340(567), *398*

Cotter, J. L., **1**: 18(148), 368(348), *422*

Cottis, **1**: 255(508), 259(508), *302*

McGuire, **1**: 187(43), *288*
McIlwain, H., **4**: 457(123,124), *515*
McIntosh, J. M., **1**: 357(260), *133, 420*
McKay, D. L., **2**: 334(538), *397*
McKay, J. F., **1**: 155(135), *179*
MacKeller, F. A., **3**: 827(758), *1163*
McKennis, **1**: 185(3,4), 186(33), 187(33, 53), *286, 287, 288;* **4**: 9(123), 17(223), 66(223), *103, 106*
McKennis, H., **4**: 477(226), *518*
McKennis, H., Jr., **3**: 332(1,22,23,24), 343 (23), 346(23), 347(22), 400(1), 418(23, 24), 431(22), 432(23), 439(22), 449(22, 266), 451(1,23,24), 567(1,23,24), 568(1), *586, 593*
McKenzie, **1**: 394(445), *425*
McKenzie, D., **4**: 446, *511*
McKillop, A., **2**: 58-59(304), *251;* **3**: 96 (484), 183(484), 777(631), 864(631), 937 (996), *247, 1160, 1169*
McKinney, W. P., **1**: 115(613), *134*
McLane, T. K., **4**: *523*
MacLean, **1**: 253(488), 255(521), *302, 303*
McLean, G. W., **1**: 155(126), *179*
McLeister, E., **1**: 15(134), *122*
McMahon, D. H., **1**: 173(222), *182*
McMahon, R. E., **4**: 466(182), *516*
McManus, E., **1**: 409(595), *430*
McMillan, F. H., **2**: 371(725), *402*
McNeil Laboratories, Inc., **4**: 6(90,93), 26-27(90), 31-32(90), 59-60(93), *102*
McOmie, J. F. W., **2**: 7(49), 11(49), 130 (49), 148(521c), 463-464(375), 466-469 (375), 473(375), *244, 257, 486;* **3**: 349 (130), 459(130), 528(130), 535(130), 716(371), *589, 1153*
McRae, J. A., **1**: 17(144,146), *123*
McWinnie, W. R., **3**: 62(241), *240;* **4**: 182 (110), *186*
Maccagnani, G., **1**: 8(51), 9(70), 119(51), *120, 121;* **2**: 42(25), *249*
Macchia, **4**: 3(10), 14(10), 33(10), 38(10), 56(10), *100*
Maccoll, A., **1**: 323(55), 330(55), *414*
Mach, **1**: 237(433), *300*
Machleidt, H., **2**: *405*
Machon, Z., **3**: 246(116), 430-431(116), 433(116), 458-459(275), 460(116), 463 (116), 466(116), 540(275), 556(116), 560-561(116), *589, 593;* **4**:

159-160(63), *185*
Macioci, F., **2**: 3(13), 8(13), 93(13), *243;* **3**: 939(1028), *1170*
Maciulus, A., **3**: 81(425,426), *245*
Mackay, M., **4**: 502(390), *523*
Mackay, R. A., **1**: 8(53a), 340(145), *120, 416*
Mackellar, F. A., **1**: 327(72), *414*
Madden, R. J., **4**: 456(119), *514*
Madronero, **1**: 256(527), 262(527), *303*
Maeda, **1**: 195(130), 196(130), *290*
Maeda, M., **2**: 50(297), *251*
Maeda, R., **2**: 274(45), 437(290), 464(400), *382, 483, 486;* **3**: 103(515), 163(515), *248*
Maeda, T., **2**: 612(190), *620*
Maekawa, H., **2**: 334(543), *397*
Maeno, Y., **3**: 335(76), 372(76), 383(76), 456(76), *588*
Maerkl, G., **2**: 25(176), 27(176), 140(176), *247*
Maffi, G., **4**: 507(428), *524*
Magidson, O. Y., **3**: 343(105), 367(105), 396(105), 428(105), 457(105), 574(105), 728(443), 800(443), *589, 1155*
Magne, F. C., **3**: 71(334,338), *242, 243*
Magnus, G., **2**: 342-343(594), *399*
Mahadevan, S., **1**: 404(556), *428*
Mahan, **1**: 274(585,587), 276(591,592), 279(625), 280(638), *304, 305, 306*
Mahan, J. E., **2**: 309(279), 335(851), 340 (571), *389, 398, 406;* **3**: 268(50), *323*
Mahendran, M., **1**: 139(3), 160(3), *176*
Maheshwari, N., **4**: 344(123), 402(123), *429*
Maheshwari, S. C., **4**: 344(123), 402(123), *429*
Mahieu, **1**: 191(128), *290*
Mahler, H. R., **1**: 154(99), *178*
Mahon, J. E., **1**: 172(237), *182*
Mahon, J. J., **3**: 847(815), *1164*
Mahrotra, N. K., **4**: 182(113), *186*
Maibaum, R., **2**: 558(200,201b), *620, 620b*
Maienthal, M., **4**: 402(141), *430*
Maier, D. V., **2**: 9(66), 16(66), 19(66), 225 (66), *244;* **3**: 65(288), *241*
Mailey, E. A., **1**: 75(439), *130;* **2**: 171(552), 467(447), 475(447), *258, 488*
Mailey, E. M., **2**: 458(328), 466(328), *485*
Maine, F. W., **3**: 919(983), *1168*

186(494), *233, 237, 247*
Rossi, L., **4**: 346(292), *435*
Rossiter, E. D., **1**: 394(441), *425*
vanRossum, J. M., **4**: 487(314), *520*
Rostafinska, **1**: 197(186,187), 203(186, 187), 208(270), 212(270), *292, 295*
Roszkowski, **4**: 6(92), 25(92), 27(92), 31 (92), *102*
Rotermel, I. A., **3**: 980(1039), *1170*
Roth, **1**: 197(199), 199(199), *292;* **4**: 12 (176), 16(215), 56(215), 57(215), *104, 105*
Roth, H. J., **1**: 351(225-227), 359(269, 270), *419, 420*
Roth, J. L., **4**: 457(128), *515*
Roth, L. J., **4**: 447(7), 476(223), *511, 518*
Roth, W., **3**: 90(457), 152(457), *246*
Rothwell, K., **1**: 4(13), *119;* **3**: 801(705), *1161*
Rouaiz, A., **2**: 12(110), 97(110), 113(110), 139-140(110), *245*
Roukema, **4**: 5(39), 24(39), 25(39), 55(39), *101*
Roukema, P. A., **3**: 120(587), *251*
Rousseau, R. J., **3**: 49(87), 94(87), 100(87), 135(87), 138(87), 173(87), 180(87), 186 (87), 188(87), *235*
Roussel UCLAF, **3**: 361(160), 584-585 (160), *590;* **4**: 453(87), *514*
Rout, **1**: 223(359), *298*
Rowbotham, J. G., **1**: 13(117), *122*
Rowe, J. D., **1**: 108(573), *133;* **3**: 685(241), 731(455,241), 735(241), 736(241), 737 (455), 738(455), 744(455), 873(455), 884 (241), 885(241), 906-907(241), *1150, 1155*
Rowe, J. M., **1**: 142(16), *176*
Rowlands, **1**: 221(345,349), *297*
Roy, **1**: 197(244,259), 207(244,259), *294*
Roy, A. B., **3**: 779(648), 864(648), *1160*
Roy, J., **3**: 458(272,273), 510(273), *593*
Roy, S. K., **1**: 107(566), 108(573), *133;* **3**: 5(26), 15(26), 22(26), 33(26), 48(73), 167(73), 685(241), 731(241,456), 735 (241,456), 736(241), 906-907(241), *38, 234, 2250*
Roy, S. R., **2**: 25(174), 196(174), *247*
Roy-Chowdhury, P., **1**: 398(496), *426*
Royer, R., **1**: 312(14,15), *412;* **2**: 302(171), *386*

Rozenberg, A. N., **2**: 328(475), 337(475), 338-339(475), *395*
Rozmarynowicz, M., **2**: 373(749), *403*
Rozum, Yu., S., **1**: 9(65), *121*
Ruan, H. T., **4**: 255(467), *440*
Rubbo, S. D., **4**: 447(18,19), *511*
Rubershtein, A. M., **4**: 123(22), 146-147 (22), *184*
Rubicek, **1**: 208(273,298), 212(273), 213 (298), 215(298), 217(273,298), 218(298), *295, 296*
Rubin, E. H., **4**: 453(82), *513*
Rubinstein, H., **3**: 3, 14(11), *38;* **4**: 177(81), *185*
Rubinstein, K., **3**: 150(647,649), *252*
Rubstov, **1**: 246(501), *302;* **4**: 5(59), 50 (59), *101*
Rubstov, M., **1**: 164(195), *181*
Rubtsov, M. U., **2**: 308(269), *389*
Rubtsov, M. V., **2**: 435(442), 437(435), 443 (435), *487; 488;* **3**: 48(62), 56(150), 102 (57), 129(62), 132(62), 145(62), 162 (150,663), 183(150), 213(507), 405(210), 452(210), 466(210), 735(491-493), 764 (589,590), 787(589), 792(589), *234, 237, 248, 253, 592, 1156, 1159*
Ruch, M., **1**: 348(207), *418*
Ruch, W., **3**: 604(19), *1144*
Ruchardt, C., **2**: 83(399), 84(399-401), *253;* **3**: 722(407), *1154*
RudKo, A. P., **1**: 9(65), *121*
Rudner, B., **3**: 53(127), *236*
Rudolf, S., **4**: 273(473), 350(473), 354 (473), *440*
Ruetgerswerke, A. G., **4**: 12(188), *105*
Ruetgerswerke and Teerverwertung, A. G., **2**: 309(275), *389*
Ruetman, S. H., **2**: 409(18a), 413(57a), 415 (72a), 427(153a), *476a, 477a, 480a*
Ruff, O., **2**: 419(98a), *478a*
Ruffin, J. M., **4**: 456(121), *515*
Ruider, G., **4**: 118(9), 124(9), 128(9), *183*
Rumon, K. A., **1**: 379(381), *423*
Rumon, K. W., **1**: 103(541), *133*
Rumpf, P., **1**: 365(316a), *421*
Runge, F., **2**: 333(536), 334(536), *397*
Runti, C., **3**: 410(223), 527-528(223), 556 (223), *592*
Ruoff, **4**: 10(132), *103*
Rus, J., **3**: 103(512), *248*

706   Cumulative Author Index

Uskokovic, M., **1**: 163(189), *181*
Usorskaya, V., **1**: 164(195), *181*
Usovskaya, V. S., **4**: 486(302), *520*
Ustavschchikov, **1**: 276(602), *295, 304, 305;*
    4: 4(28), 38(28), *100*
Ustavschchikov, B. F., **2**: 308(271), 309
    (284), 317(396), *389, 393;* **3**: 42(1), 56
    (1), 268(41,46,48,49), 269(60), 270(89),
    289(208), *232, 323, 324, 328*
Utebacu, M. V., **2**: 105(440), *254*
Utebaev, M. U., **1**: 54(340), *128*
Utermohlen, W. P., **3**: 78(401), *244*
Utke, A. R., **2**: 302(174), *386*
Utsui, Y., **3**: 988(1069), *1171*
Utsumi, **1**: 279(628), *306;* **2**: 346(637), *400*
Uvarov, B. A., **2**: 375(800), *404*

Vaartaja, O., **4**: 256(342), 262(342), *436*
Vaculik, P., **2**: 276(62), 289(62), 308(272),
    383, 389
Vahldieck, J., **2**: 328-329(483), 332(483),
    335(483), *395*
Vajda, T., **1**: 55(348), 56(357), *128;* **3**: 54
    (128-131), 148(128,129), 150(131), 155
    (128,131), 194(128), *236*
Valette, R., **4**: 448(34), *512*
Val'Kova, **4**: 4(30), 12(30), 14(30), 58(30),
    100
Val'Kova, A. K., **2**: 328-329(481,482,487,
    489), 332(481,482,487,489), 334(539),
    335(481), *395-397;* **3**: 115(1571), *250*
Vallance, D. K., **4**: 499(376), *522*
Valthuijsen, J. A., **3**: 76(384), 113(384),
    244
Vampilova, V. V., **2**: 277(79), *383*
Van Allan, J. A., **1**: 314(24,25), 329(84a),
    413, 414; **2**: 334(545), *397*
Van Ammers, M., **1**: 46(29), *233*
Van Bergen, T. J., **3**: 297(242), 319(242),
    329
Van Bragt, J., **3**: 343(102), 347-348(102),
    380(102), 417(102), 438(102), 451(102),
    569(102), *589*
Van Campen, M. G., **4**: 156(61), *185*
Vanden Berghe, A. L., **3**: 929(1019), *1170*
Van der Haak, P. J., **3**: 744(537), 816(537),
    1157
Van der Kerk, G. J. M., **4**: 261(311), *435*
Van der Meeren, **1**: 197(159), 199(159), *291*
Van der Plas, H. C., **4**: 194-196(267), 205

(267), 223(267), 228(267), 230(103), 232
    (267), 233-234(267), 286(266), 299
    (267), 305(267), 319(267), 343(267), 344
    (255,267), 345(267), 346(267,103), *429,*
    434
Van der Wal, **4**: 18(237,239), 68(239), 78
    (239), 79(237), 81(239), 97(237,239),
    106
Van Dormael, A., **1**: 112(597), *134*
Van Dijk, J., **3**: 148-149(644), *252*
Van Fossen, R. U., **3**: 298(245), *330*
Van Hell, **4**: 10(124), *103*
Van Heyningen, **4**: 5(43), 28(43), 29(43),
    35-36(43), 48(43), 53(43), *101*
Van Iderstine, **1**: 409(595), *430*
Van Swieten, P. A., **3**: 76(384), 133(384),
    244
Van Tuyle, G. C., **3**: 644(112), *1146*
Van Veelen, G. F., **3**: 929(1019), *1170*
Van Velthuijsen, J. A., **2**: 3(14), 9(14), *243;*
    4: 193(373), 204-205(373), 247(373),
    327(373), 396(373), *437*
Van Zwieten, P. A., **4**: 193(373), 204-205
    (373), 224(371), 247(373), 257(371,372),
    274(371,372), 278(371,372), 327(373),
    333-335(371,372), 396(373), *437*
Van Zyl, G., **3**: 345(112), 422-423(112),
    436(112), 471(112), *589*
Vanags, **1**: 222(350-352,354), *297*
Vanags, G., **1**: 156(146), *180;* **2**: 272-273
    (32-35), 306(35), *382;* **3**: 296(236), 347
    (121), 421(240), 448(121), *329, 592, 589*
Vanderbilt Company Inc., R. T., **4**: 256
    (343), 259(425), 261(343), 327(343), *436*
Vanderleen, W., **3**: 61(221), *239*
Vander Werf, C. A., **2**: 433(282), *483*
Vandoni, I., **1**: 7(43), *120*
Vane, F. M., **4**: 182(99), *186*
Vanecek, **4**: 17(229), *106*
Vannesland, B., **1**: 404(558), *428*
Van Orden, L. S., **4**: 453(85), *513*
Varagic, V., **4**: 510(445), *524*
Varcoe, G. L., **2**: 374(773), *404*
Vardanyan, **1**: 197(240), 205(240), *294*
Varma, **4**: 9(111), *103*
Varma, B. K., **2**: 138(511), *257;* **3**: 101
    (501), 190-191(501), 193(501), *248*
Varsel, C. J., **4**: 182(93), *186*
Varszegi, M., **4**: 478(237), *518*
Vartanyan, S. A., **3**: 645(113), *1146*

(136), 357(132), 365(164), 368(87,173), 398(87,173), 410(132), 415(132), 466 (132), 471-472(164), 488(132), 497(87, 173), 504(87), 527(164), 529(164), 531 (164), 544(42), 546(42), 560(42,164, 333), 561(164), 620-622(53), 784(53), 791(53), 795-796(53), 869(53), 877-878 (53), 880(53), *234, 248, 587, 588, 589, 590, 591, 595, 1145*

Walker, J., 4: 497(364), *522*

Walker, P. P., 3: 850(840), *1165*

Walker, S., 1: 6(27,32), *120*

Walker, S. J., 2: 30(190), *247*

Wall, M. E., 3: 332(16), 357(16), 495(16), 546(16), 560(116), *586*

Wallace, T. J., 3: 847(815), *1164*

Wallenfels, K., 1: 59(365), 145(35,39,42), 149(42,235), 152(88), 154(101), 155 (122), 160(165), 336(119), 347(197,200, 203b,c,204,205), 362(300), 366(300), 370(352,353), *128, 177, 178, 179, 180, 182, 415, 418, 421, 422;* 2: 369(716), *402*

Wallenstein, S. L., 4: 505(405), *523*

Waller, 1: 185(1), *286*

Waller, G. R., 1: 346(191,195), *418;* 3: 855 (850,851), *1165*

Wallsgrove, E. R., 2: 327(464), *395*

Walsh, E. J., 4: 143(41), 146(41), 148(41), *184*

Walsh, R. J. A., 1: 392(439), *425*

Walter, 4: 21(459), *113*

Walter, L. A., 1: 322(440), *394;* 3: 667 (191), 767(191), 797(191), *1148;* 4: 153 (59), *185*

Walter, L. W., 3: 487(285), 563(285), *594*

Walter, R., 3: 408(216), 412(216), 414 (216), *592*

Walter, W., 2: 19(169), *247;* 4: 199(438, 439), 224(438), 227(438,439), 243(438, 439), 259(439), 264(438), 327(438,439), 335(438), 347(436), 367(438), 369(439), 388(438), *439*

Waltz, 1: 188-189(66), *288*

Wampler, 4: 21(470,471), *113*

Wandrey, P., 3: 332(19), 346(19), 380(19), 422(19), 430(19), 454(19), 461-462(19), 474(19), *586*

Wang, 1: 208(300), 215(300), 243(455), *296, 301*

Wang, B. J. S., 1: 155(117), *179*

Wang, C., 3: 737(502), 785-786(502), 792 (502), 801-802(502), 808(502), 890-891 (502), *1161*

Wang, C. H., 1: 367(339), *180, 422*

Wang, C. S., 1: 46(300), 108(571), *133;* 3: 655(153), *1147*

Wang, J. H., 1: 144(31), *127, 177*

Wang, K-C., 2: 275(59), 294(59), *383*

Wang, L., 2: 275(57), 294(57), *383*

Wang, N., 1: 367(339), *180, 422*

Wang, P. J. S., 1: 366(334), *422*

Wantanabe, H., 3: 919(984), 921(984), *1169*

Warawa, E. J., 4: 456(116,117), *514*

Warburg, O., 1: 365(319), *421*

Ward, F. E., 2: 368(705), *402*

Ward, J. W., 3: 332(50), 336(50), 366(50), 390(50), 393(50), *587*

Ward, T. J., 2: 438-439(236), 442(236), 452 (236), 511-512(232), 554(232), *482, 621;* 3: 832(778), 907(778), *1163*

Waring, A. J., 3: 83(434), 149(434), *245*

Warner, C. R., 4: 143(41), 146(41), 148(41), *184*

Warner, G. H., 1: 361(282), *420*

Warner, P. F., 2: 340(563), *397*

Warner-Lambert Co., 4: 387(561), *442*

Warnoff, E. W., 1: 375(398), 383(398), *424*

Warren, B. T., 2: 522-523(61), 539-540(61), 544(61), 562-563(61), *616;* 3: 398(199), 400(199), 449(199), 760(581), 871(581), 894(581), 897(581), *591, 1158;* 4: 211(2), 223-224(2), 280(2), 335(2), *426*

Warren, E. W., 4: 238(10), 257(10), 356(10), *426*

Warren, R. J., 3: 62(236), *240*

Warrington, J. V., 2: 370(720), *402*

Wasiak, 1: 208(310), 211(310), *296*

Wasiak, J., 2: 305(201), *387*

Wass, M. N., 3: 683-684(235), 850(235), *1150*

Wasson, J. R., 3: 847(824), *1165*

Watanabe, 1: 188(61,91,92), 189(61), *288, 289*

Watanabe, H., 2: 31(191), 32(201), *247, 248*

Watanabe, M., 2: 346(626), 377(826,827), *399, 405*

Waters, W. A., 3: 806(713), *1162*

Watkins, D. A. M., 2: 7(49), 11(49), 130